De Gruyter Studies in Mathematics 50

Zoltán Sasvári

Multivariate Characteristic and Correlation Functions

De Gruyter

Mathematical Subject Classification 2010: 42B10, 43A35, 60F05, 60G10, 60G50, 60G51, 60E07, 60E10.

ISBN 978-3-11-022398-9
e-ISBN 978-3-11-022399-6
Set-ISBN 978-3-11-174044-7
ISSN 0179-0986

Library of Congress Cataloging-in-Publication Data

A CIP catalog record for this book has been applied for at the Library of Congress.

Bibliographic information published by the Deutsche Nationalbibliothek

The Deutsche Nationalbibliothek lists this publication in the Deutsche Nationalbibliografie; detailed bibliographic data are available in the internet at http://dnb.dnb.de.

Typesetting: PTP-Berlin Protago-TEX-Production GmbH, www.ptp-berlin.de
Printing and binding: Hubert & Co. GmbH & Co. KG, Göttingen
♾ Printed on acid-free paper

Printed in Germany

www.degruyter.com

Preface

Our goal in writing this book is to make the basic concepts and results on multivariate characteristic and correlation functions easily accessible to both students and researchers in a comprehensive manner. Especially in the interest of students, every attempt has been made to make the book as self-contained as possible. This led to a relatively large appendix containing topics like infinite products, functional equations, special functions or compact operators. We hope that this appendix will make the usage of the book much easier, for example in seminars. Chapters 1, 2 and parts of Chapter 3 have been 'tested' in lectures and seminars given at the Technical University Dresden, Wroclaw University of Technology, and Swansea University.

In a certain sense, characteristic functions and correlation functions are the same, the common underlying concept is positive definiteness. Many results in probability theory, mathematical statistics and stochastic processes can be derived by using these functions. While there are books on characteristic functions of one variable, books devoting some sections to the multivariate case, and books treating the general case of locally compact groups, interestingly there is no book devoted entirely to the multidimensional case which is extremely important for applications. This book is intended to fill this gap at least partially. Due to the abundance of results and of their applications a single book cannot treat all of them.

Most of the presented results are fairly well known. For this reason only few references are included, rather we cite papers and monographs where such references can be found. A brief outline of the book is as follows. The first chapter presents basic results and should be read carefully since it is essential for the understanding of the subsequent chapters. The second chapter is devoted to correlation functions, their applications to stationary processes and some connections to harmonic analysis. In Chapter 3 we deal with several special properties, Chapter 4 is devoted to the extension problem while Chapter 5 contains a few applications. Many results in the Appendix are presented with proofs. The reader might find some sections of the Appendix interesting on their own, e.g., the section *Solutions of certain functional equations* or the section *Linear independence of exponential functions.*

The numbering of chapters, sections, theorems, etc. is traditional. Equations are numbered consecutively within each paragraph, theorem, definition, and so forth. Equation (i) stands for the i-th equation within the current paragraph while equation (l.k.j.i) stands for the i-th equation within the Paragraph (l.k.j).

My warmest thanks go to Georg Berschneider who read the whole manuscript, gave valuable remarks, helped to find literature to special topics and solved several TeX problems. Björn Böttcher read substantial parts of the manuscript, I am grateful for his comments. I thank my colleagues at the Institut für Stochastik, Technische Universität Dresden, for many helpful discussions, in particular Christiane Weber for her help in preparing the figures, the index, and the bibliography, and Willi Schenk for the proof idea of Lemma B.1.2. I record my gratitude to the publishers, especially to the series editor Niels Jacob for his interest and encouragement.

Last, but not least, I would like to thank my wife, Evelyne Sasvári, for her support and understanding.

Dresden, November 2012 Zoltán Sasvári

Contents

Chapter 1

Characteristic functions

In this very first chapter we prove several classical results due to, among others, Herglotz, Bochner, Lévy, and Plancherel. We start with basic properties of characteristic functions and continue with the investigation of positive definite functions. As a byproduct, we obtain basic results on the Fourier transformation which will be needed in subsequent chapters.

Before starting we suggest to have a short look at Appendix A which contains basic notation used throughout the book.

1.1 Basic properties

This is probably the easiest section of the book. We first define characteristic functions and then show several simple properties. All the proofs are short and not difficult. In spite of their simplicity, these properties are extremely useful and frequently applied.

Definition 1.1.1. Let X be a d-dimensional random vector and denote by μ the distribution of X. The *characteristic function* $f = f_X$ of X is defined by

$$f(t) := \mathbb{E}\big(e^{i(t,X)}\big) = \int_{\mathbb{R}^d} e^{i(t,x)}\, d\mu(x), \quad t \in \mathbb{R}^d.$$

The existence of the expectation above is guaranteed by the fact that the complex-valued random variable $e^{i(t,X)}$ is bounded for each $t \in \mathbb{R}^d$. Sometimes we call f the characteristic function of the distribution μ, in which case we also write f_μ. As we will see in Section 1.3, every distribution is uniquely determined by its characteristic function.

If X has a density p, then

$$f(t) = \int_{\mathbb{R}^d} e^{i(t,x)} p(x)\, dx, \quad t \in \mathbb{R}^d.$$

If X is discrete with values $x_1, x_2, \dots$ and corresponding probabilities $p_1, p_2, \dots$, then

$$f(t) = \sum_j p_j \cdot e^{i(t,x_j)}, \quad t \in \mathbb{R}^d.$$

Especially, the functions

$$t \mapsto e^{i(t,x)} \quad \text{and} \quad t \mapsto \cos((t,x)) = \frac{1}{2} \cdot e^{i(t,x)} + \frac{1}{2} \cdot e^{i(t,-x)}$$

are characteristic functions for every $x \in \mathbb{R}^d$. As a further example, assume that X is uniformly distributed on the parallelepiped

$$K = [a_1, b_1] \times \cdots \times [a_d, b_d] \subset \mathbb{R}^d,$$

where $a_j < b_j$. We then have

$$\begin{aligned} f_X(t) &= \frac{1}{\lambda(K)} \int_K \mathrm{e}^{\mathrm{i}(t,x)}\, \mathrm{d}x = \frac{1}{\lambda(K)} \int_K \prod_{j=1}^d \mathrm{e}^{\mathrm{i}t_j x_j}\, \mathrm{d}x \\ &= \frac{1}{\lambda(K)} \cdot \prod_{j=1}^d \int_{a_j}^{b_j} \mathrm{e}^{\mathrm{i}t_j x_j}\, \mathrm{d}x_j = \prod_{j=1}^d \frac{\mathrm{e}^{\mathrm{i}b_j t_j} - \mathrm{e}^{\mathrm{i}a_j t_j}}{\mathrm{i}(b_j - a_j)t_j} \end{aligned}$$

where the factors of the last product are equal to 1 (by definition) if $t_j = 0$.

In the last example, the characteristic function f_X is real-valued whenever $a_j = -b_j$ for all j. We will show later (cf. Theorem 1.3.13) that f_X is real-valued if and only if X and $-X$ have the same distribution, i.e., if μ is symmetric: $\mu = \mu^{\triangle}$. The proof of the "if part" is very simple, but for the "only if part" we need the fact that every distribution is uniquely determined by its characteristic function.

Theorem 1.1.2. *Every characteristic function f on $\mathbb{R}^d$ has the following properties:*

(i) $f(0) = 1$;

(ii) $|f(t)| \le 1, \quad t \in \mathbb{R}^d$;

(iii) $f(-t) = \overline{f(t)}, \quad t \in \mathbb{R}^d$;

(iv) *f is uniformly continuous on $\mathbb{R}^d$;*

(v) *the inequality*

$$\sum_{i,j=1}^n f(t_i - t_j) c_i \overline{c_j} \ge 0$$

holds for every positive integer n, for all $t_1, \dots, t_n \in \mathbb{R}^d$, and for all $c_1, \dots, c_n \in \mathbb{C}$.

Proof. The first three properties follow immediately from Definition 1.1.1. To prove (iv) assume that f is the characteristic function of a random vector X. Then

$$|f(t+h) - f(t)| = |\mathbb{E}\,[\mathrm{e}^{\mathrm{i}(t,X)} \cdot (\mathrm{e}^{\mathrm{i}(h,X)} - 1)]| \le \mathbb{E}(|\mathrm{e}^{\mathrm{i}(h,X)} - 1|)$$

where the upper bound is independent of t and, by the dominated convergence theorem, tends to zero as $h \to 0$. This shows that f is uniformly continuous on $\mathbb{R}^d$.

The last statement follows from

$$\sum_{i,j=1}^{n} f(t_i - t_j)c_i\overline{c_j} = \sum_{i,j=1}^{n} \mathbb{E}(\mathrm{e}^{\mathrm{i}(t_i - t_j, X)})c_i\overline{c_j} = \mathbb{E}\Big(\sum_{i,j=1}^{n} c_i \mathrm{e}^{\mathrm{i}(t_i,X)}\,\overline{c_j \mathrm{e}^{\mathrm{i}(t_j,X)}}\Big)$$
$$= \mathbb{E}\Big(\Big|\sum_{i=1}^{n} c_i \mathrm{e}^{\mathrm{i}(t_i,X)}\Big|^2\Big) \geq 0. \qquad \square$$

We will see in Section 1.7 that inequality (1.1.2.v) is a characteristic property of characteristic functions.

The next theorem is very useful in the study of sums of independent random vectors.

Theorem 1.1.3. *If X and Y are independent d-dimensional random vectors, then the characteristic function of their sum is*

$$f_{X+Y} = f_X \cdot f_Y.$$

Proof. Let $t \in \mathbb{R}^d$ be arbitrary. Since X and Y are independent, the complex-valued random variables $\mathrm{e}^{\mathrm{i}(t,X)}$ and $\mathrm{e}^{\mathrm{i}(t,Y)}$ are independent as well. Thus,

$$f_{X+Y}(t) = \mathbb{E}(\mathrm{e}^{\mathrm{i}(t,X+Y)}) = \mathbb{E}(\mathrm{e}^{\mathrm{i}(t,X)} \cdot \mathrm{e}^{\mathrm{i}(t,Y)})$$
$$= \mathbb{E}(\mathrm{e}^{\mathrm{i}(t,X)}) \cdot \mathbb{E}(\mathrm{e}^{\mathrm{i}(t,Y)}) = f_X(t) \cdot f_Y(t). \qquad \square$$

For arbitrary distributions μ and ν on $\mathbb{R}^d$ there exist independent d-dimensional random vectors X and Y such that μ and ν are the distributions of X and Y, respectively.[1] From this and Theorem 1.1.3 we obtain:

Corollary 1.1.4. *The product of two characteristic functions of d variables is again a characteristic function.*

Noting that $\overline{f}$ is the characteristic function of $-X$, we obtain the following corollary.

Corollary 1.1.5. *If X and Y are independent identically distributed random vectors with characteristic function f, then $|f|^2$ is the characteristic function of $X - Y$.*

Next we present further simple methods to construct characteristic functions from given ones.

---

[1] Note that the distribution of $X + Y$ is $\mu * \nu$ so that the equation in Theorem 1.1.3 can also be written as $f_{\mu*\nu} = f_\mu \cdot f_\nu$.

Theorem 1.1.6. *Let R be a random variable and X be a d-dimensional random vector. If R and X are independent, then*[2]

$$f_{RX}(t) = \int_{-\infty}^{\infty} f_X(st)\,\mathrm{d}\mu_R(s), \quad t \in \mathbb{R}^d.$$

Proof. Applying equation (F.1.3) we obtain

$$\begin{aligned} f_{RX}(t) &= \mathbb{E}\big[\mathrm{e}^{\mathrm{i}(t,RX)}\big] = \mathbb{E}\big[\mathrm{e}^{\mathrm{i}R(t,X)}\big] \\ &= \int_{-\infty}^{\infty} \mathrm{e}^{\mathrm{i}s}\,\mathrm{d}\mu_{R(t,X)}(s) \\ &= \int_{-\infty}^{\infty} \mathrm{e}^{\mathrm{i}s}\,\mathrm{d}\mu_R \star \mu_{(t,X)}(s) \\ &= \int_{-\infty}^{\infty}\int_{-\infty}^{\infty} \mathrm{e}^{\mathrm{i}su}\,\mathrm{d}\mu_{(t,X)}(u)\,\mathrm{d}\mu_R(s) \\ &= \int_{-\infty}^{\infty}\int_{\mathbb{R}^d} \mathrm{e}^{\mathrm{i}s(t,x)}\,\mathrm{d}\mu_X(x)\,\mathrm{d}\mu_R(s) \\ &= \int_{-\infty}^{\infty} f_X(st)\,\mathrm{d}\mu_R(s)\,. \end{aligned}$$ □

Theorem 1.1.7. *Let X be a d-dimensional random vector. Then the equation*

$$f_{AX+b}(t) = \mathrm{e}^{\mathrm{i}(t,b)} \cdot f_X(A^{\mathrm{T}}t), \quad t \in \mathbb{R}^n$$

holds for every linear mapping $A: \mathbb{R}^d \to \mathbb{R}^n$ and $b \in \mathbb{R}^n$.
In particular, if $X = (X_1, X_2)$ where X_1 and X_2 are random variables, then

$$f_{a_1X_1+a_2X_2}(t) = f_X(a_1t, a_2t), \quad a_1, a_2, t \in \mathbb{R}.$$

Proof. We have

$$f_{AX+b}(t) = \mathbb{E}[\mathrm{e}^{\mathrm{i}(t,AX+b)}] = \mathbb{E}\left[\mathrm{e}^{\mathrm{i}(t,b)} \cdot \mathrm{e}^{\mathrm{i}(A^T t,X)}\right] = \mathrm{e}^{\mathrm{i}(t,b)} \cdot f_X(A^{\mathrm{T}}t).$$ □

Theorem 1.1.8. *Let X_1 and X_2 be independent d_1- and d_2-dimensional random vectors, respectively, and write $d = d_1 + d_2$. Then the characteristic function of the d-dimensional random vector $X = (X_1, X_2)$ is given by*

$$f_X(t) = f_{X_1}(t_1) \cdot f_{X_2}(t_2)$$

where $t_1 \in \mathbb{R}^{d_1}$, $t_2 \in \mathbb{R}^{d_2}$, and $t = (t_1, t_2) \in \mathbb{R}^d$.[3]

---

[2] See also Lemma 1.5.7.

[3] See also Theorem 1.3.10.

Proof. The theorem follows from

$$f_X(t) = \mathbb{E}(e^{i(t,X)}) = \mathbb{E}(e^{i(t_1,X_1)} \cdot e^{i(t_2,X_2)})$$

using the independence of X_1 and X_2. □

Applying the same argument as in the proof of Corollary 1.1.4, we obtain the following corollary.

Corollary 1.1.9. *Let f_i be a characteristic function of d_i variables ($i = 1, 2$). Then the function f defined by*

$$f(t) = f_1(t_1) \cdot f_2(t_2)$$

where $t_1 \in \mathbb{R}^{d_1}$, $t_2 \in \mathbb{R}^{d_2}$, and $t = (t_1, t_2) \in \mathbb{R}^d$, is a characteristic function of $d = d_1 + d_2$ variables.

Remark 1.1.10. It is a useful fact that the characteristic function f_X of a random vector $X = (X_1, \ldots, X_d)$ is determined by the characteristic functions f_{L_y} of all linear combinations

$$L_y = y_1 X_1 + \cdots + y_d X_d = (y, X), \quad y \in \mathbb{R}^d$$

and vice versa. Indeed,

$$f_{L_y}(s) = \mathbb{E}(e^{is \cdot L_y}) = \mathbb{E}(e^{is \cdot (y,X)}) = f_X(s \cdot y), \quad y \in \mathbb{R}^d, \ s \in \mathbb{R}. \tag{1}$$

In particular, $f_X(y) = f_{L_y}(1)$. From (1) we also see that $s \mapsto f_X(s \cdot y)$ is a characteristic function of one variable for all $y \in \mathbb{R}^d$.

Remark 1.1.11. Let $X = (X_1, \ldots, X_d)$ be a random vector and $i_1, \ldots, i_k$ be integers such that $1 \le i_1 < \cdots < i_k \le d$. It follows immediately from Definition 1.1.1 that the characteristic function of the random vector $(X_{i_1}, \ldots, X_{i_k})$ can be obtained by setting $t_i = 0$ for $i \notin \{i_1, \ldots, i_k\}$ in $f_X(t_1, \ldots, t_d)$. The distribution of $(X_{i_1}, \ldots, X_{i_k})$ is usually called a *marginal distribution*.

Lemma 1.1.12. *Let μ be a distribution on $\mathbb{R}^d$. The real part*

$$\operatorname{Re} f_\mu(t) = \int_{\mathbb{R}^d} \cos((t, x))\, d\mu(x), \quad t \in \mathbb{R}^d$$

of its characteristic function f_μ is the characteristic function of the distribution $(\mu + \mu^\triangle)/2$.

Proof. The statement follows immediately from the fact that

$$\int h(t)\, d\mu^\triangle(t) = \int h(-t)\, d\mu(t)$$

for every bounded continuous function h on $\mathbb{R}^d$. □

In the rest of this section we present some simple examples of characteristic functions of one variable. These functions are shown in Figures 1.1–1.5.

Example 1.1.13.

(a) The *exponential distribution* with parameter $\alpha > 0$ has the density function $x \mapsto \alpha \mathbf{1}_{[0,\infty)}(x) \cdot \mathrm{e}^{-\alpha x}$, $x \in \mathbb{R}$, and its characteristic function is given by

$$\int_0^\infty \alpha \mathrm{e}^{-\alpha x} \cdot \mathrm{e}^{\mathrm{i}tx}\,\mathrm{d}x = \frac{\alpha \cdot \mathrm{e}^{(\mathrm{i}t-\alpha)\cdot x}}{\mathrm{i}t-\alpha}\Big|_{x=0}^{x=\infty} = \frac{\alpha}{\alpha - \mathrm{i}t}$$

(see Figure 1.1).

(b) The *Laplace distribution* with parameter $\beta > 0$ has the density function $x \mapsto \frac{1}{2\beta} \cdot \mathrm{e}^{-\frac{|x|}{\beta}}$, $x \in \mathbb{R}$. Its characteristic function is given by

$$\begin{aligned}\int_{-\infty}^\infty \frac{1}{2\beta} \cdot \mathrm{e}^{-\frac{|x|}{\beta}} \cdot \mathrm{e}^{\mathrm{i}tx}\,\mathrm{d}x &= \int_0^\infty \frac{1}{2\beta} \cdot \mathrm{e}^{(-\frac{1}{\beta}+\mathrm{i}t)\cdot x}\,\mathrm{d}x \\ &\quad + \int_{-\infty}^0 \frac{1}{2\beta} \cdot \mathrm{e}^{(\frac{1}{\beta}+\mathrm{i}t)\cdot x}\,\mathrm{d}x \\ &= \frac{1}{2\beta}\left[-\frac{1}{\mathrm{i}t - \frac{1}{\beta}} + \frac{1}{\mathrm{i}t + \frac{1}{\beta}}\right] = \frac{1}{1+\beta^2 t^2}\end{aligned}$$

(see Figure 1.2). Using Corollary 1.1.5 and the last two examples we see that the difference of two independent identically distributed exponential random variables is Laplace distributed.

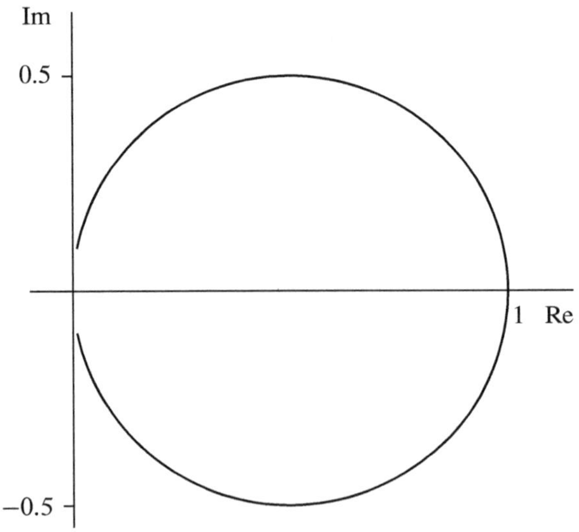

Figure 1.1. The function $t \mapsto \frac{1}{1-\mathrm{i}t}$, $t \in [-10, 10]$, from Example 1.1.13(a).

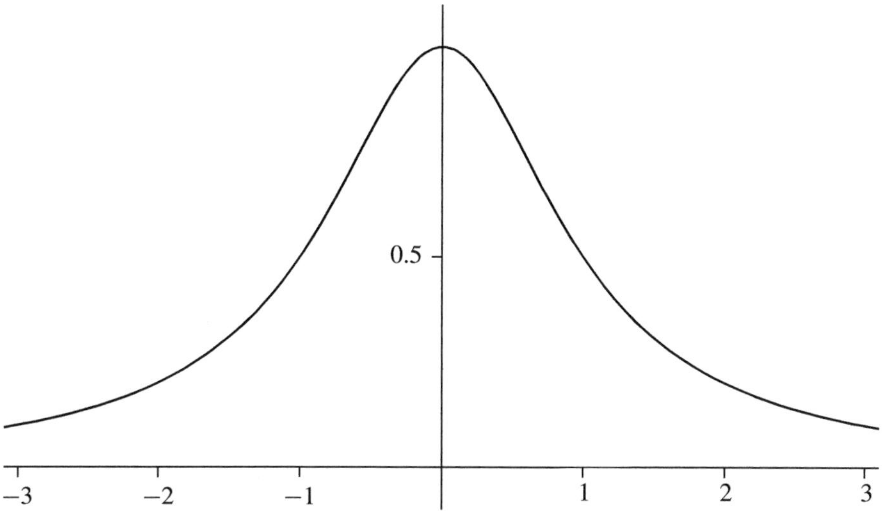

Figure 1.2. The function $t \mapsto \frac{1}{1+t^2}$ from Example 1.1.13(b).

(c) To compute the characteristic function of the *binomial distribution*

$$\mu = \sum_{k=0}^{n} \binom{n}{k} p^k (1-p)^{n-k} \delta_k$$

with parameters $n \geq 1$ and $p \in [0, 1]$, let $X_1, \ldots, X_n$ be independent random variables such that $P(X_j = 0) = 1 - p$, $P(X_j = 1) = p$ for all j. Then μ is the distribution of $X = X_1 + \cdots + X_n$. Applying Theorem 1.1.3 we obtain

$$f_\mu(t) = f_X(t) = [f_{X_1}(t)]^n = [1 - p + p \cdot \mathrm{e}^{\mathrm{i}t}]^n$$

(see Figure 1.3 and Figure 1.4).

(d) The characteristic function of the *Poisson distribution*

$$\mathrm{e}^{-\lambda} \cdot \sum_{k=0}^{\infty} \frac{\lambda^k}{k!} \cdot \delta_k$$

with parameter $\lambda \geq 0$ is

$$\mathrm{e}^{-\lambda} \cdot \sum_{k=0}^{\infty} \frac{\lambda^k}{k!} \cdot \mathrm{e}^{\mathrm{i}tk} = \mathrm{e}^{-\lambda} \cdot \sum_{k=0}^{\infty} \frac{(\lambda \mathrm{e}^{\mathrm{i}t})^k}{k!} = \mathrm{e}^{\lambda(\mathrm{e}^{\mathrm{i}t}-1)}$$

(see Figure 1.5).

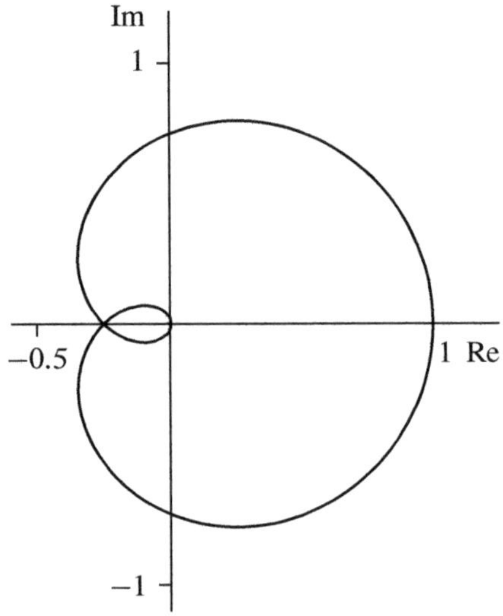

Figure 1.3. The function $t \mapsto [1 - p + p \cdot e^{it}]^n$ from Example 1.1.13(c) with $p = 1/2$ and $n = 4$.

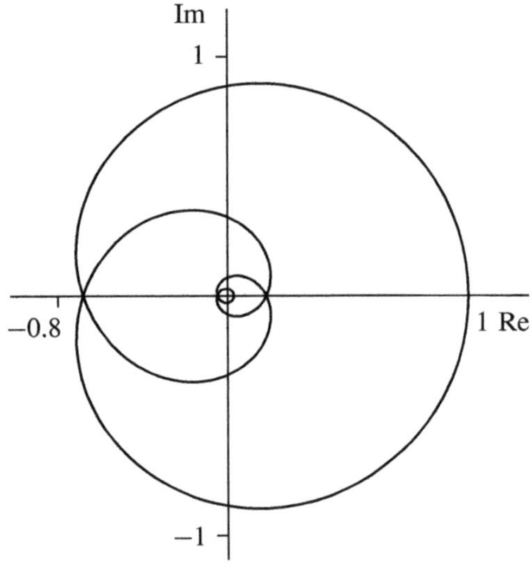

Figure 1.4. The function $t \mapsto [1 - p + p \cdot e^{it}]^n$ from Example 1.1.13(c) with $p = 0.7$ and $n = 4$.

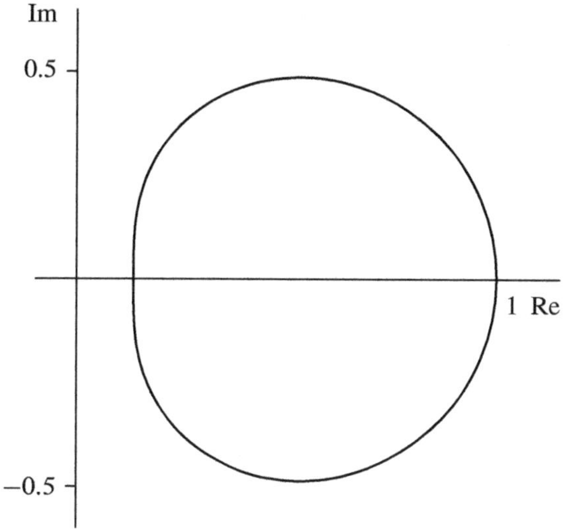

Figure 1.5. The function $t \mapsto \exp(e^{it} - 1)$ from Example 1.1.13(d).

1.2 Differentiability

The characteristic functions in Example 1.1.13 are infinitely often differentiable (even analytic) on $\mathbb{R}$. We will show in Example 1.3.7 that the characteristic function of the Cauchy distribution with parameter $\alpha = 1$ is given by $f(t) = e^{-|t|}$. This function is not differentiable at 0. Example B.1.19 shows that characteristic functions may be nowhere differentiable.

Throughout this section, μ denotes a probability distribution on $\mathbb{R}^d$ with characteristic function f. We will see that differentiability of f is closely related to the existence of moments of μ.

Theorem 1.2.1. *Let $\alpha \in \mathbb{N}_0^d$ be such that the moment M_α of μ exists. Then the partial derivative $D^\alpha f$ exists and we have*

(i) $D^\alpha f(t) = i^{|\alpha|} \int_{\mathbb{R}^d} x^\alpha \cdot e^{i(t,x)} \, d\mu(x), \quad t \in \mathbb{R}^d;$

(ii) $D^\alpha f(0) = i^{|\alpha|} \cdot M_\alpha;$

(iii) *$D^\alpha f$ is uniformly continuous and $|D^\alpha f|$ is bounded by the absolute moment A_α of μ.*

Proof. Suppose that $\alpha_j > 0$ for some j and let $e_1, \dots, e_d$ be the standard basis of $\mathbb{R}^d$. For $h \in \mathbb{R} \setminus \{0\}$ we have

$$\frac{f(t + he_j) - f(t)}{h} = \int_{\mathbb{R}^d} x_j \cdot \frac{e^{ihx_j} - 1}{hx_j} \cdot e^{i(t,x)} \, d\mu(x).$$

Theorem B.1.5 with $n = 0$ shows that the integrand is dominated by $|x_j|$. Taking the limit $h \to 0$ and applying Lebesgue's theorem on dominated convergence we obtain

$$\frac{\partial}{\partial t_j} f(t) = \mathrm{i} \int_{\mathbb{R}^d} x_j \cdot \mathrm{e}^{\mathrm{i}(t,x)} \, \mathrm{d}\mu(x), \quad t \in \mathbb{R}^d.$$

Repeating this argument we obtain (i). Equation (ii) is obtained from (i) by setting $t = 0$.

From (i) we infer that

$$|D^\alpha f(t)| \le \int_{\mathbb{R}^d} |x^\alpha| \, \mathrm{d}\mu(x) = A_\alpha.$$

The uniform continuity of $D^\alpha f$ follows from

$$\begin{aligned} |D^\alpha f(t+h) - D^\alpha f(t)| &= \left| \int_{\mathbb{R}^d} x^\alpha \cdot (\mathrm{e}^{\mathrm{i}(h,x)} - 1) \cdot \mathrm{e}^{\mathrm{i}(t,x)} \, \mathrm{d}\mu(x) \right| \\ &\le \int_{\mathbb{R}^d} |x^\alpha| \cdot |\mathrm{e}^{\mathrm{i}(h,x)} - 1| \, \mathrm{d}\mu(x) \end{aligned}$$

by dominated convergence. □

The previous theorem has the following immediate consequences.

Corollary 1.2.2. *Let $X = (X_1, \dots, X_d)$ be a d-dimensional random vector with characteristic function f. If X_j is square integrable for all j, then the covariance matrix* $\operatorname{cov}(X) = (c_{jk})_{j,k=1}^d$ *is given by*

$$c_{jk} = -\frac{\partial^2}{\partial t_j \partial t_k} f(0) + \frac{\partial}{\partial t_j} f(0) \cdot \frac{\partial}{\partial t_k} f(0).$$

Corollary 1.2.3. *Assume that $\mu \ne \delta_0$ and that for some $\alpha \in \mathbb{N}_0^d$ the moment $M_{2\alpha}$ of μ exists. Then the function g defined by*

$$g(t) = \frac{(-1)^{|\alpha|}}{M_{2\alpha}} \cdot D^{2\alpha} f(t), \quad t \in \mathbb{R}^d$$

is a characteristic function. The corresponding distribution ν is given by

$$\mathrm{d}\nu(x) = \frac{1}{M_{2\alpha}} \cdot x^{2\alpha} \, \mathrm{d}\mu(x).$$

Example 1.2.4. As an application of Theorem 1.2.1 we compute the mean and the variance of the distributions from Example 1.1.13. Let f denote the characteristic function of a random variable X with finite variance. Then, in view of (1.2.1.ii),

$$\mathbb{E}(X) = -\mathrm{i} f'(0), \quad \mathbb{E}(X^2) = -f''(0)$$

and hence

$$\operatorname{Var}(X) = \mathbb{E}(X^2) - (\mathbb{E}\,X)^2 = -f''(0) - f'(0)^2.$$

(a) If X is exponentially distributed with parameter $\alpha > 0$ then we have

$$f'(t) = \frac{\mathrm{d}}{\mathrm{d}t}\frac{\alpha}{\alpha - \mathrm{i}t} = \frac{\alpha \cdot \mathrm{i}}{(\alpha - \mathrm{i}t)^2}, \quad f''(t) = -\frac{2\alpha}{(\alpha - \mathrm{i}t)^3}$$

and hence $\mathbb{E}(X) = \frac{1}{\alpha}$, $\mathbb{E}(X^2) = \frac{2}{\alpha^2}$ and $\operatorname{Var}(X) = \frac{1}{\alpha^2}$.

(b) In the case of the Laplace distribution with parameter $\beta > 0$ we have

$$f'(t) = \frac{\mathrm{d}}{\mathrm{d}t}\frac{1}{1+\beta^2 t^2} = -\frac{2\beta^2 t}{(1+\beta^2 t^2)^2}, \quad f''(t) = \frac{2\beta^2(3\beta^2 t^2 - 1)}{(1+\beta^2 t^2)^3}.$$

Thus, $\mathbb{E}(X) = 0$ and $\mathbb{E}(X^2) = \operatorname{Var}(X) = 2\beta^2$.

(c) Assume that X has the binomial distribution with parameters n and p. Writing $q := 1 - p$ we have

$$f'(t) = \frac{\mathrm{d}}{\mathrm{d}t}(q + p\mathrm{e}^{\mathrm{i}t})^n = \mathrm{i}np \cdot (q + p\mathrm{e}^{\mathrm{i}t})^{n-1} \cdot \mathrm{e}^{\mathrm{i}t}$$
$$f''(t) = -np \cdot (q + p\mathrm{e}^{\mathrm{i}t})^{n-2} \cdot (q + pn\mathrm{e}^{\mathrm{i}t}) \cdot \mathrm{e}^{\mathrm{i}t}.$$

Consequently, $\mathbb{E}(X) = np$, $\mathbb{E}(X^2) = np(np+q)$ and $\operatorname{Var}(X) = npq$.

(d) If X has the Poisson distribution with parameter $\lambda > 0$, then

$$f'(t) = \frac{\mathrm{d}}{\mathrm{d}t}\mathrm{e}^{\lambda(\mathrm{e}^{\mathrm{i}t}-1)} = \mathrm{i}\lambda \cdot \mathrm{e}^{\mathrm{i}t} \cdot \mathrm{e}^{\lambda(\mathrm{e}^{\mathrm{i}t}-1)}, \quad f''(t) = -\lambda \cdot \mathrm{e}^{\mathrm{i}t} \cdot (1 + \lambda \mathrm{e}^{\mathrm{i}t}) \cdot \mathrm{e}^{\lambda(\mathrm{e}^{\mathrm{i}t}-1)}$$

and therefore $\mathbb{E}(X) = \lambda$, $\mathbb{E}(X^2) = \lambda + \lambda^2$ and $\operatorname{Var}(X) = \lambda$.

The next theorem follows immediately from Theorem 1.2.1 and Theorem B.6.1 with $a = 0$.

Theorem 1.2.5. *If for some positive integer n all moments M_α, $\alpha \in \mathbb{N}_0^d$, $|\alpha| \le n$, of μ exist, then*

$$f(t) = \sum_{|\alpha| \le n} \frac{M_\alpha}{\alpha!} \cdot (\mathrm{i}t)^\alpha + \sum_{|\alpha| = n} R_\alpha(t) \cdot t^\alpha, \quad t \in \mathbb{R}^d$$

where

$$R_\alpha(t) = \frac{n}{\alpha!} \cdot \int_0^1 (D^\alpha f(xt) - D^\alpha f(0))(1-x)^{n-1}\,\mathrm{d}x.$$

Moreover, $|R_\alpha(t)| \le \frac{2A_\alpha}{\alpha!}$ and $\lim_{t\to 0} R_\alpha(t) = 0$.

Sometimes it is more convenient to use the following expansion which is obtained by using Theorem B.6.2.

Theorem 1.2.6. *Under the conditions of the previous theorem there exist $\theta, \eta \in (0,1)$ depending on t such that*

$$f(t) = \sum_{|\alpha|\le n-1} \frac{M_\alpha}{\alpha!} \cdot (\mathrm{i}t)^\alpha + \sum_{|\alpha|=n} \frac{\operatorname{Re} D^\alpha f(\theta t) + \mathrm{i} \cdot \operatorname{Im} D^\alpha f(\eta t)}{\alpha!} \cdot t^\alpha.$$

The next special case is frequently used.

Corollary 1.2.7. *Let $X = (X_1, \ldots, X_d)$ be a random vector such that $\mathbb{E}(X_j^2) < \infty$ for all j. For the characteristic function f of X we then have*

$$f(t) = 1 + \mathrm{i} \cdot \sum_{j=1}^{d} \mathbb{E}(X_j) t_j - \frac{1}{2} \sum_{i,j=1}^{d} [\mathbb{E}(X_i X_j) + R_{i,j}(t)]\, t_i t_j, \quad t \in \mathbb{R}^d$$

where $|R_{i,j}(t)| \le 2\mathbb{E}(|X_i X_j|)$ and $\lim_{t\to 0} R_{i,j}(t) = 0$. Moreover, there exist $\theta, \eta \in (0,1)$ depending on t such that

$$f(t) = 1 + \mathrm{i} \cdot \sum_{j=1}^{d} \mathbb{E}(X_j) t_j + \frac{1}{2} \sum_{i,j=1}^{d} [\operatorname{Re} f_{t_i,t_j}(\theta t) + \mathrm{i} \cdot \operatorname{Im} f_{t_i,t_j}(\eta t)]\, t_i t_j.$$

Lemma 1.2.8. *Let $t \in \mathbb{R}^d$ be such that the inequality*

$$1 - \operatorname{Re} f(s \cdot t) \le C_t \cdot s^2, \quad s \in \mathbb{R} \tag{1}$$

holds with some $C_t \ge 0$. Then,

$$\int_{\mathbb{R}^d} (t,x)^2 \,\mathrm{d}\mu(x) \le 2 \cdot C_t. \tag{2}$$

In particular, if (1) holds for all $t \in \mathbb{R}^d$, then all moments of second order of μ are finite and hence all partial derivatives of order two of f exist.[4]

Proof. Let $s \in \mathbb{R} \setminus \{0\}$, $t \in \mathbb{R}^d$ and $T > 0$ be arbitrary. We then have

$$\begin{aligned} C_t \ge \frac{1 - \operatorname{Re} f(s \cdot t)}{s^2} &= \int_{\mathbb{R}^d} \frac{1 - \cos(s(t,x))}{s^2} \,\mathrm{d}\mu(x) \\ &\ge \int_{-T}^{T} \cdots \int_{-T}^{T} \frac{1 - \cos(s(t,x))}{s^2} \,\mathrm{d}\mu(x). \end{aligned}$$

Using the inequality (B.1.6.2) and Lebesgue's theorem on dominated convergence we see that the right-hand side tends to

$$\frac{1}{2} \int_{-T}^{T} \cdots \int_{-T}^{T} (t,x)^2 \,\mathrm{d}\mu(x)$$

as $s \to 0$. Since $T > 0$ is arbitrary, we find that (2) holds.

4 Compare this result with Theorem 1.11.4.

The second statement is obtained by choosing special t's with $t_j = 1$ and $t_i = 0$, $i \neq j$, in (2). □

The existence of $D^\alpha f$ does not imply the existence of the moment M_α in general.[5] However, the following is true.

Theorem 1.2.9. *If for some $\alpha \in \mathbb{N}_0^d \setminus \{0\}$ and for $g = \operatorname{Re} f$ all partial derivatives $D^\beta g$, $\beta < 2\alpha$, exist in an open neighborhood of zero and if $D^{2\alpha} g$ exists at zero, then the moment $M_{2\alpha}$ of μ exists. Moreover, the partial derivative $D^{2\alpha} f$ exists on all of $\mathbb{R}^d$.*

Proof. The second statement follows from the first one using Theorem 1.2.1. To prove the first statement assume first that $|\alpha| = 1$. By our assumption the function

$$h(s) = h_\alpha(s) = \operatorname{Re} f(s \cdot \alpha), \quad s \in \mathbb{R}$$

is differentiable in a neighborhood of 0 and $h''(0)$ exists. Since h is even, $h'(0) = 0$. Applying (1.1.2.ii) and the mean value theorem we see that

$$0 \leq \frac{1 - h(s)}{s^2} = \frac{h'(\eta s)}{s} \leq \frac{h'(\eta s)}{\eta s}, \quad s \neq 0$$

where $\eta = \eta(s) \in (0, 1)$. Since $h'(0) = 0$, the right-hand side tends to $h''(0)$ as $s \to 0$. Consequently, there exists a constant $C = C_\alpha$ such that

$$\frac{1 - h(s)}{s^2} \leq C, \quad s \neq 0$$

(note that $|h(s)| \leq 1$ for all s). Thus, by Lemma 1.2.8, the moment $M_{2\alpha}$ exists.

The general case $|\alpha| > 1$ follows from Corollary 1.2.3 using induction on $|\alpha|$. □

We continue with a simple application to conditional expectations.

Definition 1.2.10. Let X_0 be a random variable with finite expectation and let $X = (X_1, \dots, X_d)$ be a random vector. The random variable X_0 is said to have *polynomial regression* of order k on X if there exists a polynomial

$$Q(t) = \sum_{\alpha \in \mathbb{N}_0^d,\ |\alpha| \leq k} b_\alpha \cdot t^\alpha, \quad t \in \mathbb{R}^d \tag{1}$$

of order k such that

$$\mathbb{E}(X_0 | X) = Q(X). \tag{2}$$

If $k = 0, 1$ or 2 we speak of *constant, linear* or *quadratic regression*, respectively.

[5] See Remark 1.10.6 in [6] for an example.

Whether X_0 has polynomial regression on X can be answered in terms of the characteristic function of the random vector $(X_0, X_1, \dots, X_d)$.

Theorem 1.2.11. *Let $k \in \mathbb{N}_0$ and let $X_0, X_1, \dots, X_d$ be random variables such that all moments of order at most k of the random vector $Y = (X_0, X_1, \dots, X_d)$ exist. Denote by g the characteristic function of Y. Then the equation* (1.2.10.2) *holds if and only if*[6]

$$\mathrm{i}D^{(1,0,\dots,0)}g(0,t) = \sum_{\alpha\in\mathbb{N}_0^d,\ |\alpha|\le k} (-\mathrm{i})^{|\alpha|} b_\alpha D^{(0,\alpha_1,\dots,\alpha_d)} g(0,t), \quad t \in \mathbb{R}^d.$$

Proof. Write $X = (X_1, \dots, X_d)$. By Lemma F.4.2, equation (1.2.10.2) holds if and only if

$$\mathbb{E}\left(X_0 \cdot \mathrm{e}^{\mathrm{i}(t,X)}\right) = \sum_\alpha b_\alpha \cdot \mathbb{E}\left(X^\alpha \cdot \mathrm{e}^{\mathrm{i}(t,X)}\right), \quad t \in \mathbb{R}^d.$$

Hence, the statement follows immediately from Theorem 1.2.1. □

Next we prove an interesting result: If the product of two characteristic functions is $2n$-times differentiable, then so are the factors.[7]

Theorem 1.2.12. *Suppose that $f = f_1 f_2$ where f_1, f_2 are characteristic functions on $\mathbb{R}$ and let μ_1, μ_2 be the corresponding distributions. If f is $2n$-times differentiable, then so is f_1 and*

$$|f_1^{(k)}(0)| \le 2\left(A_k^{1/k} + |m_2|\right)^k, \quad 1 \le k \le 2n \tag{1}$$

where m_2 is an arbitrary median[8] *of μ_2 and A_k is the absolute moment of order k of $\mu := \mu_1 * \mu_2$. Analogous statements hold for f_2.*

Proof. Note first that f is the characteristic function of μ. Since f is $2n$-times differentiable, the moments A_k, $k = 1, \dots, 2n$, exist by Theorem 1.2.9. For every $a \in \mathbb{R}$ we have

$$\infty > A_k(a) := \int_{\mathbb{R}} |x-a|^k \,\mathrm{d}(\mu_1 * \mu_2)(x) = \int_{\mathbb{R}}\int_{\mathbb{R}} |t+s-a|^k \,\mathrm{d}\mu_1(t)\,\mathrm{d}\mu_2(s).$$

If t and $s-a$ have the same sign, then $|t+s-a|^k \ge |t|^k$. Using this inequality and the fact that the integrand above is zero if $s = a$ and $t = 0$, we obtain

$$\begin{aligned} A_k(a) &\ge \int_{[a,\infty)}\int_{[0,\infty)} |t|^k \,\mathrm{d}\mu_1(t)\,\mathrm{d}\mu_2(s) + \int_{(-\infty,a]}\int_{(-\infty,0)} |t|^k \,\mathrm{d}\mu_1(t)\,\mathrm{d}\mu_2(s) \\ &= \mu_2\big([a,\infty)\big)\int_{[0,\infty)} |t|^k \,\mathrm{d}\mu_1(t) + \mu_2\big((-\infty,a]\big)\int_{(-\infty,0)} |t|^k \,\mathrm{d}\mu_1(t). \end{aligned}$$

---

[6] Further results of this type can be found in [33].

[7] See also Theorem 3.3.9.

[8] This means that $\mu_2((-\infty, m_2]) \ge \frac{1}{2}$ and $\mu_2([m_2, \infty)) \ge \frac{1}{2}$.

These inequalities show that if $a = m_2$ is a median of μ_2, then

$$\int_{-\infty}^{\infty} |t|^k \, d\mu_1(t) \leq 2A_k(m_2), \quad k = 1, \ldots, 2n. \tag{2}$$

By virtue of Theorem 1.2.1, the function f_1 is $2n$-times differentiable. Moreover, by (2) and (1.2.1.iii), we have

$$|f_1^{(k)}(0)| \leq 2A_k(m_2). \tag{3}$$

On the other hand,

$$\begin{aligned} A_k(m_2) &= \int_{-\infty}^{\infty} |x - m_2|^k \, d\mu(x) \\ &\leq \int_{-\infty}^{\infty} \sum_{j=0}^{k} \binom{k}{j} |x|^j |m_2|^{k-j} \, d\mu(x) = \sum_{j=0}^{k} \binom{k}{j} A_j \cdot |m_2|^{k-j}. \end{aligned}$$

The last sum is, in view of inequality (F.1.5), majorized by

$$\sum_{j=0}^{k} \binom{k}{j} \left(A_k^{1/k}\right)^j |m_2|^{k-j} = \left(A_k^{1/k} + |m_2|\right)^k$$

which, together with (3), proves the inequality (1).

Interchanging the roles of f_1 and f_2, we obtain the last statement of the theorem. □

The next theorem generalizes the differentiability statement of Theorem 1.2.12.[9]

Theorem 1.2.13. *Let $\delta > 0$ and f be a $2n$-times differentiable function on the interval $(-\delta, \delta)$ such that $f(-t) = \overline{f(t)}$, $|t| < \delta$. Further, let $f_1, \ldots, f_r$ be characteristic functions on $\mathbb{R}$ and $\alpha_1, \ldots, \alpha_r$ be positive real numbers such that*

$$\prod_{j=1}^{r} |f_j(t)|^{\alpha_j} = |f(t)|, \quad -\delta < t < \delta.$$

Then the f_j's are $2n$-times differentiable on $\mathbb{R}$.

Proof. Let

$$g_j(t) = f_j(t) f_j(-t) = |f_j(t)|^2, \quad t \in \mathbb{R}$$

and $g(t) = f(t)f(-t) = |f(t)|^2$, $|t| < \delta$. The g_j's are nonnegative characteristic functions, the corresponding distributions will be denoted by μ_j. The function g is even, nonnegative and $2n$-times differentiable. Moreover,

$$\prod_{j=1}^{r} g_j^{\alpha_j}(t) = g(t), \quad -\delta < t < \delta. \tag{1}$$

[9] See also Theorem 3.3.14.

We prove by induction on k that g_j is $2k$-times differentiable for all $k = 1, \dots, n$. The statement for f_j follows then from Theorem 1.2.12.

Assume first that $k = 1$. Since $g_j(t) \le 1$ we have $g(t) \le g_j^{\alpha_j}(t)$. Using this and the mean value theorem for the function $x \mapsto x^{\alpha_j}$ we see that

$$\frac{1-g(t)}{t^2} \ge \frac{1-g_j^{\alpha_j}(t)}{t^2} = \frac{1-g_j(t)}{t^2} \cdot \alpha_j \cdot s(t)^{\alpha_j - 1}$$

where $g_j(t) \le s(t) \le 1$. As g is even $g'(0) = 0$ and hence the left side tends to $-\frac{1}{2}g''(0)$ if t tends to zero. On the other hand, $s(t)$ tends to 1 and therefore

$$\frac{1-g_j(t)}{t^2} \le C, \quad t \neq 0$$

for some constant C. Lemma 1.2.8 shows that g_j is twice differentiable.

Suppose now that g_j is $2k$-times differentiable for some $k < n$. Then

$$g_j^{(2l-1)}(0) = 0, \quad l = 1, \dots, k; \; j = 1, \dots, r.$$

We choose $\delta_0 \in (0, \delta]$ such that $g(t) \neq 0$ if $|t| < \delta_0$. Then we have

$$\sum_{j=1}^{r} \alpha_j \log \frac{1}{g_j(t)} = -h(t), \quad -\delta_0 < t < \delta_0 \tag{2}$$

where $h = \log g$. This equation and Faà di Bruno's formula (B.1.14) with $m = 2k$, $g = -\log$ and $f = g_j$ show that for $t \in (-\delta_0, \delta_0)$ we have

$$-h^{(2k)}(t) = \sum_{j=1}^{r} \alpha_j \sum c(b) \left(\frac{g_j'(t)}{g_j(t)}\right)^{b_1} \left(\frac{g_j''(t)}{g_j(t)}\right)^{b_2} \cdots \left(\frac{g_j^{(2k)}(t)}{g_j(t)}\right)^{b_{2k}} \tag{3}$$

where the last sum is over all $b \in \mathbb{N}_0^{2k}$ satisfying $b_1 + 2b_2 + \dots + 2kb_{2k} = 2k$ and

$$c(b) = \frac{(2k)!(-1)^{|b|} \cdot (|b|-1)!}{b_1! b_2! \dots b_{2k}! \, (1!)^{b_1} (2!)^{b_2} \dots ((2k)!)^{b_{2k}}}.$$

Setting $t = 0$ in (3), subtracting the resulting expression from (3) and using that $c(b) = -1$ if $b = (0, 0, \dots, 1)$, we obtain

$$\sum_{j=1}^{r} \alpha_j \left(\frac{g_j^{(2k)}(t)}{g_j(t)} - g_j^{(2k)}(0) \right) =$$

$$h^{(2k)}(t) - h^{(2k)}(0) \sum_{j=1}^{r} \alpha_j \sum c(b) \cdot [S_{j,b}(t) - S_{j,b}(0)] \tag{4}$$

where the last sum is over all $b \in \mathbb{N}_0^{2k}$ satisfying $b_{2k} = 0$, $b_1 + 2b_2 + \cdots + (2k-1)b_{2k-1} = 2k$ and

$$S_{j,b}(t) = \left(\frac{g_j'(t)}{g_j(t)}\right)^{b_1} \left(\frac{g_j''(t)}{g_j(t)}\right)^{b_2} \cdots \left(\frac{g_j^{(2k-1)}(t)}{g_j(t)}\right)^{b_{2k-1}}.$$

Using that g_j is $2k$-times continuously differentiable and $g_j(0) = 1$, $g_j^{(2l-1)}(0) = 0$, $1 \leq l \leq k$, we obtain

$$\frac{1}{g_j(t)} = 1 + O(t^2), \quad t \to 0 \tag{5}$$

and

$$g_j^{(2l-1)}(t) = O(t), \quad g_j^{(2l-2)}(t) = g_j^{(2l-2)}(0) + O(t^2), \quad t \to 0.$$

These relations show that $S_{j,b}(t) - S_{j,b}(0) = O(t^2)$ if $b \neq (0, 0, \ldots, 1)$. Since h is $2k$-times differentiable we conclude that the left-hand side of (4) is $O(t^2)$. Applying this and (5) we get

$$\sum_{j=1}^{r} \alpha_j \left(g_j^{(2k)}(t) - g_j^{(2k)}(0)\right) = O(t^2).$$

Since $g_j^{(2k)}/g_j^{(2k)}(0)$ is a characteristic function (cf. Corollary 1.2.3), inequality (1.1.2.ii) shows that the terms in this sum have the same sign and hence they are $O(t^2)$. By Lemma 1.2.8 the function $g_j^{(2k)}$ is twice differentiable, completing the induction on k. □

Remark 1.2.14. If all the numbers α_k are integers or rational numbers, then Theorem 1.2.13 follows easily from Theorem 1.2.12 using the fact that f_j^{α} is a characteristic function for all $\alpha \in \mathbb{N}$. If g is a nonnegative characteristic function and α is not an integer, then g^{α} is not necessarily a characteristic function. To see a simple example, let $g(t) = \cos^2 t$, $t \in \mathbb{R}$. Then $g^{1/2}(t) = |\cos t|$ and $g^{1/2}$ is not a characteristic function since it is infinitely differentiable on $(-\pi/2, \pi/2)$ but it is not differentiable at $\pi/2$ (cf. Theorem 1.2.9).

Remark 1.2.15. The set of all points where a given characteristic function is differentiable can be quite complicated. To see examples, let $b \geq 2$ be an integer and $\{\varepsilon_n\}_1^{\infty}$ be a sequence of nonnegative numbers tending to zero. Denote by $\mathcal{D}$ the set of all $t \in \mathbb{R}$ where the characteristic function g defined by

$$g(t) = C \sum_{n=1}^{\infty} \varepsilon_n b^{-n} \cos b^n t, \quad t \in \mathbb{R}$$

is differentiable (the constant C is determined by the condition $g(0) = 1$). Then g is differentiable at 0. Moreover:

(i) For every open interval $I \subset \mathbb{R}$ the set $\mathcal{D} \cap I$ has the same cardinality as $\mathbb{R}$.

(ii) If $\sum_1^\infty \varepsilon_n < \infty$ then $\mathcal{D} = \mathbb{R}$ and g' is continuous.

(iii) If $\sum_1^\infty \varepsilon_n^2 = \infty$ then $\lambda(\mathcal{D}) = 0$.

This result is due to A. Zygmund, a proof can be found in Section 1.10 of [6].

1.3 Inversion theorems

The main result of this section is Theorem 1.3.3, which states that distributions are uniquely determined by their characteristic functions. This result allows us to work with functions instead of measures, which is very useful for example in the study of limit theorems. At the end of this section we present some simple applications of the uniqueness theorem.

Theorem 1.3.1. *Let f be the characteristic function of a d-dimensional distribution μ and let $a, b \in \mathbb{R}^d$ such that $a_j < b_j$, $j = 1, \dots, d$. Then* [10]

$$\lim_{T\to\infty} \frac{1}{(2\pi)^d} \int_{-T}^{T} \cdots \int_{-T}^{T} \prod_{j=1}^{d} \frac{e^{-ia_j t_j} - e^{-ib_j t_j}}{it_j} f(t)\, dt = \int_{\mathbb{R}^d} \prod_{j=1}^{d} \chi_j(x)\, d\mu(x)$$

where

$$\chi_j(x) = \begin{cases} 1 & \text{if } x_j \in (a_j, b_j) \\ \frac{1}{2} & \text{if } x_j = a_j \text{ or } x_j = b_j \\ 0 & \text{if } x_j \notin [a_j, b_j]. \end{cases}$$

Proof. For $T > 0$ write

$$\begin{aligned} I(T) &:= \frac{1}{(2\pi)^d} \int_{-T}^{T} \cdots \int_{-T}^{T} \prod_{j=1}^{d} \frac{e^{-ia_j t_j} - e^{-ib_j t_j}}{it_j} f(t)\, dt \\ &= \frac{1}{(2\pi)^d} \int_{-T}^{T} \cdots \int_{-T}^{T} \left(\prod_{j=1}^{d} \frac{e^{-ia_j t_j} - e^{-ib_j t_j}}{it_j} \int_{\mathbb{R}^d} e^{i(x,t)}\, d\mu(x) \right) dt. \end{aligned}$$

In view of Theorem B.1.5,

$$\left| \frac{e^{-ia_j t_j} - e^{-ib_j t_j}}{it_j} \cdot e^{i(x,t)} \right| = \left| \frac{1 - e^{i(b_j - a_j)t_j}}{t_j} \right| \le b_j - a_j,$$

[10] The value of $(e^{-ia_j t_j} - e^{-ib_j t_j})/it_j$ at $t_j = 0$ is defined by continuity. It is equal to $b_j - a_j$.

and hence we may apply Fubini's theorem:

$$\begin{aligned} I(T) &= \frac{1}{(2\pi)^d}\int_{\mathbb{R}^d}\int_{-T}^{T}\cdots\int_{-T}^{T}\prod_{j=1}^{d}\frac{e^{i(x_j-a_j)t_j}-e^{i(x_j-b_j)t_j}}{it_j}\,dt\,d\mu(x) \\ &= \int_{\mathbb{R}^d}\prod_{j=1}^{d}\left(\frac{1}{2\pi}\int_{-T}^{T}\left[\frac{\sin(x_j-a_j)t_j}{t_j}-\frac{\sin(x_j-b_j)t_j}{t_j}\right]dt_j\right)d\mu(x) \\ &=: \int_{\mathbb{R}^d}\prod_{j=1}^{d}K_j(T,x)\,d\mu(x). \end{aligned}$$

It follows from Corollary B.1.8 that $\lim_{T\to\infty} K_j(T,x) = \chi_j(x)$. By Theorem B.1.9, we have $|K_j(T,x)| < 2$. Hence, the statement of the theorem follows by dominated convergence. □

Corollary 1.3.2. *If the μ-measure of the boundary of the set*

$$J = (a_1, b_1) \times \cdots \times (a_d, b_d)$$

is equal to zero, then

$$\lim_{T\to\infty}\frac{1}{(2\pi)^d}\int_{-T}^{T}\cdots\int_{-T}^{T}\prod_{j=1}^{d}\frac{e^{-ia_jt_j}-e^{-ib_jt_j}}{it_j}f(t)\,dt = \mu(J).$$

From this corollary we immediately obtain the following theorem.

Theorem 1.3.3 (Uniqueness Theorem). *Every distribution μ is uniquely determined by its characteristic function f.*

Proof. It suffices to show that the μ-measure of sets of the form

$$J = (a_1, b_1) \times \cdots \times (a_d, b_d)$$

is uniquely determined by f. If the boundary ∂J of J has zero μ-measure, then we can apply the previous corollary. Otherwise we consider the sets

$$J_\epsilon = (a_1+\epsilon, b_1-\epsilon) \times \cdots \times (a_d+\epsilon, b_d-\epsilon), \quad \epsilon > 0.$$

If $\epsilon_1 \neq \epsilon_2$ then $\partial J_{\epsilon_1} \cap \partial J_{\epsilon_2} = \emptyset$. Since μ is a finite measure the set

$$\{\epsilon > 0 : \mu(\partial J_\epsilon) \neq 0\}$$

is denumerable. Consequently, there exists a sequence $\{\epsilon_n\}_1^\infty$ such that $\mu(\partial J_{\epsilon_n}) = 0$ and

$$J = \bigcup_{n=1}^{\infty} J_{\epsilon_n}$$

from which the theorem follows. □

The next lemma deals with a special case of Gaussian distributions which will be investigated in more detail in Sections 1.10 and 3.5.

Lemma 1.3.4. *The function φ_σ, $\sigma > 0$, given by*

$$\varphi_\sigma(x) = \frac{1}{(\sqrt{2\pi}\cdot\sigma)^d}\cdot \mathrm{e}^{-\frac{\|x\|^2}{2\sigma^2}}, \quad x \in \mathbb{R}^d$$

is a probability density[11] *with characteristic function*

$$g_\sigma(t) = \mathrm{e}^{-\frac{\sigma^2\|t\|^2}{2}}, \quad t \in \mathbb{R}^d.$$

Moreover,

$$\varphi_\sigma = \frac{1}{(\sqrt{2\pi}\cdot\sigma)^d}\cdot g_{1/\sigma}. \tag{1}$$

Proof. We consider first the case $d = 1$. By Corollary B.1.11, the function φ_σ is a probability density. The corresponding characteristic function is

$$g_\sigma(t) = \frac{1}{\sqrt{2\pi}\cdot\sigma}\cdot\int_{-\infty}^{\infty} \mathrm{e}^{-\frac{y^2}{2\sigma^2}}\cdot \mathrm{e}^{ity}\,\mathrm{d}y.$$

Differentiating (cf. Theorem 1.2.1) and integrating by parts we get

$$\begin{aligned} g'_\sigma(t) &= \frac{\mathrm{i}}{\sqrt{2\pi}\cdot\sigma}\cdot\int_{-\infty}^{\infty} y\mathrm{e}^{-\frac{y^2}{2\sigma^2}}\cdot \mathrm{e}^{ity}\,\mathrm{d}y \\ &= -\frac{\sigma^2 t}{\sqrt{2\pi}\cdot\sigma}\cdot\int_{-\infty}^{\infty} \mathrm{e}^{-\frac{y^2}{2\sigma^2}}\cdot \mathrm{e}^{ity}\,\mathrm{d}y = -\sigma^2 t g_\sigma(t). \end{aligned}$$

Every solution of this differential equation has the form $t \mapsto C\mathrm{e}^{-\frac{\sigma^2 t^2}{2}}$. Since $g_\sigma(0) = 1$ we must have $C = 1$.

The case $d > 1$ is obtained now from the equation

$$\varphi_\sigma(x) = \prod_{j=1}^{d} \frac{1}{\sqrt{2\pi}\cdot\sigma}\cdot \mathrm{e}^{-\frac{1}{2\sigma^2}\cdot x_j^2}, \quad x \in \mathbb{R}^d.$$

The last statement is trivial. □

The next lemma is a special form of *Parseval's identity* (see also equation (i) in Plancherel's Theorem 1.8.9).

Lemma 1.3.5. *Let μ and ν be distributions on $\mathbb{R}^d$ with characteristic functions f_μ and f_ν, respectively. Then*

$$\int_{\mathbb{R}^d} \mathrm{e}^{\mathrm{i}(t,y)} f_\mu(t)\,\mathrm{d}\nu(t) = \int_{\mathbb{R}^d} f_\nu(x+y)\,\mathrm{d}\mu(x)$$

holds for all $y \in \mathbb{R}^d$.

[11] The distribution corresponding to the case $\sigma = 1$ is called *standard Gaussian* (see also Figure B.2).

Proof. By definition of f_μ

$$e^{i(t,y)} f_\mu(t) = \int_{\mathbb{R}^d} e^{i(t,x+y)} \, d\mu(x).$$

Integrating both sides with respect to ν and using Fubini's theorem we obtain the desired equation. □

Throughout the rest of this section f denotes the characteristic function of a probability distribution μ on $\mathbb{R}^d$.

The most frequently applied inversion formula is presented in the next theorem.

Theorem 1.3.6. *If $f \in \mathcal{L}^1(\mathbb{R}^d)$ then μ is absolutely continuous and the corresponding density p is given by*

$$p(y) = \frac{1}{(2\pi)^d} \int_{\mathbb{R}^d} f(t) \cdot e^{-i(t,y)} \, dt, \quad y \in \mathbb{R}^d.$$

Moreover, p is bounded and uniformly continuous.

Proof. Since f is integrable it can be written as a (complex) linear combination of four nonnegative integrable functions. Thus, the left-hand side of the equation above represents a linear combination of four characteristic functions. As characteristic functions are bounded and uniformly continuous, it remains to prove the first part of the theorem.

Let X and Y be independent d-dimensional random vectors such that μ is the distribution of X and Y is standard Gaussian (cf. Lemma 1.3.4). For any $\sigma > 0$ the density and the characteristic function of σY are given by the functions φ_σ and g_σ of Lemma 1.3.4, respectively. Applying equation (F.1.2) and Lemma 1.3.4 we see that the random vector $S_n = X + \frac{1}{n}Y$, $n \in \mathbb{N}$, has a density p_n given by

$$p_n(y) = \int_{\mathbb{R}^d} \varphi_{1/n}(y-x) \, d\mu(x) = \left(\frac{n}{\sqrt{2\pi}}\right)^d \cdot \int_{\mathbb{R}^d} g_n(y-x) \, d\mu(x), \quad y \in \mathbb{R}^d.$$

Lemma 1.3.5 shows that the integral on the right-hand side is equal to

$$\int_{\mathbb{R}^d} e^{-i(x,y)} f(x) \varphi_n(x) \, dx$$

and therefore

$$p_n(y) = \frac{1}{(2\pi)^d} \int_{\mathbb{R}^d} e^{-i(t,y)} f(t) e^{-\frac{1}{2n^2}\|t\|^2} \, dt. \tag{1}$$

The sequence $\{S_n\}_1^\infty$ converges pointwise to X. Therefore, the corresponding sequence of distributions converges weakly to μ. By Theorem E.1.7,

$$\lim_{n\to\infty} \int_{\mathbb{R}^d} h(y) p_n(y) \, dy = \int_{\mathbb{R}^d} h(y) \, d\mu(y) \tag{2}$$

for every bounded continuous function h on $\mathbb{R}^d$. Since the modulus of the integrand on the right-hand side of (1) is bounded by $|f|$, Lebesgue's dominated convergence theorem can be applied to show that $p_n(y)$ tends to

$$p(y) := \frac{1}{(2\pi)^d} \int_{\mathbb{R}^d} f(t) \cdot \mathrm{e}^{-\mathrm{i}(t,y)}\, \mathrm{d}t, \quad y \in \mathbb{R}^d$$

for all $y \in \mathbb{R}^d$. From (1) we also see that the sequence $\{p_n\}_{n=1}^{\infty}$ is uniformly bounded. Applying Lebesgue's theorem once more and using (2) we obtain

$$\begin{aligned} \int_{\mathbb{R}^d} h(y) p(y)\, \mathrm{d}y &= \int_{\mathbb{R}^d} \lim_{n\to\infty} h(y) p_n(y)\, \mathrm{d}y \\ &= \lim_{n\to\infty} \int_{\mathbb{R}^d} h(y) p_n(y)\, \mathrm{d}y = \int_{\mathbb{R}^d} h(y)\, \mathrm{d}\mu(y). \end{aligned}$$

From this we conclude that μ is absolutely continuous with density p. □

Example 1.3.7. Let

$$f(t) = \frac{1}{1+\alpha^2 t^2}, \quad t \in \mathbb{R}$$

be the characteristic function of the Laplace distribution with density

$$p(y) = \frac{1}{2\alpha} \cdot \mathrm{e}^{-\frac{|y|}{\alpha}}, \ y \in \mathbb{R}$$

(see (1.1.13.b)). By the previous theorem,

$$\frac{1}{2\alpha} \cdot \mathrm{e}^{-\frac{|y|}{\alpha}} = \frac{1}{2\pi} \int_{-\infty}^{\infty} \frac{\mathrm{e}^{-ity}}{1+\alpha^2 t^2}\, \mathrm{d}t.$$

Replacing y by $-y$, this equation shows that $y \mapsto \mathrm{e}^{-\frac{|y|}{\alpha}}$ is the characteristic function of the *Cauchy distribution* with density $t \mapsto \frac{1}{\pi} \cdot \frac{\alpha}{1+\alpha^2 t^2}$.

Theorem 1.3.8. *For every $a \in \mathbb{R}^d$ we have*

$$\lim_{T\to\infty} \frac{1}{(2T)^d} \int_{-T}^{T} \cdots \int_{-T}^{T} \mathrm{e}^{-\mathrm{i}(a,t)} f(t)\, \mathrm{d}t = \mu(\{a\}).$$

Proof. Applying Fubini's theorem we obtain

$$\begin{aligned} &\frac{1}{(2T)^d} \int_{-T}^{T} \cdots \int_{-T}^{T} \mathrm{e}^{-\mathrm{i}(a,t)} f(t)\, \mathrm{d}t \\ &= \frac{1}{(2T)^d} \int_{-T}^{T} \cdots \int_{-T}^{T} \mathrm{e}^{-\mathrm{i}(a,t)} \left(\int_{\mathbb{R}^d} \mathrm{e}^{\mathrm{i}(t,x)}\, \mathrm{d}\mu(x) \right) \mathrm{d}t \\ &= \int_{\mathbb{R}^d} \left(\frac{1}{(2T)^d} \int_{-T}^{T} \cdots \int_{-T}^{T} \mathrm{e}^{\mathrm{i}(t,x-a)}\, \mathrm{d}t \right) \mathrm{d}\mu(x) \end{aligned}$$

$$= \int_{\mathbb{R}^d} \left(\prod_{j=1}^{d} \frac{1}{2T} \int_{-T}^{T} e^{it_j \cdot (x_j - a_j)} \, dt_j \right) d\mu(x)$$

$$=: \int_{\mathbb{R}^d} \left(\prod_{j=1}^{d} K_j(x_j, T) \right) d\mu(x)$$

where $K_j(a_j, T) = 1$ and

$$\lim_{T \to \infty} K_j(x_j, T) = \lim_{T \to \infty} \frac{\sin T(x_j - a_j)}{T(x_j - a_j)} = 0, \quad x_j \neq a_j.$$

Hence, the assertion of the theorem follows by dominated convergence. □

Theorem 1.3.9. *Assume that* $\mu = \mu_d + \mu_c$ *where* μ_c *is a continuous nonnegative measure and*

$$\mu_d = \sum_n p_n \delta_{x_n}$$

is a nonnegative discrete measure. Then

$$\lim_{T \to \infty} \frac{1}{(2T)^d} \int_{-T}^{T} \cdots \int_{-T}^{T} |f(t)|^2 \, dt = \sum_n p_n^2.$$

Consequently, μ *is continuous if and only if the limit on the left-hand side is equal to zero.*

Proof. Let X_1 and X_2 be independent random vectors with characteristic function f. Then $|f|^2$ is the characteristic function of $X = X_1 - X_2$ and

$$P(X = 0) = \sum_n P([X_1 = x_n] \cap [X_2 = x_n]) = \sum_n p_n^2.$$

Applying Theorem 1.3.8 with $a = 0$ and with f replaced by $|f|^2$ we obtain the assertion of the theorem. □

We close this section with some simple applications of the uniqueness Theorem 1.3.3.

Theorem 1.3.10. *The random variables* $X_1, \ldots, X_d$ *with characteristic functions* $f_1, \ldots, f_d$ *are independent if and only if*

$$f_X(t) = \prod_{j=1}^{d} f_j(t_j), \quad t \in \mathbb{R}^d \tag{1}$$

where f_X *denotes the characteristic function of the random vector* $X = (X_1, \ldots, X_d)$.

Proof. Denoting by μ_j the distribution of X_j, the random variables X_j's are independent if and only if the distribution of X is equal to the product measure $\mu_1 \times \cdots \times \mu_d$. The statement of the theorem follows from the uniqueness Theorem 1.3.3 and from the fact that the right-hand side of equation (1) is equal to the characteristic function of the product measure $\mu_1 \times \cdots \times \mu_d$. □

Theorem 1.3.11. *A probability measure μ on $\mathbb{R}^d$ is uniquely determined by its values assigned to the half spaces*

$$H_{y,r} := \{x \in \mathbb{R}^d : (y, x) \le r\}, \quad y \in \mathbb{R}^d,\ r \in \mathbb{R}.$$

Proof. Let X be the identity mapping on $\mathbb{R}^d$ and consider X as a random vector on the probability space $(\Omega, \mathcal{A}, P) := (\mathbb{R}^d, \mathcal{B}_d, \mu)$. Then μ is the distribution of X and

$$H_{y,r} = \{\omega \in \Omega : (y, X(\omega)) \le r\}.$$

Thus, the values $\mu(H_{y,r})$, $r \in \mathbb{R}$, determine the distribution (and hence the characteristic function) of the random variable $\omega \mapsto (y, X(\omega))$ for all $y \in \mathbb{R}^d$. On the other hand, the characteristic functions of these random variables determine the characteristic function of X (cf. Remark 1.1.10). This fact together with Theorem 1.3.3 proves the result. □

The next result can be proved in the same way as the previous one. We omit the proof.

Theorem 1.3.12. *The distribution of a random vector X is uniquely determined by the distributions of all orthogonal projections of X on lines going through the origin.*

Recall that a probability distribution μ on $\mathbb{R}^d$ is called *symmetric* if $\mu = \mu^{\triangle}$, i.e., $\mu(B) = \mu(-B)$ for all Borel subsets of $\mathbb{R}^d$. The distribution of a random vector X is symmetric if and only if X and $-X$ have the same distribution.

Theorem 1.3.13. *For a random vector X the following conditions are equivalent.*

(i) *The distribution μ of X is symmetric.*

(ii) *The characteristic function f of X is real-valued.*

(iii) $f(t) = \int_{\mathbb{R}^d} \cos((t, x))\, \mathrm{d}\mu(x), \quad t \in \mathbb{R}^d.$

Proof. If μ is symmetric then X and $-X$ have the same distribution and hence

$$f(t) = \mathbb{E}\left(\mathrm{e}^{\mathrm{i}(t,X)}\right) = \mathbb{E}\left(\mathrm{e}^{\mathrm{i}(t,-X)}\right) = f(-t) = \overline{f(t)}, \quad t \in \mathbb{R}^d.$$

Thus, f is real-valued.

Suppose that f is real-valued. Then $f(t) = \overline{f(-t)} = f(-t)$ for all $t \in \mathbb{R}^d$, hence μ and $\mu^{\triangle}$ have the same characteristic function. Therefore $\mu = \mu^{\triangle}$, by Theorem 1.3.3.

The equivalence of (ii) and (iii) follows from

$$f(t) = \mathbb{E}\left[\cos(t, X)\right] + \mathrm{i}\mathbb{E}\left[\sin(t, X)\right]. \qquad \square$$

1.4 Basic properties of positive definite functions

An essential property of characteristic functions and covariance functions is that they are positive definite. In the present section we investigate positive definiteness on its own. In the main part of the book we deal with positive definite functions on the group $\mathbb{R}^d$, but occasionally we need also the group $\mathbb{Z}^d$. For this reason it is practical to treat the basic properties of these functions in a more general setting. Throughout this section the symbol G denotes an arbitrary commutative group. We use additive notation for the group operation.[12]

We have already shown in Theorem 1.1.2 that every characteristic function is positive definite in the following sense.

Definition 1.4.1. A complex-valued function f on G is said to be *positive definite* if the inequality

$$\sum_{i,j=1}^{n} f(t_i - t_j) c_i \overline{c_j} \geq 0$$

holds for every positive integer n, for all $t_1, \dots, t_n \in G$, and for all $c_1, \dots, c_n \in \mathbb{C}$.

We denote by $P(G)$ the set of all positive definite functions on G while $P^c(\mathbb{R}^d)$ and $P^m(\mathbb{R}^d)$ denote the set of all continuous and Lebesgue measurable positive definite functions on $\mathbb{R}^d$, respectively.

Remark 1.4.2. By the definition above, f is positive definite if and only if the matrix

$$A = \big(f(t_i - t_j)\big)_{i,j=1}^{n}$$

is positive semidefinite[13] for an arbitrary choice of $n \in \mathbb{N}$ and $t_1, \dots, t_n \in G$.

The inequality in Definition 1.4.1 can be formulated in terms of finitely supported complex measures as follows. Let t_j, c_j be as in Definition 1.4.1 and define the meas-

[12] Note that most of the results on positive definite functions presented in this book have their analogues on locally compact groups (see for example [49]). The restriction of considering only commutative groups without topology makes the treatment simpler. On the other hand, the results obtained are sufficient for many applications in probability theory.

[13] Section D.2 contains some basic results on positive semidefinite matrices.

ure ν by

$$\nu = \sum_{j=1}^{n} c_j \delta_{t_j}.$$

Then $\tilde{\nu} = \sum_{j=1}^{n} \overline{c_j} \delta_{-t_j}$ and

$$\nu * \tilde{\nu} * f(t) = \int_G f(t-x)\, \mathrm{d}(\nu * \tilde{\nu})(x) = \sum_{i,j=1}^{n} f(t - t_i + t_j) c_i \overline{c_j}, \quad t \in G.$$

Setting here $t = 0$ and replacing t_i by $-t_i$ we obtain the sum in Definition 1.4.1. We have thus proved the following lemma.

Lemma 1.4.3. *A complex-valued function f on G is positive definite if and only if the inequality*

$$\nu * \tilde{\nu} * f(0) \geq 0$$

holds for all finitely supported complex measures ν on G.

Next we present two examples where we show the positive definiteness of a given function by proving that the matrix A in Remark 1.4.2 is positive semidefinite.

Example 1.4.4. A simple example of a positive definite function on G is given by the function $\mathbf{1}_{\{0\}}$. In this case the matrix A after Definition 1.4.1 is the $n \times n$ identity matrix whenever the t_i's are mutually different.

More generally, if G_0 is a subgroup of G, then $\mathbf{1}_{G_0}$ is positive definite. To see this let F be a non-void finite subset of G. The set F can be written as $F = \cup_{k=1}^{n} F_k$ where the sets F_k are contained in mutually different cosets of G_0. Let n_k denote the number of elements of F_k and choose a numbering $t_1, t_2, \ldots, t_m$ of the elements of F such that $t_i \in F_k$ if and only if

$$\sum_{j=1}^{k-1} n_j < i \leq \sum_{j=1}^{k} n_j$$

where the sum on the left-hand side is interpreted as 0 if $k = 1$. Then

$$\big(f(t_i - t_j)\big)_{i,j=1}^{m} = \begin{bmatrix} A_1 & 0 & & 0 \\ 0 & A_2 & & 0 \\ & & \ddots & \\ 0 & 0 & & A_n \end{bmatrix}$$

where all entries of A_k are equal to 1. It is easy to check that this matrix is positive semidefinite.

We will see in Section 1.7 that the 'building stones' of positive definite functions are the characters defined below.

Definition 1.4.5. A complex-valued function $\gamma \neq 0$ on G is called a *character* of G if $\gamma(x+y) = \gamma(x)\gamma(y)$ and $\gamma(-x) = \overline{\gamma(x)}$ for all $x, y \in G$.

It is easy to check that $\gamma(0) = 1$ and $|\gamma(x)| = 1$ for all $x \in G$ if γ is a character of G.

Example 1.4.6. Every character γ of G is positive definite. Indeed,

$$\sum_{i,j=1}^{n} \gamma(t_i - t_j) c_i \overline{c_j} = \sum_{i,j=1}^{n} \gamma(t_i)\overline{\gamma(t_j)} c_i \overline{c_j} = \left| \sum_{i=1}^{n} c_i \gamma(t_i) \right|^2 \geq 0.$$

Lemma 1.4.7. *Every continuous character of* $\mathbb{R}^d$ *has the form*

$$\gamma(t) = \mathrm{e}^{\mathrm{i}(x,t)}, \quad t \in \mathbb{R}^d$$

with some $x \in \mathbb{R}^d$. *If* χ *is a character of* $\mathbb{Z}^d$ *then there exists* $z \in \mathbb{T}^d$ *such that*

$$\chi(n) = z^n, \quad n \in \mathbb{Z}^d.$$

Proof. The first statement follows immediately from Theorem C.9.2 using the fact that $|\gamma| = 1$.

To prove the second one let $e_1, \ldots, e_d$ be the standard basis of $\mathbb{Z}^d$. If χ is a character of $\mathbb{Z}^d$ and $n = (n_1, \ldots, n_d) \in \mathbb{Z}^d$ then

$$\chi(n) = \chi(n_1 \cdot e_1 + \cdots + n_d \cdot e_d) = \chi(e_1)^{n_1} \cdot \ldots \cdot \chi(e_d)^{n_d} = z^n$$

where $z_i = \chi(e_i) \in \mathbb{T}$. □

Next we prove several simple properties of positive definite functions. From Theorem D.2.3 we obtain the following lemma:

Lemma 1.4.8. *Let* $f \in P(G)$. *Then* f *is* Hermitian, *i.e.,* $f(-x) = \overline{f(x)}$ *holds for all* $x \in G$.

Theorem 1.4.9. *Let* $f_1, f_2 \in P(G)$. *Then the functions* $\overline{f_1}$, $\operatorname{Re} f_1$, $|f_1|^2$, *and* $f_1 \cdot f_2$ *are positive definite. Moreover,* $p_1 f_1 + p_2 f_2$ *is positive definite for all* $p_1, p_2 \geq 0$.

Proof. The positive definiteness of $\overline{f_1}$ and $p_1 f_1 + p_2 f_2$ follow at once from Definition 1.4.1. We therefore have $\operatorname{Re} f_1 = \frac{1}{2}(f_1 + \overline{f_1}) \in P(G)$. The fact that $f_1 f_2$ is positive definite is an immediate consequence of Schur's theorem stating that the pointwise product of nonnegative definite matrices is nonnegative definite (cf. Theorem D.2.12). It follows that $|f_1|^2 = f_1 \overline{f_1} \in P(G)$. □

The next theorem follows immediately from Definition 1.4.1.

Theorem 1.4.10. *If a sequence (or net) of positive definite functions converges pointwise to a function f then f is positive definite.*

As a consequence of Theorems 1.4.9 and 1.4.10 we obtain the following corollary:

Corollary 1.4.11. *The set $P(G)$ is a convex cone closed in the topology of pointwise convergence.*

Next we prove some frequently used inequalities.[14]

Theorem 1.4.12. *Let $f \in P(G)$. Then*

(i) $|f(x)| \le f(0)$ *for all* $x \in G$;

(ii) $|f(x) - f(y)|^2 \le 2f(0)[f(0) - \operatorname{Re} f(x-y)]$ *for all* $x, y \in G$;

(iii) $|f(0)f(x+y) - f(x)f(y)|^2 \le [f(0)^2 - |f(x)|^2][f(0)^2 - |f(y)|^2]$ *for all* $x, y \in G$;

(iv) $\left|\sum_{i=1}^{n} a_i f(x_i)\right|^2 \le f(0) \sum_{i,j=1}^{n} f(x_i - x_j) a_i \overline{a_j}$ *for an arbitrary choice of* $n \in \mathbb{N}$, $x_1, \dots, x_n \in G$, *and* $a_1, \dots, a_n \in \mathbb{C}$;

(v) $$\left|\sum_{i=1}^{n}\sum_{j=1}^{m} f(x_i - y_j) a_i \overline{b_j}\right|^2 \le \left(\sum_{i,j=1}^{n} f(x_i - x_j) a_i \overline{a_j}\right)\left(\sum_{i,j=1}^{m} f(y_i - y_j) b_i \overline{b_j}\right)$$
for all $n, m \in \mathbb{N}$, $x_1, \dots, x_n, y_1, \dots, y_m \in G$, *and for all complex numbers* $a_1, \dots, a_n, b_1, \dots, b_m$.

Proof. We prove (v) first. Denote by $\mathcal{L}$ the complex linear space of all functions of the form $g = \sum_1^n a_i \mathbf{1}_{\{x_i\}}$, $a_i \in \mathbb{C}$, $x_i \in G$. For such g and $h = \sum_1^m b_i \mathbf{1}_{\{y_i\}} \in \mathcal{L}$ we define the inner product (g, h) by

$$(g, h) := \sum_{i=1}^{n}\sum_{j=1}^{m} f(x_i - y_j) a_i \overline{b_j}.$$

It is easy to check that this definition is correct, i.e., (g, h) does not depend on the particular representations of g and h. As $f \in P(G)$, this inner product is nonnegative definite. Thus, the Cauchy–Schwarz inequality $|(g, h)|^2 \le (g, g) \cdot (h, h)$ holds, and this is the same as (v).
Setting $m = 1$, $y_1 = 0$, and $b_1 = 1$ in (v), we get (iv).

The inequality (ii) follows readily from (iv) by setting $n = 2$, $x_1 = x$, $x_2 = y$, $a_1 = 1, a_2 = -1$. Putting $n = 1$, $x_1 = x$, and $a_1 = 1$ in (iv), we get $|f(x)|^2 \le f(0)^2$ and, since $f(0) = (\mathbf{1}_{\{0\}}, \mathbf{1}_{\{0\}}) \ge 0$, this gives $|f(x)| \le f(0)$.

---

[14] See also Theorem 1.11.6.

To prove (iii) we first remark that if a matrix of the form

$$\begin{pmatrix} 1 & a & b \\ \overline{a} & 1 & c \\ \overline{b} & \overline{c} & 1 \end{pmatrix}$$

is positive semidefinite, then its determinant is nonnegative (cf. Theorem D.2.9). Computing this determinant we obtain

$$1 + a\overline{b}c + \overline{a}b\overline{c} \geq |a|^2 + |b|^2 + |c|^2$$

or equivalently,

$$|c - \overline{a}b|^2 \leq (1 - |a|^2)(1 - |b|^2). \tag{1}$$

Assume first that $f(0) = 1$. Applying (1) to the (Hermitian) matrix

$$\big(f(x_i - x_j)\big)_{i,j=1}^3 = \begin{pmatrix} 1 & f(-x) & f(y) \\ f(x) & 1 & f(x+y) \\ f(-y) & f(-x-y) & 1 \end{pmatrix}$$

where $x_1 = 0$, $x_2 = x$, and $x_3 = -y$, we obtain (iii). If $f(0) = 0$, then $f = 0$ by (i) and so (iii) is satisfied. The case $f(0) \neq 0$ can be reduced to the case $f(0) = 1$ by considering the positive definite function $f/f(0)$. □

Corollary 1.4.13. *Let $f \in P(G)$ be such that $f(0) = 1$ and write*

$$G_0 := \{x \in G : |f(x)| = 1\}.$$

Then G_0 is a subgroup of G and

$$f(x + y) = f(x)f(y)$$

for all $x \in G_0$ and $y \in G$.

Proof. This follows immediately from inequality (1.4.12.iii) and Lemma 1.4.8. □

Corollary 1.4.14. *Let $f \in P(\mathbb{R})$ be such that $f(0) = 1$. If there exist $n \in \mathbb{N}$ and $h > 0$ such that $f(h)^n = 1$ then f is periodic with period nh.*

Proof. The case $n = 1$ follows readily from Corollary 1.4.13. Using this, the general case is obtained by noting that $|f(h)| = 1$ and therefore, by Corollary 1.4.13, $f(nh) = f(h)^n = 1$. □

The functions

$$t \mapsto \mathrm{e}^{\mathrm{i}(t,x)}, \quad t \mapsto \cos((t,x)) \quad \text{and} \quad t \mapsto \mathrm{e}^{-\|t\|^2}, \quad t, x \in \mathbb{R}^d$$

provide examples for the subgroup G_0 in the previous corollary. Example 1.4.4 shows that for an arbitrary subgroup G_0 of G there exists $f \in P(G)$ such that $G_0 = \{x \in G : |f(x)| = 1\}$.

A simple corollary of inequality (1.4.12.iii) is the following:

Corollary 1.4.15. *Let $f_n \in P(G)$ be such that $f_n(0) = 1$, $n \in \mathbb{N}$. Then the set of all $x \in G$ with $f_n(x) \to 1$, $n \to \infty$, is a subgroup of G. In particular, if $G = \mathbb{R}^d$ and $f_n \to 1$ on a neighborhood of 0 then $f_n \to 1$ on $\mathbb{R}^d$.*

Next we list some simple results that can be used to construct positive definite functions.

Lemma 1.4.16. *Let $f_i \in P(G_i)$ where G_i, $i = 1, 2$, is a commutative group. Then the function*

$$(x_1, x_2) \mapsto f_1(x_1) f_2(x_2)$$

is positive definite on the product group $G_1 \times G_2$.

Proof. The lemma follows immediately from the fact that the entrywise product of positive semidefinite matrices is positive semidefinite (see Schur's Theorem D.2.12). □

Lemma 1.4.17. *Let g be a given positive definite function on G and let a_n, $n \in \mathbb{N}_0$, be nonnegative numbers such that the series*

$$\varphi(z) := \sum_{n=0}^{\infty} a_n z^n, \quad z \in \mathbb{C}$$

is convergent whenever $|z| \le g(0)$. Then the function $f(t) := \varphi(g(t))$, $t \in G$, is positive definite.

Proof. By (1.4.12.i), we have $|g(t)| \le g(0)$, and hence the definition of f is correct. The positive definiteness of f follows from Theorems 1.4.9 and 1.4.10. □

For example, the functions e^g and $1/\,(2g(0) - g)$, are positive definite if $g \in P(G)$ and $g(0) \neq 0$.

Lemma 1.4.18. *Let $f \in P(G)$ and let*

$$\mu = \sum_{j=1}^{n} c_j \delta_{t_j}, \quad n \in \mathbb{N},\ c_j \in \mathbb{C},\ t_j \in G$$

be an arbitrary complex measure with finite support. Then the function

$$g(t) = \mu * \tilde{\mu} * f(t) = \sum_{i,j=1}^{n} f(t - t_i + t_j) c_i \overline{c_j}, \quad t \in G$$

is positive definite. In particular, the function f_c^y defined by

$$f_c^y(t) = (1 + |c|^2) f(t) + c f(t + y) + \overline{c} f(t - y), \quad t \in G$$

is positive definite for all $y \in G$ and $c \in \mathbb{C}$.

Proof. Let ν be an arbitrary finitely supported complex measure on G. The first statement of the lemma follows from the equation

$$\nu * \tilde{\nu} * g(0) = (\nu * \mu) * (\nu * \mu)^{\sim} * f(0),$$

using Lemma 1.4.3 and the positive definiteness of f. Setting $\mu = \delta_0 + \overline{c}\delta_y$, we obtain the second statement. □

It follows from Corollary 1.4.11 that the set

$$P_0(G) := \{f \in P(G) : f(0) = 1\}$$

is convex. Using the previous lemma we are now able to identify the extreme points (cf. D.4.1) of this set. First we prove a simple but useful statement.

Lemma 1.4.19. *If f is an extreme point of $P_0(G)$ and $f = f_1 + f_2$ where $f_i \in P(G)$, then $f_i = f_i(0) \cdot f$, $i = 1, 2$.*

Proof. If $f_1(0) = 0$, then $f_1 = 0$ (cf. (1.4.12.i)) and hence $f_2 = f$. The case $f_2(0) = 0$ can be treated in the same way. If $f_1(0)$ and $f_2(0)$ are different from zero then $f_i / f_i(0) \in P_0(G)$ and

$$f = f_1(0) \cdot \frac{f_1}{f_1(0)} + f_2(0) \cdot \frac{f_2}{f_2(0)}.$$

Since f is extremal and $1 = f(0) = f_1(0) + f_2(0)$ we conclude that $f_i / f_i(0)$ is equal to f. □

Theorem 1.4.20. *A complex-valued function γ on G is an extreme point of $P_0(G)$ if and only if it is a character of G.*

Proof. Let $\gamma \in P_0(G)$ be extremal. For every $y \in G$ and $c \in \mathbb{C}$, the function

$$\gamma_c^y(x) = (1 + |c|^2)\gamma(x) + c\gamma(x + y) + \bar{c}\gamma(x - y)$$

is positive definite by Lemma 1.4.18. Moreover,

$$\gamma = \frac{1}{2(1+|c|^2)} \cdot (\gamma_c^y + \gamma_{-c}^y) .$$

Since γ is extremal the previous lemma shows that $\gamma_c^y(x) = g(c,y)\gamma(x)$ with a certain function g. Thus,

$$\gamma_c^y(x) - \gamma_{-c}^y(x) = 2[c\gamma(x+y) + \bar{c}\gamma(x-y)] = g_1(c,y)\gamma(x), \tag{1}$$

where $g_1(c,y) = g(c,y) - g(-c,y)$. Setting $c = 1$ and $c = \mathrm{i}$ in (1) , we get

$$\gamma(x+y) = l(y)\gamma(x) \tag{2}$$

with

$$l(y) = \frac{1}{4} \cdot [g_1(1,y) - \mathrm{i}g_1(\mathrm{i},y)].$$

Setting $x = 0$ in (2) we obtain $\gamma(y) = l(y)$, and therefore

$$\gamma(x+y) = \gamma(x)\gamma(y), \quad x, y \in G.$$

Noting that positive definite functions are Hermitian we see that γ is a character of G.

We now prove that every character γ is an extreme point of $P_0(G)$. We have already seen in Example 1.4.6 that γ is positive definite. Assume that $\gamma = p\gamma_1 + (1-p)\gamma_2$, where $\gamma_1, \gamma_2 \in P_0(G)$ and $0 < p < 1$. It follows immediately from the relations $|\gamma| = 1$, $|\gamma_1| \le 1$ and $|\gamma_2| \le 1$ that $\gamma = \gamma_1 = \gamma_2$, i.e., γ is extremal. □

1.5 Further properties of positive definite functions on $\mathbb{R}^d$

We continue the investigation of positive definite functions, but now we make use of the special structure of $\mathbb{R}^d$.

Lemma 1.5.1. *Let $f \in P^c(\mathbb{R}^d)$ such that $|f(t)| = 1$ holds for all t from a neighborhood of $0 \in \mathbb{R}^d$. Then*

$$f(t) = \mathrm{e}^{\mathrm{i}(x,t)}, \quad t \in \mathbb{R}^d$$

for some $x \in \mathbb{R}^d$.

Proof. If a subgroup of $\mathbb{R}^d$ contains a neighborhood of the zero, then it is equal to $\mathbb{R}^d$. The statement follows therefore from Corollary 1.4.13 and Theorem C.9.2. □

Corollary 1.5.2. *Let $f \in P(\mathbb{R}^d)$. If* $\operatorname{Re} f$ *is continuous at* 0, *then f is uniformly continuous on $\mathbb{R}^d$.*

Proof. This is a consequence of (1.4.12.ii). □

The next two results can be applied to construct positive definite functions.

Lemma 1.5.3. *Let $f \in P^c(\mathbb{R}^d)$ and let μ be an arbitrary complex measure on $\mathcal{B}_d$. Then the function*

$$\mu * \tilde{\mu} * f(t) = \int_{\mathbb{R}^d} \int_{\mathbb{R}^d} f(t-x-y)\,\mathrm{d}\mu(x)\,\mathrm{d}\tilde{\mu}(y), \quad t \in \mathbb{R}^d$$

is positive definite. In particular, the function

$$t \mapsto \int_{\mathbb{R}^d} \int_{\mathbb{R}^d} f(t+x-y)h(x)\overline{h(y)}\,\mathrm{d}x\,\mathrm{d}y, \quad t \in \mathbb{R}^d$$

is positive definite for all $h \in \mathcal{L}^1(\mathbb{R}^d)$.

Proof. By Lemma 1.4.18, the statement is true if μ has finite support. Let μ now be arbitrary, and let $\{\mu_\alpha\}$ be a net of finitely supported complex measures converging weakly to μ (cf. Theorem E.1.8). By Theorem E.2.5, the net $\{\mu_\alpha * \tilde{\mu}_\alpha\}$ converges weakly to $\mu * \tilde{\mu}$, and hence $\{\mu_\alpha * \tilde{\mu}_\alpha * f\}$ converges pointwise to $\mu * \tilde{\mu} * f$. Thus, the first assertion follows from Theorem 1.4.10. The second assertion follows from the first one by setting $\mathrm{d}\mu(x) = h(-x)\,\mathrm{d}x$. □

Lemma 1.5.4. *Let $g \in \mathcal{L}^2(\mathbb{R}^d)$. Then the function*

$$f(t) := g * \tilde{g}(t) = \int_{\mathbb{R}^d} g(t+y)\overline{g(y)}\,\mathrm{d}y, \quad t \in \mathbb{R}^d$$

belongs to $P^c(\mathbb{R}^d)$ and it vanishes at infinity.

Proof. For all $t_1, \dots, t_n \in \mathbb{R}^d$, and for all $c_1, \dots, c_n \in \mathbb{C}$ we have

$$\begin{aligned}
\sum_{i,j=1}^{n} f(t_i - t_j)c_i\overline{c_j} &= \int_{\mathbb{R}^d} \sum_{i,j=1}^{n} g(t_i - t_j + y)\overline{g(y)}c_i\overline{c_j}\,\mathrm{d}y \\
&= \int_{\mathbb{R}^d} \sum_{i,j=1}^{n} g(t_i + y)\overline{g(y + t_j)}c_i\overline{c_j}\,\mathrm{d}y \\
&= \int_{\mathbb{R}^d} \left| \sum_{i=1}^{n} g(t_i + y)c_i \right|^2 \mathrm{d}y \ge 0
\end{aligned}$$

i.e., f is positive definite. The remaining statements follow from Theorem E.2.4. □

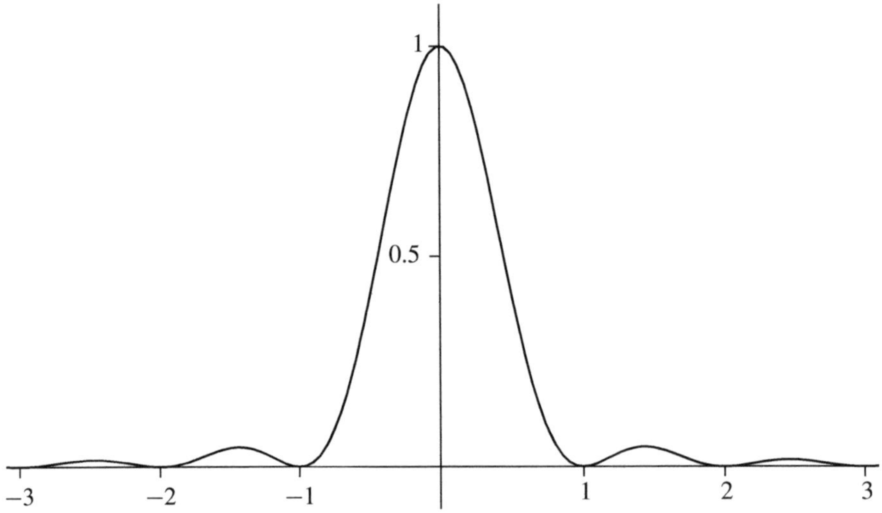

Figure 1.6. The function p_a from Remark 1.5.5 with $a = 2\pi$.

Remark 1.5.5. Setting $d = 1$ and $g = \frac{1}{\sqrt{a}} \cdot \mathbf{1}_{[-a/2,a/2]}$, $a > 0$, in the previous lemma we see that the function

$$\psi_a(t) := \begin{cases} 1 - \frac{|t|}{a} & \text{if } |t| \le a \\ 0 & \text{if } |t| > a \end{cases}, \qquad t \in \mathbb{R}$$

is positive definite. We show that the function

$$p_a(t) = \frac{a}{2\pi} \cdot \left[\frac{\sin(ta/2)}{ta/2}\right]^2, \qquad t \in \mathbb{R}$$

is a density and the corresponding characteristic function is ψ_a.
Indeed, let X and Y be independent random variables having $\frac{1}{a} \cdot \mathbf{1}_{[-a/2,a/2]}$ as their density. The characteristic functions of X and Y are given by

$$f_X(t) = f_Y(t) = \frac{\sin(ta/2)}{ta/2}.$$

In view of Theorem 1.1.3, the characteristic function of their sum is

$$f_{X+Y}(t) = f_X(t) \cdot f_Y(t) = \left[\frac{\sin(ta/2)}{ta/2}\right]^2 = \frac{2\pi}{a} \cdot p_a(t).$$

Applying (F.1.2), a short computation shows that the density of $X+Y$ is equal to $\frac{1}{a} \cdot \psi_a$. From Theorem 1.3.6 with f replaced by f_{X+Y} we conclude that p_a is a density and the corresponding characteristic function is ψ_a.[15]

[15] The function ψ_a is an example of the so-called *Pólya-type* characteristic functions, which we will treat in more detail in Section 3.9.

Lemma 1.5.6. *The following statements hold.*[16]

(i) *There exists a sequence* $\{g_n\}_1^\infty$ *of functions* $g_n \in P^c(\mathbb{R}^d)$ *of the form* $g_n = h_n * \tilde{h}_n$ *where* $h_n \in C_{00}(\mathbb{R}^d)$ *such that* $g_n(0) = 1$ *and* $\{g_n\}_1^\infty$ *tends to* 1 *uniformly on compact sets.*

(ii) *There exists a sequence* $\{g_n\}_1^\infty$ *of functions* $g_n \in P(\mathbb{Z}^d)$ *of the form* $g_n = h_n * \tilde{h}_n$ *where* h_n *is a complex-valued function on* $\mathbb{Z}^d$ *with finite support such that* $g_n(0) = 1$ *and* $\{g_n\}_1^\infty$ *tends pointwise to* 1.

Proof. (i) For $n \in \mathbb{N}$ let $S_n := \{t \in \mathbb{R}^d : 0 \le t_j \le n,\ j = 1, \dots, d\}$ and

$$h_n := \frac{1}{\sqrt{\lambda(S_n)}} \cdot \mathbf{1}_{S_n}.$$

For all $r \in [0, n]$ we have

$$\lambda(S_n \cap (S_n + x)) \ge (n - r)^d, \quad x \in \mathbb{R}^d,\ |x_i| \le r.$$

Using this inequality we see that the sequence $\{h_n * \tilde{h}_n\}_1^\infty$ has the desired properties.

(ii) We define $S_n := \{t \in \mathbb{Z}^d : 0 \le t_j \le n,\ j = 1, \dots, d\}$ and

$$h_n := \frac{1}{\sqrt{|S_n|}} \cdot \mathbf{1}_{S_n}$$

where $|\cdot|$ denotes the counting measure. For all $r \in [0, n]$ we then have

$$|S_n \cap (S_n + x)| \ge (n + 1 - r)^d, \quad x \in \mathbb{Z}^d,\ |x_i| \le r$$

and hence the sequence $\{h_n * \tilde{h}_n\}_1^\infty$ has the desired properties. □

The next lemma follows immediately from the definition of positive definiteness, we omit the proof.

Lemma 1.5.7. *Let* $f \in P(\mathbb{R}^d)$ *be Borel measurable and let* ν *be a finite, nonnegative measure on* $\mathcal{B}_1$. *Then the function* g *defined by*

$$g(t) = \int_{\mathbb{R}} f(xt)\, \mathrm{d}\nu(x), \quad t \in \mathbb{R}^d$$

is positive definite. In particular, the function $t \mapsto f(xt),\ t \in \mathbb{R}^d$, *is positive definite for all* $x \in \mathbb{R}$.[17]

---

[16] See also Lemma 3.2.3 and Lemma 3.2.4.

[17] See also Theorem 1.1.6.

Lemma 1.5.8. *A continuous function f on $\mathbb{R}^d$ is positive definite if and only if the inequality*

$$\int_{\mathbb{R}^d}\int_{\mathbb{R}^d} f(x-y)h(x)\overline{h(y)}\,\mathrm{d}x\,\mathrm{d}y \ge 0$$

holds for arbitrary functions $h \in C_{00}(\mathbb{R}^d)$.

Proof. The 'only if part' follows from Lemma 1.5.3 with $t = 0$ and $\mathrm{d}\mu(x) = h(x)\,\mathrm{d}x$, using the fact that positive definite functions are nonnegative at 0.

The assumption of the 'if part' means that $h * \tilde{h} * f(0) \ge 0$ for all $h \in C_{00}(\mathbb{R}^d)$. Let μ be an arbitrary finitely supported complex measure. It follows from Theorems E.1.9 and E.2.5 that there exists a sequence $\{h_n\}_1^\infty$ of functions $h_n \in C_{00}(\mathbb{R}^d)$ such that $\mu_n * \tilde{\mu}_n$ converges weakly to $\mu * \tilde{\mu}$, where $\mathrm{d}\mu_n(x) = h_n(x)\,\mathrm{d}x$. From this we conclude that $\mu * \tilde{\mu} * f(0) \ge 0$. Thus, the positive definiteness of f follows from Lemma 1.4.3. □

Theorem 1.5.9. *If $f \in P^c(\mathbb{R}^d) \cap \mathcal{L}^1(\mathbb{R}^d)$, then*[18]

$$\int_{\mathbb{R}^d} f(x)\mathrm{e}^{\mathrm{i}(t,x)}\,\mathrm{d}x \ge 0, \quad t \in \mathbb{R}^d.$$

If $f \in P(\mathbb{Z}^d) \cap \mathcal{L}^1(\mathbb{Z}^d)$, then

$$\sum_{n\in\mathbb{Z}^d} f(n)\mathrm{e}^{\mathrm{i}(t,n)} \ge 0, \quad t \in \mathbb{R}^d.$$

Proof. Since the integrand is positive definite for all fixed t, it suffices to prove that

$$\int_{\mathbb{R}^d} f(x)\,\mathrm{d}x \ge 0$$

for all $f \in P^c(\mathbb{R}^d) \cap \mathcal{L}^1(\mathbb{R}^d)$. To show this inequality we choose a sequence $\{g_n\}$ as in Lemma 1.5.6. Then $|fg_n| \le |f|$ and Lebesgue's theorem on dominated convergence shows that

$$\int_{\mathbb{R}^d} f(x)\,\mathrm{d}x = \lim_{n\to\infty}\int_{\mathbb{R}^d} f(x)g_n(x)\,\mathrm{d}x \ge 0.$$

Applying Fubini's theorem and Lemma 1.5.8 we obtain

$$\begin{aligned}\int_{\mathbb{R}^d} f(x)g_n(x)\,\mathrm{d}x &= \int_{\mathbb{R}^d} f(x)\int_{\mathbb{R}^d} h_n(x+y)\overline{h_n(y)}\,\mathrm{d}y\,\mathrm{d}x\\ &= \int_{\mathbb{R}^d}\int_{\mathbb{R}^d} f(x)h_n(x+y)\overline{h_n(y)}\,\mathrm{d}x\,\mathrm{d}y\\ &= \int_{\mathbb{R}^d}\int_{\mathbb{R}^d} f(x-y)h_n(x)\overline{h_n(y)}\,\mathrm{d}x\,\mathrm{d}y \ge 0.\end{aligned}$$

The second statement can be proved in the same way by replacing the integrals by sums. □

[18] See Theorems 1.8.13 and 1.9.6 for closely related results.

The next lemma will be used in the proof of Theorem 1.5.11.

Lemma 1.5.10. *Let $K \subset \mathbb{R}^d$ be compact and let $\{g_n\}_1^\infty$ be a uniformly bounded sequence of Lebesgue measurable functions on $\mathbb{R}^d$ converging λ-almost everywhere to a function g. Then the sequence $\{\varphi_n\}_1^\infty$ with*

$$\varphi_n(t) = \int_K g_n(t+x)\,\mathrm{d}x, \quad t \in \mathbb{R}^d$$

converges uniformly on every compact set to the function φ defined by

$$\varphi(t) = \int_K g(t+x)\,\mathrm{d}x, \quad t \in \mathbb{R}^d.$$

Proof. We may assume that $\lambda(K) > 0$, $g = 0$ and $|g_n| \le 1$. Suppose, on the contrary, that for some compact set $C \subset \mathbb{R}^d$ the sequence $\{\varphi_n\}$ does not converge uniformly to 0 on C. Then there exist a positive number δ, a sequence $\{t_n\}_1^\infty \subset C$, and a subsequence $\{\varphi_{k_n}\}_1^\infty$ such that $|\varphi_{k_n}(t_n)| \ge \delta$. We claim that the Lebesgue measure of the set

$$S_n = \left\{x \in K : |g_{k_n}(t_n + x)| \ge \frac{\delta}{2\lambda(K)}\right\}$$

is at least $\frac{\delta}{2}$. Indeed, the inequality $\lambda(S_n) < \frac{\delta}{2}$ would imply that

$$\begin{aligned} |\varphi_{k_n}(t_n)| &\le \int_{S_n} |g_{k_n}(t_n+x)|\,\mathrm{d}x + \int_{K\setminus S_n} |g_{k_n}(t_n+x)|\,\mathrm{d}x \\ &\le \lambda(S_n) + \lambda(K)\cdot\frac{\delta}{2\lambda(K)} < \delta \end{aligned}$$

contradicting our choice of t_n and k_n. Thus, $|g_{k_n}(y)| \ge \frac{\delta}{2\lambda(K)} > 0$ for all $y \in \tilde{S}_n := S_n + t_n \subset K + C$ and $\lambda(\tilde{S}_n) = \lambda(S_n) \ge \frac{\delta}{2}$. Consequently,

$$\lambda(\limsup \tilde{S}_n) = \lambda\left(\bigcap_{n=1}^\infty \bigcup_{m=n}^\infty \tilde{S}_m\right) \ge \frac{\delta}{2}$$

which is a contradiction since the sequence $\{g_n\}_1^\infty$ does not converge to 0 on the set $\limsup \tilde{S}_n$ which is contained in the compact set $K + C$. □

We close this section with an important result concerning the convergence of positive definite functions.

Theorem 1.5.11. *If a sequence $\{f_n\}_1^\infty$ of functions $f_n \in P^c(\mathbb{R}^d)$ converges pointwise to a continuous function f, then the convergence is uniform on every compact subset of $\mathbb{R}^d$.*

Proof. Let $C \subset \mathbb{R}^d$ be an arbitrary compact subset. For a bounded complex-valued function h on C write $\|h\|_C := \sup\{\,|h(t)| : t \in C\}$. Using this notation we have to show that $\|f_n - f\|_C \to 0$.

We may assume that $f_n(0) = f(0) = 1$. Since f is continuous at 0, for every $\varepsilon > 0$ there exists a compact set K with positive Lebesgue measure satisfying

$$\frac{2}{\lambda(K)} \int_K [1 - \operatorname{Re} f(x)]\,\mathrm{d}x < \frac{\varepsilon^2}{9}. \tag{1}$$

Since $f_n(x) \to f(x)$, this inequality also holds for f_n if n is sufficiently large, say $n \geq N_0$. Define φ and φ_n as in Lemma 1.5.10 with g and g_n replaced by f and f_n, respectively. We show that

$$\left\| \frac{1}{\lambda(K)}\varphi - f \right\|_C < \frac{\varepsilon}{3} \quad \text{and} \quad \left\| \frac{1}{\lambda(K)}\varphi_n - f_n \right\|_C < \frac{\varepsilon}{3}. \tag{2}$$

Indeed, using the Cauchy–Schwarz inequality, (1.4.12.ii), and (1) we obtain

$$\begin{aligned}
\left| f(t) - \frac{1}{\lambda(K)}\varphi(t) \right|^2 &= \left| \frac{1}{\lambda(K)} \int_K f(t)\,\mathrm{d}x - \frac{1}{\lambda(K)} \int_K f(t+x)\,\mathrm{d}x \right|^2 \\
&= \frac{1}{\lambda(K)^2} \left| \int_K [f(t) - f(t+x)]\,\mathrm{d}x \right|^2 \\
&\leq \frac{1}{\lambda(K)^2} \left(\int_K 1\,\mathrm{d}x \right) \cdot \left(\int_K |f(t) - f(t+x)|^2\,\mathrm{d}x \right) \\
&\leq \frac{2}{\lambda(K)} \int_K [1 - \operatorname{Re} f(x)]\,\mathrm{d}x \leq \frac{\varepsilon^2}{9}.
\end{aligned}$$

The second inequality in (2) can be proved in the same way.

By Lemma 1.5.10 (note that $|f_n(t)| \leq f_n(0) = 1$), there exists $N \geq N_0$ such that

$$\left\| \frac{1}{\lambda(K)}\varphi_n - \frac{1}{\lambda(K)}\varphi \right\|_C < \frac{\varepsilon}{3} \quad \text{if} \quad n \geq N. \tag{3}$$

From (2) and (3) we infer that $\|f - f_n\|_C < \varepsilon$, $n \geq N$. □

1.6 Lévy's continuity theorem

We now give necessary and sufficient conditions for the weak convergence of probability distributions in terms of their characteristic functions.

Theorem 1.6.1. *If a sequence $\{\mu_n\}_1^\infty$ of probability distributions on $\mathbb{R}^d$ converges weakly to a probability distribution μ, then the sequence $\{f_n\}_1^\infty$ of the corresponding characteristic functions converges uniformly on every compact set to the characteristic function of μ.*

Proof. By weak convergence (cf. Theorem E.1.7),

$$\lim_{n\to\infty} f_n(t) = \lim_{n\to\infty} \int_{\mathbb{R}^d} \mathrm{e}^{\mathrm{i}(t,x)}\, \mathrm{d}\mu_n(x) = \int_{\mathbb{R}^d} \mathrm{e}^{\mathrm{i}(t,x)}\, \mathrm{d}\mu(x), \quad t \in \mathbb{R}^d.$$

That the convergence is uniform on compact sets follows from Theorem 1.5.11. □

Lemma 1.6.2. *Let f be the characteristic function of a distribution μ on $\mathbb{R}^d$. For $a > 0$ we then have*

$$\mu(K_a) \geq 1 - 7 \cdot \sup_{t \in K_{1/a}} [1 - \operatorname{Re} f(t)]$$

where $K_a = [-a, a]^d$.

Proof. Using Fubini's theorem and the example after Definition 1.1.1 we obtain

$$\begin{aligned}
\left(\frac{2}{a}\right)^d \cdot \sup_{t \in K_{1/a}} [1 - \operatorname{Re} f(t)] &\geq \int_{K_{1/a}} [1 - \operatorname{Re} f(t)]\, \mathrm{d}t \\
&= \int_{K_{1/a}} \int_{\mathbb{R}^d} [1 - \cos((t, y))]\, \mathrm{d}\mu(y)\, \mathrm{d}t \\
&= \int_{\mathbb{R}^d} \int_{K_{1/a}} [1 - \cos((t, y))]\, \mathrm{d}t\, \mathrm{d}\mu(y) \\
&= \left(\frac{2}{a}\right)^d \cdot \int_{\mathbb{R}^d} \left[1 - \prod_{j=1}^d \frac{\sin y_j/a}{y_j/a}\right] \mathrm{d}\mu(y) \\
&\geq \left(\frac{2}{a}\right)^d \cdot \int_{\mathbb{R}^d \setminus K_a} \left[1 - \prod_{j=1}^d \frac{\sin y_j/a}{y_j/a}\right] \mathrm{d}\mu(y)
\end{aligned}$$

If $y \in \mathbb{R}^d \setminus K_a$, then $|y_j/a| \geq 1$ for at least one j. Since

$$\frac{\sin x}{x} < \frac{101}{120} < \frac{6}{7} \quad \text{if} \quad |x| \geq 1 \quad \text{and} \quad \frac{\sin x}{x} < 1 \quad \text{if} \quad x \in \mathbb{R} \setminus \{0\}$$

(cf. (B.1.6.5) and (B.1.6.1)), the last integrand is not less than $\frac{1}{7}$ on the set $\mathbb{R}^d \setminus K_a$. We thus obtain

$$\sup_{t \in K_{1/a}} [1 - \operatorname{Re} f(t)] \geq \frac{1}{7} \cdot [1 - \mu(K_a)],$$

from which the lemma follows. □

The next theorem is frequently applied in probability theory.

Theorem 1.6.3 (Lévy's continuity theorem). *Let $\{\mu_n\}_1^\infty$ be a sequence of probability distributions on $\mathbb{R}^d$ such that the sequence $\{f_n\}_1^\infty$ of the corresponding characteristic* *functions converges pointwise to a continuous function f. Then f is the characteristic function of a probability distribution μ and the sequence $\{\mu_n\}_1^\infty$ converges weakly to μ.*

Proof. In view of Theorem E.1.13, the sequence $\{\mu_n\}_{n=1}^\infty$ contains a subsequence $\{\mu_{n_k}\}_{k=1}^\infty$ converging vaguely to some finite nonnegative measure μ satisfying $\mu(\mathbb{R}^d) \le 1$. We show that $\mu(\mathbb{R}^d) = 1$.

By Lemma 1.6.2, the inequality

$$\mu_{n_k}(K_m) \ge 1 - 7 \cdot \sup_{t \in K_{1/m}} [1 - \operatorname{Re} f_{n_k}(t)]$$

holds for all k and $m \in \mathbb{N}$. Taking lim sup with respect to k and using Theorem E.1.11 we obtain

$$\mu(K_m) \ge \limsup_k \mu_{n_k}(K_m) \ge 1 - 7 \cdot \sup_{t \in K_{1/m}} [1 - \operatorname{Re} f(t)].$$

Now letting $m \to \infty$ and noting that f is continuous and $f(0) = 1$, we see that $\mu(\mathbb{R}^d) \ge 1$. Thus, $\mu(\mathbb{R}^d) = 1$ and hence $\{\mu_{n_k}\}_{k=1}^\infty$ converges weakly to μ (cf. Theorem E.1.12).

From Theorem 1.6.1 we conclude that f is the characteristic function of μ. Since μ is uniquely determined by f, we see that every weakly convergent subsequence of $\{\mu_n\}_1^\infty$ has the same limit μ. This shows that the whole sequence converges weakly to μ. □

Remark 1.6.4. In view of Theorem 1.4.10 and Corollary 1.5.2, it suffices to assume continuity of $\operatorname{Re} f$ at 0 in the previous theorem. This assumption cannot be dropped. Indeed, with the notation introduced in Remark 1.5.5, $\lim_{n\to\infty} \psi_{1/n} = \mathbf{1}_{\{0\}}$. The limit is not a characteristic function though it is continuous at every point different from zero.

The next theorem is usually called the *Cramér–Wold device*.

Theorem 1.6.5. *A sequence $\{X_n\}$ of d-dimensional random vectors converges in distribution to a random vector X if and only if for all $a \in \mathbb{R}^d$ the sequence $\{(a, X_n)\}$ of random variables converges in distribution to (a, X).*

Proof. Suppose that $X_n \to X$ in distribution. For any $a \in \mathbb{R}^d$ the function $x \mapsto (a, x)$ is continuous on $\mathbb{R}^d$. Applying Corollary E.1.16 we see that $(a, X_n) \to (a, X)$ in distribution.

Conversely, suppose that $(a, X_n) \to (a, X)$ in distribution for all $a \in \mathbb{R}^d$. Denoting by $f_{a,n}$ and f_a the corresponding characteristic functions, Theorem 1.6.1 shows that

$$\lim_n f_{a,n}(s) = f_a(s), \quad s \in \mathbb{R}.$$

Using Remark 1.1.10 we obtain

$$f_{X_n}(a) = f_{a,n}(1) \to f_a(1) = f_X(a), \quad a \in \mathbb{R}^d.$$

Hence, by Lévy's continuity theorem, $X_n \to X$ in distribution. □

The next result will be used in the proof of Schoenberg's Theorem 3.8.5.

Proposition 1.6.6. *Let p, p_n, $n \in \mathbb{N}$, be probability densities on $\mathbb{R}^d$ with corresponding characteristic functions f_n and f, respectively. If $p_n \to p$ pointwise, then $f_n \to f$ uniformly on $\mathbb{R}^d$.*

Proof. We have

$$\begin{aligned} |f_n(rt) - f(rt)| &= \left| \int_{\mathbb{R}^d} e^{i(rt,x)} p_n(x)\,dx - \int_{\mathbb{R}^d} e^{i(rt,x)} p(x)\,dx \right| \\ &\le \int_{\mathbb{R}^d} |p_n(x) - p(x)|\,dx. \end{aligned}$$

By Scheffe's Lemma E.1.19 the integral above tends to zero as $n \to \infty$. This completes the proof. □

1.7 The theorems of Bochner and Herglotz

We already know that functions of the form

$$f(t) = \sum_{j=1}^{n} p_j \gamma_j(t), \quad t \in G$$

where G is a commutative group, $p_j \ge 0$, and γ_j is a character of G, are positive definite. We will show that every positive definite function on G can be obtained if we replace the weighted sum above by an integral with respect to a nonnegative measure.

First we consider the case $G = \mathbb{R}^d$.

Lemma 1.7.1. *Let μ be a probability measure on $\mathbb{R}^d$ and for $n \in \mathbb{N}$ define the probability measure μ_n by $\mu_n(B) = \mu(nB)$, $B \in \mathcal{B}_d$. Then*

$$\lim_{n\to\infty} \mu_n * g(t) = g(t), \quad t \in \mathbb{R}^d$$

holds for any bounded, continuous function g on $\mathbb{R}^d$.

Proof. Let $\epsilon > 0$ and $t \in \mathbb{R}^d$ be arbitrary and choose a bounded, open neighborhood U of 0 such that

$$|g(t) - g(t - x)| < \epsilon, \quad x \in U.$$

We have

$$\begin{aligned}|g(t)-\mu_n * g(t)| &= \left|\int_{\mathbb{R}^d} g(t)-g(t-x)\,\mathrm{d}\mu_n(x)\right| \\ &\le \epsilon\cdot\mu_n(U)+2\|g\|_\infty\cdot\mu_n(\mathbb{R}^d\setminus U) \\ &\le \epsilon+2\|g\|_\infty\cdot\mu(\mathbb{R}^d\setminus nU).\end{aligned}$$

The lemma follows now from the fact that $\lim_{n\to\infty}\mu(\mathbb{R}^d\setminus nU)=0$. □

Applying the previous lemma to an absolutely continuous measure μ we immediately obtain the following corollary:

Corollary 1.7.2. *Let p be a probability density on $\mathbb{R}^d$ and for $n\in\mathbb{N}$ put*

$$p_n(t)=n^d\cdot p(nt),\quad t\in\mathbb{R}^d.$$

Then

$$\lim_{n\to\infty} p_n * g(t)=g(t),\quad t\in\mathbb{R}^d$$

holds for any bounded, continuous function g on $\mathbb{R}^d$.

We are now able to characterize continuous positive definite functions on $\mathbb{R}^d$. We formulate the result in terms of characteristic functions.

Theorem 1.7.3 (Bochner). *A continuous complex-valued function f on $\mathbb{R}^d$ is a characteristic function if and only if $f(0)=1$ and f is positive definite.*

Proof. We have already seen in Theorem 1.1.2 that characteristic functions are positive definite. Conversely, let f be a continuous positive definite function with $f(0)=1$ and assume first that f is integrable. Further, let g_σ and φ_σ be as in Lemma 1.3.4. Applying Fubini's theorem we obtain

$$\begin{aligned}f * g_\sigma(t) &= \int_{\mathbb{R}^d} f(x)g_\sigma(t-x)\,\mathrm{d}x \\ &= \int_{\mathbb{R}^d} f(x)\cdot\left[\int_{\mathbb{R}^d}\mathrm{e}^{\mathrm{i}(t-x,y)}\varphi_\sigma(y)\,\mathrm{d}y\right]\mathrm{d}x \\ &= \int_{\mathbb{R}^d}\varphi_\sigma(y)\cdot\left[\int_{\mathbb{R}^d} f(x)\mathrm{e}^{-\mathrm{i}(y,x)}\,\mathrm{d}x\right]\cdot\mathrm{e}^{\mathrm{i}(t,y)}\,\mathrm{d}y \\ &=: \int_{\mathbb{R}^d}\varphi_\sigma(y)\cdot q(y)\cdot\mathrm{e}^{\mathrm{i}(t,y)}\,\mathrm{d}y.\end{aligned}$$

Theorem 1.5.9 shows that $q(y)\ge 0$. Since $q(y)\le\int|f(x)|\,\mathrm{d}x$, the function $y\mapsto\varphi_\sigma(y)q(y)$ is integrable. From this, using (1.3.4.1), we conclude that

$$\frac{f * g_\sigma}{f * g_\sigma(0)}=\frac{f * \varphi_{1/\sigma}}{f * \varphi_{1/\sigma}(0)}=: h_{1/\sigma}$$

is a characteristic function. By Lemma 1.7.2, the sequence $\{h_n\}_1^\infty$ converges pointwise to f. Application of Lévy's continuity Theorem 1.6.3 shows that f is a characteristic function.

In the general case where f is not supposed to be integrable, we consider the functions $f \cdot g_{1/n}$ which are positive definite and integrable. Moreover, $f(t) = \lim_n f \cdot g_{1/n}(t)$, $t \in \mathbb{R}^d$. A second application of Lévy's continuity theorem completes the proof. □

An alternative formulation of Bochner's theorem is the following.

Theorem 1.7.4. *A continuous complex-valued function f on $\mathbb{R}^d$ is positive definite if and only if it can be represented in the form*

$$f(t) = \int_{\mathbb{R}^d} \mathrm{e}^{\mathrm{i}(t,x)}\, \mathrm{d}\mu(x), \quad t \in \mathbb{R}^d,$$

with some nonnegative finite Borel measure μ on $\mathbb{R}^d$. The measure μ is uniquely determined by f.

In the case $d = 1$ the next theorem is due to G. Herglotz. We will deduce it from the more general Theorem 1.7.8.[19]

Theorem 1.7.5. *A complex-valued function f on $\mathbb{Z}^d$ is positive definite if and only if it can be represented in the form*

$$f(n) = \int_{[0,2\pi)^d} \mathrm{e}^{\mathrm{i}(n,t)}\, \mathrm{d}\mu(t), \quad n \in \mathbb{Z}^d$$

with some nonnegative finite Borel measure μ on $[0, 2\pi)^d$. The measure μ is uniquely determined by f.

Remark 1.7.6. Let G be an arbitrary commutative group and denote by $L = L(G)$ the complex linear space of all bounded complex-valued functions on G. Equipped with the topology of pointwise convergence, L becomes a locally convex linear space (cf. Section D.4). Notice that for fixed $x \in G$ the function $h \mapsto h(x)$ is a continuous linear functional on L. The convex set

$$P_0(G) = \{f \in P(G) : f(0) = 1\}$$

is a closed subset of L. Since $|f(x)| \leq 1$, $x \in G$, this set is compact (cf. Theorem B.2.8). We have already seen (cf. Theorem 1.4.20) that the set of all extreme points of $P_0(G)$ is equal to the set of all characters of G which we will denote by $\hat{G}$. The set $\hat{G}$ is a closed subset of $P_0(G)$. Hence it is compact.

[19] The formulations of Theorems 1.7.4, 1.7.5, and 1.7.8 can be unified within the framework of Fourier transformation on locally compact commutative groups, see for example [49].

Remark 1.7.7. If $G = \mathbb{Z}^d$ then we can identify $\hat{G}$ with the set $\mathbb{T}^d$ (see Lemma 1.4.7). Moreover, the topology of pointwise convergence on $\hat{G}$ is the same as the metric topology of $\mathbb{T}^d$. Since every $z \in \mathbb{T}$ can uniquely be written as $z = \mathrm{e}^{\mathrm{i}t}$ with some $t \in [0, 2\pi)$, we can identify $\hat{G}$ also with $[0, 2\pi)^d$. This is sometimes more convenient. However, in this case the topology of pointwise convergence is not the same as the metric topology (note that $[0, 2\pi)^d$ is not compact). Nevertheless, both topologies generate the same Borel sets.

Theorem 1.7.8. *A complex-valued function f on a commutative group G is positive definite if and only if it can be represented in the form*

$$f(x) = \int_{\hat{G}} \gamma(x)\, \mathrm{d}\mu(\gamma), \qquad x \in G$$

with some nonnegative finite Radon measure μ on $\hat{G}$. The measure μ is uniquely determined by f.[20]

Proof. If f has the form above, then

$$\sum_{i,j=1}^{n} f(x_i - x_j) c_i \overline{c_j} = \int_{\hat{G}} \left| \sum_{i=1}^{n} c_i \gamma(x_i) \right|^2 \mathrm{d}\mu(\gamma) \ge 0,$$

showing that f is positive definite. To prove the other direction we may assume that $f \in P_0(G)$. By Theorem D.4.6 (see also Remark 1.7.6), there exists a probability measure $\mu = \mu_f$ on $\hat{G}$ such that

$$l(f) = \int_{\hat{G}} l(\gamma)\, \mathrm{d}\mu(\gamma)$$

holds for every continuous linear functional l on $L(G)$. Taking here the linear functionals l_x defined by $l_x(h) = h(x)$, $h \in L(G)$, we obtain the desired representation of f.

To prove uniqueness, assume that ν is a finite Radon measure on $\hat{G}$ such that

$$\int_{\hat{G}} \gamma(x)\, \mathrm{d}\mu(\gamma) = \int_{\hat{G}} \gamma(x)\, \mathrm{d}\nu(\gamma), \qquad x \in G.$$

Denote by $\mathcal{A}$ the linear space of (continuous) functions on $\hat{G}$ spanned by all functions of the form $\gamma \mapsto \gamma(x)$, $\gamma \in \hat{G}$ where $x \in G$. By the equation above, $\int g\, \mathrm{d}\mu = \int g\, \mathrm{d}\nu$ for all $g \in \mathcal{A}$. It is easy to check that $\mathcal{A}$ satisfies the first two conditions of Theorem B.2.4 (Stone–Weierstraß theorem). To show that the last condition is satisfied, let $x, y \in G$ be such that $x \neq y$. Applying the first part of the proof to the positive

[20] The measure μ is sometimes called a *representing measure* or *spectral measure* of f.

definite function $\mathbf{1}_{\{0\}}$ we see that there exists a probability measure λ on $\hat{G}$ such that

$$0 = \mathbf{1}_{\{0\}}(x-y) = \int_{\hat{G}} \gamma(x-y)\,\mathrm{d}\lambda(\gamma).$$

This implies the existence of $\chi \in \hat{G}$ such that $\chi(x-y) \neq 1$ from which $\chi(x) \neq \chi(y)$ follows.

Thus, $\mathcal{A}$ satisfies the conditions of the Stone–Weierstraß theorem (cf. Theorem B.2.4) and therefore an arbitrary function $g \in \mathrm{C}(\hat{G})$ can be uniformly approximated by functions from $\mathcal{A}$. This shows that $\int g\,\mathrm{d}\mu = \int g\,\mathrm{d}\nu$ for all $g \in \mathrm{C}(\hat{G})$, and hence $\mu = \nu$. □

Applying Theorem E.1.8[21] we obtain the following corollary:

Corollary 1.7.9. *Every positive definite function f on G is the pointwise limit of a net $\{f_\alpha\}$ of positive definite functions of the form*

$$f_\alpha = \sum_{i=1}^{n_\alpha} p_i^\alpha \gamma_i^\alpha$$

where $n_\alpha \in \mathbb{N}$, $p_i^\alpha \geq 0$, $\sum_{i=1}^{n_\alpha} p_i^\alpha = f(0)$ and $\gamma_i^\alpha \in \hat{G}$.

The next statement has been show in the proof of Theorem 1.7.8. We state it explicitly because of its importance.

Corollary 1.7.10. *If x and y are distinct elements of G, then there exists a character γ of G such that $\gamma(x) \neq \gamma(y)$.*

Another nice corollary of Theorem 1.7.8 is the existence and uniqueness of a *translation invariant* probability measure on $\hat{G}$, which is called the *normalized Haar measure*.

Corollary 1.7.11. *There exists a unique Radon probability measure λ on $\hat{G}$ such that*

$$\int_{\hat{G}} f(\chi\gamma)\,\mathrm{d}\lambda(\gamma) = \int_{\hat{G}} f(\gamma)\,\mathrm{d}\lambda(\gamma), \quad f \in \mathrm{C}(\hat{G}),\ \chi \in \hat{G}.$$

Proof. Let λ and $\mathcal{A}$ be as in the proof of Theorem 1.7.8. It follows from the definition of λ that the equation above holds for all $f \in \mathcal{A}$, while the density property of $\mathcal{A}$ implies that it holds for all $f \in \mathrm{C}(\hat{G})$.

Assume that μ is a Radon probability measure satisfying

$$\int_{\hat{G}} f(\chi\gamma)\,\mathrm{d}\mu(\gamma) = \int_{\hat{G}} f(\gamma)\,\mathrm{d}\mu(\gamma), \quad f \in \mathrm{C}(\hat{G}),\ \chi \in \hat{G}.$$

[21] The corollary also follows from Kreĭn–Milman's Theorem D.4.5.

Setting $f(\gamma) := \gamma(x)$ where $x \in G$ we obtain

$$\chi(x) \int_{\hat{G}} \gamma(x)\,\mathrm{d}\mu(\gamma) = \int_{\hat{G}} (\chi\gamma)(x)\,\mathrm{d}\mu(\gamma) = \int_{\hat{G}} \gamma(x)\,\mathrm{d}\mu(\gamma), \qquad x \in G,\ \chi \in \hat{G}.$$

We have shown in the proof of Theorem 1.7.8 that for all $x \neq 0$ there exists $\chi \in \hat{G}$ such that $\chi(x) \neq \chi(0) = 1$. Using this and the relation above we conclude that

$$\int_{\hat{G}} \gamma(x)\,\mathrm{d}\mu(\gamma) = \mathbf{1}_{\{0\}}(x), \quad x \in G$$

and hence $\mu = \lambda$. □

Example 1.7.12. Identifying the character group of $\mathbb{Z}^d$ with $[0, 2\pi)^d$ the restriction of the Lebesgue measure to $[0, 2\pi)^d$ is translation invariant in the sense of the previous corollary. Thus, dividing this restriction by $(2\pi)^d$ we obtain the normalized Haar measure.

Corollary 1.7.13. *If $\nu \in \mathcal{M}^b(G)$ is such that*

$$\int_G \gamma(x)\,\mathrm{d}\nu(x) = 0, \quad \gamma \in \hat{G}$$

then $\nu = 0$.

Proof. For all $y \in G$ we have

$$\begin{aligned}\nu(\{y\}) = \int_G \mathbf{1}_{\{0\}}(x-y)\,\mathrm{d}\nu(x) &= \int_G \left(\int_{\hat{G}} \gamma(x-y)\,\mathrm{d}\lambda(\gamma)\right)\mathrm{d}\nu(x)\\ &= \int_{\hat{G}} \left(\gamma(-y)\int_G \gamma(x)\,\mathrm{d}\nu(x)\right)\mathrm{d}\lambda(\gamma) = 0.\end{aligned}$$ □

Proof of Theorem 1.7.5. We identify the character group of $\mathbb{Z}^d$ with $[0, 2\pi)^d$ (see Remark 1.7.7). Since the topology of pointwise convergence generates the same Borel σ-algebra as the metric topology of $[0, 2\pi)^d$, Theorem 1.7.5 follows immediately from Theorem 1.7.8. □

Alternative proof of Theorem 1.7.5 in the case $d = 1$. We show here only the existence of μ. Definition 1.4.1 with $c_j = \mathrm{e}^{-\mathrm{i}jx}$, $x \in \mathbb{R}$, and $t_j = j$ gives

$$0 \le P_N(x) := \frac{1}{N}\sum_{j=1}^{N}\sum_{k=1}^{N} f(j-k)\mathrm{e}^{-\mathrm{i}(j-k)x} = \sum_{m=-N}^{N}\left(1 - \frac{|m|}{N}\right) f(m)\mathrm{e}^{-\mathrm{i}mx}$$

and therefore

$$\frac{1}{2\pi}\int_0^{2\pi} e^{inx} P_N(x)\,dx = \frac{1}{2\pi}\sum_{m=-N}^{N}\left(1-\frac{|m|}{N}\right) f(m)\int_0^{2\pi} e^{i(n-m)x}\,dx$$
$$= \left(1-\frac{|n|}{N}\right)_+ \cdot f(n)$$

for all $n \in \mathbb{Z}$. Consequently, the equation

$$d\mu_N(x) := \frac{1}{2\pi} P_N(x)\,dx, \quad x \in [0, 2\pi)$$

defines a nonnegative measure on $[0, 2\pi)$, such that

$$\int_{[0,2\pi)} e^{inx}\,d\mu_N(x) = \left(1-\frac{|n|}{N}\right)_+ f(n). \tag{1}$$

In particular, $\mu_N([-\pi, \pi)) = f(0)$. By Theorem E.1.13, the sequence $\{\mu_N\}$ contains a subsequence converging weakly to some nonnegative measure μ.[22] From (1) we conclude that

$$\int_{[0,2\pi)} e^{inx}\,d\mu(x) = f(n). \quad \square$$

1.8 Fourier transformation on $\mathbb{R}^d$

The aim of the present section is to prove some basic and frequently used results on the Fourier transform on $\mathbb{R}^d$. Our previous results on inversion theorems, on positive definite functions and Bochner's theorem enable us to give short proofs.

Definition 1.8.1. For $g \in \mathcal{L}^1(\mathbb{R}^d)$ or $g \in L^1(\mathbb{R}^d)$ let

$$\hat{g}(t) = \frac{1}{(2\pi)^{d/2}}\int_{\mathbb{R}^d} g(x)e^{-i(t,x)}\,dx, \quad t \in \mathbb{R}^d$$

and

$$\check{g}(t) = \frac{1}{(2\pi)^{d/2}}\int_{\mathbb{R}^d} g(x)e^{i(t,x)}\,dx, \quad t \in \mathbb{R}^d.$$

The function $\hat{g}$ is called the *Fourier transform of* g while $\check{g}$ is the *inverse Fourier transform of* g. In some cases it is convenient to consider the Fourier transformation on the Banach space $L^1(\mathbb{R}^d)$, in other cases it is more natural to deal with $\mathcal{L}^1(\mathbb{R}^d)$, the elements of which are functions and not equivalence classes of functions.

[22] This follows also from Helly's selection theorem.

For a finite complex measure μ on $\mathbb{R}^d$ we define the *Fourier–Stieltjes transform* of μ by

$$\hat{\mu}(t) = \int_{\mathbb{R}^d} \mathrm{e}^{-\mathrm{i}(t,x)}\,\mathrm{d}\mu(x), \quad t \in \mathbb{R}^d$$

while

$$\check{\mu}(t) = \int_{\mathbb{R}^d} \mathrm{e}^{\mathrm{i}(t,x)}\,\mathrm{d}\mu(x), \quad t \in \mathbb{R}^d$$

is the *inverse Fourier–Stieltjes transform* of μ.

Note that if φ is a density, then the corresponding characteristic function is equal to $(2\pi)^{d/2}\check{\varphi}$. On the other hand, if μ is the distribution of a random vector, then the corresponding characteristic function is $\check{\mu}$.

To see an example of the Fourier transform, let φ_σ be the Gaussian density as defined in Lemma 1.3.4. Then, by the same lemma,

$$\hat{\varphi}_\sigma = \check{\varphi}_\sigma = \frac{1}{\sigma^d} \cdot \varphi_{1/\sigma}\,.$$

Theorem 1.8.2. *Every function $g \in L^1(\mathbb{R}^d)$ is uniquely determined by its Fourier transform $\hat{g}$ (inverse Fourier transform $\check{g}$) which is a bounded, uniformly continuous function on $\mathbb{R}^d$. Moreover, for all $g, h \in L^1(\mathbb{R}^d)$ we have*

(i) $(g + h)\hat{} = \hat{g} + \hat{h}$;

(ii) $(cg)\hat{} = c\hat{g}$ *for all* $c \in \mathbb{C}$;

(iii) $(g * h)\hat{} = (2\pi)^{d/2}\, \hat{g} \cdot \hat{h}$;

(iv) $\overline{\hat{g}} = (\overline{g})\check{} = (\tilde{g})\hat{}$;

(v) $\hat{g}$ *is real-valued if and only if* $g = \tilde{g}$;

(vi) $\sup\{|\hat{g}(t)| : t \in \mathbb{R}^d\} \le \dfrac{1}{(2\pi)^{d/2}} \displaystyle\int_{\mathbb{R}^d} |g(x)|\,\mathrm{d}x$;

(vii) $(g_y)\hat{}(t) = \mathrm{e}^{\mathrm{i}(t,y)}\hat{g}(t)$ *and*
$(g_y)\check{}(t) = \mathrm{e}^{-\mathrm{i}(t,y)}\check{g}(t)$, *where* $g_y(x) = g(x + y)$.

The analogues of (i)–(iii), (v) *and* (vi) *are also valid for the inverse Fourier transform* $\check{g}$.

Proof. The properties (i)–(iv) and (vi) are simple consequences of Definition 1.8.1, we omit the details. Since g is integrable it can be written as a linear combination of four density functions. Using this, uniform continuity of $\hat{g}$ (and of $\check{g}$) follows from (1.1.2.iv). As to uniqueness, the proof of Theorem 1.3.3 can also be applied to $\mu = g\lambda$. Property (v) follows from (iv) and from the uniqueness. □

The next theorem can be proved in the same way as Theorem 1.8.2. We omit the proof.

Theorem 1.8.3. *Every measure $\mu \in \mathcal{M}^b(\mathbb{R}^d)$ is uniquely determined by its Fourier–Stieltjes transform $\hat{\mu}$ which is a bounded, uniformly continuous function on $\mathbb{R}^d$. For all $\mu, \nu \in \mathcal{M}^b(\mathbb{R}^d)$ we have*

(i) $(\mu + \nu)\hat{} = \hat{\mu} + \hat{\nu}$;

(ii) $(c\mu)\hat{} = c\hat{\mu}$ *for all* $c \in \mathbb{C}$;

(iii) $(\mu * \nu)\hat{} = \hat{\mu} \cdot \hat{\nu}$;

(iv) $(\mu * f)\hat{} = \hat{\mu} \cdot \hat{f}$ *for all* $f \in L^1(\mathbb{R}^d)$;

(v) $(\tilde{\mu})\hat{} = \overline{\hat{\mu}}$;

(vi) $\hat{\mu}$ *is real-valued if and only if* $\mu = \tilde{\mu}$.

All these properties are also valid for $\check{\mu}$.

Next we investigate the behavior of the Fourier transform at infinity.

Theorem 1.8.4 (Riemann–Lebesgue lemma). *If $g \in L^1(\mathbb{R}^d)$ then $\hat{g} \in C_0(\mathbb{R}^d)$.*

Proof. Let g first be the indicator function of a parallelepiped

$$[a_1, b_1] \times \cdots \times [a_d, b_d] \subset \mathbb{R}^d.$$

Then

$$\hat{g}(t) = \prod_{j=1}^{d} \frac{e^{ib_j t_j} - e^{ia_j t_j}}{it_j}, \qquad t \in \mathbb{R}^d,$$

and this tends to zero as $\|t\| \to \infty$, since $|t_j| \to \infty$ for at least one j.

For each function $g \in L^1(\mathbb{R}^d)$ and $\epsilon > 0$ there exists a function h such that h is a finite linear combination of indicator functions of parallelepipeds and $\|g-h\|_1 < \epsilon/2$. From the first step we know that there exists a positive number K such that $|\hat{h}(t)| < \epsilon/2$ whenever $|t| > K$. For such t we have

$$|\hat{g}(t)| = |(g-h)\hat{}(t) + \hat{h}(t)| \le \|g-h\|_1 + |\hat{h}(t)| < \epsilon. \quad \square$$

Theorem 1.8.5. *If μ is a finite discrete measure on $\mathbb{R}^d$, then for all $t \in \mathbb{R}^d$ and $s \in \mathbb{R}$ we have*

$$\liminf_{r\to\infty} |\hat{\mu}(st) - \hat{\mu}(rt)| = 0.$$

Proof. The function $f(s) := \hat{\mu}(st)$, $s \in \mathbb{R}$, has the form

$$f(s) = \sum_{n=1}^{\infty} c_n e^{ix_n s}, \qquad s \in \mathbb{R}$$

where $x_n \in \mathbb{R}$, $c_n \in \mathbb{C}$ and $\sum_{n=1}^{\infty} |c_n| < \infty$. Therefore, the theorem follows from Corollary C.2.5 and Theorem C.2.7. $\square$

Remark 1.8.6. Applying the previous theorem with $s = 0$ and $t = 1$ to a discrete probability measure μ on $\mathbb{R}$ we see that

$$\limsup_{r\to\infty} |\hat{\mu}(r)| = 1.$$

It can be shown (see, e.g., [6], Corollary 1.11.8) that for every $p \in [0, 1]$ there exists a continuous, singular probability measure ν on $\mathbb{R}$ such that

$$\limsup_{r\to\infty} |\hat{\nu}(r)| = p.$$

To see another interesting example for the behavior at infinity of a Fourier transform, let $q \in (0, 1)$ and let X_n, $n \in \mathbb{N}_0$, be independent random variables such that

$$P(X_n = -1) = P(X_n = +1) = \frac{1}{2}.$$

The characteristic function of the random variable

$$X = \sum_{n=0}^{\infty} q^n X_n$$

does not tend to zero at infinity if and only if $\theta := \frac{1}{q}$ is a Pisot number[23] different from 2. This is the so-called Zygmund–Salem theorem; we refer to [6, Sect. 1.11] for more details.

The Fourier transform of an integrable function may not be integrable (cf. Example 1.8.14). However, we have the following result:[24]

Theorem 1.8.7. *Let f be a continuous and integrable positive definite function on $\mathbb{R}^d$. Then both $\hat{f}$ and $\check{f}$ are integrable, nonnegative and*

$$f = (\hat{f}\,)\check{} = (\check{f}\,)\hat{}.$$

Proof. We may assume that $f(0) = 1$. Then f is a characteristic function in view of Bochner's theorem. By Theorem 1.3.6, f corresponds to some density p, i.e., $f = (2\pi)^{d/2}\check{p}$. On the other hand, Theorem 1.3.6 shows that $p = (1/2\pi)^{d/2}\hat{f}$. Combining these facts we see that $\hat{f} \geq 0$ and $f = (\hat{f}\,)\check{}$. The relations $\check{f} \geq 0$ and $f = (\check{f}\,)\hat{}$ follow from (1.8.2.iv). □

We will extend the Fourier transformation to square integrable functions. For this we need the following result:

Lemma 1.8.8. *The set*

$$L := \{\hat{g} : g \in L^1(\mathbb{R}^d) \cap L^2(\mathbb{R}^d)\}$$

---

23 A real number $\theta > 1$ is said to be a *Pisot number* if it satisfies an equation

$$\theta^n + b_1\theta^{n-1} + \cdots + b_n = 0$$

where $b_1, \ldots, b_n$ are integers, and all other roots of this equation have moduli less than 1.

24 See also Theorem 1.8.13.

is a dense linear subspace of $L^2(\mathbb{R}^d)$ *and*

$$\|g\|_2 = \|\hat{g}\|_2, \quad g \in L^1(\mathbb{R}^d) \cap L^2(\mathbb{R}^d).$$

Proof. If $g \in L^1(\mathbb{R}^d) \cap L^2(\mathbb{R}^d)$, then the function $f = g * \tilde{g}$ belongs to $P^c(\mathbb{R}^d) \cap L^1(\mathbb{R}^d)$ (cf. Lemma 1.5.4 and Theorem E.2.4) and $\hat{f} = (2\pi)^{d/2} |\hat{g}|^2$ in view of (1.8.2.iii) and (1.8.2.iv). By Theorem 1.8.7, the function $\hat{f}$ is integrable and therefore $\hat{g} \in L^2(\mathbb{R})$, i.e., $L \subset L^2(\mathbb{R}^d)$. It is obvious that L is a linear subspace.

Applying the first equation in Theorem 1.8.7, we get

$$\int_{\mathbb{R}^d} |g(x)|^2 \,\mathrm{d}x = g * \tilde{g}(0) = f(0) = \frac{1}{(2\pi)^{d/2}} \int_{\mathbb{R}^d} \hat{f}(x) \,\mathrm{d}x = \int_{\mathbb{R}^d} |\hat{g}(x)|^2 \,\mathrm{d}x,$$

that is, $\|g\|_2 = \|\hat{g}\|_2$.

Suppose now that $\psi \in L^2(\mathbb{R}^d)$ and ψ is orthogonal to L. Since $L^1(\mathbb{R}^d) \cap L^2(\mathbb{R}^d)$ is translation-invariant, (1.8.2.vi) shows that L is invariant under multiplication by the function $t \mapsto \mathrm{e}^{\mathrm{i}(t,x)}$ for all $x \in \mathbb{R}^d$. Define the function ϕ by $\phi(t) = \mathrm{e}^{-(t,t)}$. Since ϕ and ψ are square integrable their product is in $L^1(\mathbb{R}^d)$. By Lemma 1.3.4, the function ϕ is in L, thus the identity

$$\int_{\mathbb{R}^d} \psi(t)\phi(t)\mathrm{e}^{-\mathrm{i}(t,x)} \,\mathrm{d}t = 0$$

holds for all $x \in \mathbb{R}^d$, i.e., $(\psi\phi)\hat{} = 0$. The first part of Theorem 1.8.2 implies that $\psi\phi = 0$ and hence also $\psi = 0$. Thus, 0 is the only element in $L^2(\mathbb{R}^d)$ that is orthogonal to L; therefore L is dense in $L^2(\mathbb{R}^d)$. □

Theorem 1.8.9 (Plancherel). *We consider the Fourier transformation, introduced in Definition 1.8.1, on the set* $L^1(\mathbb{R}^d) \cap L^2(\mathbb{R}^d)$. *This transformation can be extended in a unique way to a linear isometry from* $L^2(\mathbb{R}^d)$ *onto* $L^2(\mathbb{R}^d)$. *Denoting the extension by the same symbol, the equations*[25]

(i) $\displaystyle\int_{\mathbb{R}^d} f(x)\overline{g(x)} \,\mathrm{d}x = \int_{\mathbb{R}^d} \hat{f}(t)\overline{\hat{g}(t)} \,\mathrm{d}t$ and

(ii) $(2\pi)^{d/2} (f \cdot g)\hat{} = \hat{f} * \hat{g}$.

hold for all $f, g \in L^2(\mathbb{R}^d)$.

Proof. Write $H = L^1(\mathbb{R}^d) \cap L^2(\mathbb{R}^d)$. By Lemma 1.8.8, the Fourier transformation is a linear isometry from H into $L^2(\mathbb{R}^d)$. The first statement follows immediately from the fact that H is a dense subspace of $L^2(\mathbb{R}^d)$. Equation (i) is a consequence of the identity

$$4f\bar{g} = |f + g|^2 - |f - g|^2 + \mathrm{i}|f + \mathrm{i}g|^2 - \mathrm{i}|f - \mathrm{i}g|^2.$$

Replacing $g(x)$ by $\overline{g(x)\mathrm{e}^{\mathrm{i}(y,x)}}$, $y \in \mathbb{R}^d$, in (i) we obtain (ii). □

[25] The first equation is usually referred to as Parseval's identity.

Remark 1.8.10. If $g \in L^2(\mathbb{R}^d)$ then we call $\hat{g}$ the L^2 *Fourier transform of* g. Note that $\hat{g}$ is not a function but an equivalence class of functions and relations such as $\hat{g} = \hat{h}$ or $\hat{g} \geq 0$ are relations between equivalence classes. However, if $g \in L^1(\mathbb{R}^d) \cap L^2(\mathbb{R}^d)$ then we can regard $\hat{g}$ as a function given by

$$\hat{g}(t) = \frac{1}{(2\pi)^{d/2}} \int_{\mathbb{R}^d} g(x) e^{-i(t,x)} \, dx.$$

It will be clear from the context in every case whether we mean the fixed function $\hat{g}$ or the equivalence class in $L^2(\mathbb{R}^d)$ containing $\hat{g}$.

The inverse Fourier transformation can be extended to square integrable functions in the same way.

Theorem 1.8.11. *We consider the inverse Fourier transformation, introduced in Definition 1.8.1, on the set* $L^1(\mathbb{R}^d) \cap L^2(\mathbb{R}^d)$. *This transformation can be extended in a unique way to a linear isometry from* $L^2(\mathbb{R}^d)$ *onto* $L^2(\mathbb{R}^d)$. *Denoting the extension by the same symbol, the equations*

(i) $\phi = (\check{\phi})\hat{} = (\hat{\phi})\check{}$

(ii) $(2\pi)^{d/2} (\phi \cdot \psi)\check{} = \check{\phi} * \check{\psi}$

(iii) $(\phi * \psi)\check{} = (2\pi)^{d/2} \cdot \check{\phi} \cdot \check{\psi}$

hold for all $\phi, \psi \in L^2(\mathbb{R}^d)$.

Proof. The inverse Fourier transform of a function in $L^1(\mathbb{R}^d) \cap L^2(\mathbb{R}^d)$ is square integrable. This follows from (1.8.2.iv) and the square integrability of the Fourier transform. If $\phi, g \in L^1(\mathbb{R}^d) \cap L^2(\mathbb{R}^d)$, then

$$\begin{aligned} \int_{\mathbb{R}^d} \check{\phi}(x) \overline{g(x)} \, dx &= \frac{1}{(2\pi)^{d/2}} \int_{\mathbb{R}^d} \int_{\mathbb{R}^d} \phi(t) e^{i(t,x)} \, dt \, \overline{g(x)} \, dx \\ &= \frac{1}{(2\pi)^{d/2}} \int_{\mathbb{R}^d} \phi(t) \int_{\mathbb{R}^d} \overline{g(x) e^{i(t,x)}} \, dx \, dt \\ &= \int_{\mathbb{R}^d} \phi(t) \overline{\hat{g}(t)} \, dt. \end{aligned}$$

Using (1.8.9.i) and the equations above we see that

$$\int_{\mathbb{R}^d} (\check{\phi})\hat{}(t) \overline{\hat{g}(t)} \, dt = \int_{\mathbb{R}^d} \check{\phi}(x) \overline{g(x)} \, dx = \int_{\mathbb{R}^d} \phi(t) \overline{\hat{g}(t)} \, dt.$$

Since the set L in Lemma 1.8.8 is dense in $L^2(\mathbb{R}^d)$, we conclude that

$$\phi = (\check{\phi})\hat{}, \quad \phi \in L^1(\mathbb{R}^d) \cap L^2(\mathbb{R}^d). \tag{1}$$

It follows from (1) and (1.8.9.i) that $\|\phi\|_2 = \|\check{\phi}\|_2$. Thus, the mapping $\phi \mapsto \check{\phi}$ maps $L^1(\mathbb{R}^d) \cap L^2(\mathbb{R}^d)$, which is dense in $L^2(\mathbb{R}^d)$, isometrically into $L^2(\mathbb{R}^d)$, and therefore it extends to a unique isometric mapping from $L^2(\mathbb{R}^d)$ onto $L^2(\mathbb{R})$. From (1) we conclude that (i) holds while (ii) is obtained in the same way as (1.8.9.ii). Taking the inverse Fourier transform of both sides of (1.8.9.ii) we obtain (iii). □

Remark 1.8.12. We call $\check{\phi}$, $\phi \in L^2(\mathbb{R}^d)$, the *inverse L^2 Fourier transform of ϕ*. Remark 1.8.10 holds for this transformation, too.

Theorem 1.8.13. *Suppose that $f \in L^1(\mathbb{R}^d) \cap L^\infty(\mathbb{R}^d)$ and that $\hat{f}$ is real-valued and nonnegative. Then $\hat{f}$ is in $L^1(\mathbb{R}^d)$.*

Proof. First we note that $f \in L^2(\mathbb{R}^d)$. Let φ_σ be the Gaussian density as defined in Lemma 1.3.4. Applying (1.8.9.i) with $g = \varphi_{1/n}$ and using that $\hat{\varphi}_{1/n} = n^d \varphi_n$ we obtain

$$\begin{aligned}\frac{1}{(2\pi)^{d/2}} \int_{\mathbb{R}^d} \hat{f}(t) \mathrm{e}^{-\frac{\|t\|^2}{2n^2}} \,\mathrm{d}t &= \int_{\mathbb{R}^d} \hat{f}(t) \hat{\varphi}_{1/n}(t) \,\mathrm{d}t \\ &= \int_{\mathbb{R}^d} f(x) \varphi_{1/n}(x) \,\mathrm{d}x \le \|f\|_\infty.\end{aligned}$$

The integrand on the left is nonnegative and monotone increasing as a function of n. Taking the limit $n \to \infty$ and applying Beppo Levi's theorem on monotone convergence we obtain

$$\int_{\mathbb{R}^d} \hat{f}(t) \,\mathrm{d}t \le (2\pi)^{d/2} \|f\|_\infty < \infty.$$

Thus, $\hat{f}$ is integrable. □

Next we give two simple examples where the Fourier transform is square integrable but not integrable.

Example 1.8.14. By Example 1.1.13(a), the Fourier transform of the function $p(x) = \mathbf{1}_{[0,\infty)}(x) \cdot \mathrm{e}^{-x}$, $x \in \mathbb{R}$, is equal to $q(t) = \frac{1}{\sqrt{2\pi}} \cdot \frac{1}{1+it}$. This function is square integrable but not integrable. By (1.8.11.i), the inverse L^2 Fourier transform of q is p.

A similar example is given by the indicator function p of the set $[-1, 1]^d$. The computations at the end of Definition 1.1.1 show that

$$q(t) := \hat{p}(t) = \check{p}(t) = \left(\frac{2}{\pi}\right)^{d/2} \prod_{j=1}^{d} \frac{\sin t_j}{t_j}, \quad t \in \mathbb{R}^d.$$

The function q belongs to $L^2(\mathbb{R}^d) \setminus L^1(\mathbb{R}^d)$ and $\hat{q} = \check{q} = p$.

The representing measure of the positive definite function $\mu * \tilde{\mu} * f$ from Lemma 1.5.3 can be easily determined.

Theorem 1.8.15. *Assume that $f \in P^c(\mathbb{R}^d)$ admits the representing measure σ and let $\mu \in \mathcal{M}^b(\mathbb{R}^d)$ be arbitrary. Then the function $\mu * \tilde{\mu} * f$ is positive definite and its representing measure σ_μ is given by $\mathrm{d}\sigma_\mu(s) = |\hat{\mu}(s)|^2\, \mathrm{d}\sigma(s)$.*

Proof. Using

$$f(t) = \int e^{i(t,s)}\, \mathrm{d}\sigma(s)$$

(integration over $\mathbb{R}^d$) we obtain

$$\begin{aligned}
\mu * \tilde{\mu} * f(t) &= \int f(t-x)\, \mathrm{d}(\mu * \tilde{\mu})(x) \\
&= \iint f(t-x+y)\, \mathrm{d}\mu(x)\, \mathrm{d}\overline{\mu}(y) \\
&= \iiint e^{i(t-x+y,s)}\, \mathrm{d}\sigma(s)\, \mathrm{d}\mu(x)\, \mathrm{d}\overline{\mu}(y) \\
&= \int e^{i(t,s)} \int e^{-i(x,s)}\, \mathrm{d}\mu(x) \overline{\int e^{-i(y,s)}\, \mathrm{d}\mu(y)}\, \mathrm{d}\sigma(s) \\
&= \int e^{i(t,s)} |\hat{\mu}(s)|^2\, \mathrm{d}\sigma(s).
\end{aligned}$$

□

Using L^2 Fourier transformation we can characterize characteristic functions corresponding to absolutely continuous distributions.

Theorem 1.8.16. *Let f be the characteristic function of a distribution σ on $\mathbb{R}^d$. Then σ is absolutely continuous if and only if there exists $g \in \mathcal{L}^2(\mathbb{R}^d)$ such that*

$$f(t) = g * \tilde{g}(t) = \int_{\mathbb{R}^d} g(t+x)\overline{g(x)}\, \mathrm{d}x, \quad t \in \mathbb{R}^d.$$

For this function g we have $\mathrm{d}\sigma(x) = |\hat{g}(x)|^2\, \mathrm{d}x$.

Proof. Assume first that $\mathrm{d}\sigma(t) = p(t)\, \mathrm{d}t$ where $p \in \mathcal{L}^1(\mathbb{R}^d)$, $p \geq 0$, so that

$$f = (2\pi)^{\frac{d}{2}} \cdot \check{p}.$$

Choose an arbitrary Lebesgue measurable function q with $p = |q|^2$ (we can take $q = \sqrt{p}$). Then $q \in \mathcal{L}^2(\mathbb{R}^d)$. Setting $g := \check{q}$ we have $g \in \mathcal{L}^2(\mathbb{R}^d)$, $\hat{g} = q$ and

$$(g * \tilde{g})\hat{} = (2\pi)^{\frac{d}{2}} \cdot |\hat{g}|^2 = (2\pi)^{\frac{d}{2}} \cdot |q|^2 = (2\pi)^{\frac{d}{2}} \cdot p = \hat{f}.$$

Thus, $f = g * \tilde{g}$.

Assume now that $f = g * \tilde{g}$ with some function $g \in \mathcal{L}^2(\mathbb{R}^d)$. Parseval's identity (1.8.9.i) shows that

$$f(t) = \int_{\mathbb{R}^d} g(t+x)\overline{g(x)}\, \mathrm{d}x = \int_{\mathbb{R}^d} e^{i(t,x)} |\hat{g}(x)|^2\, \mathrm{d}x$$

i.e., $\mathrm{d}\sigma(x) = |\hat{g}(x)|^2\, \mathrm{d}x$. □

1.9 Fourier transformation on discrete commutative groups

In this section we prove some basic facts on the Fourier transformation on a commutative group G and its character group $\hat{G}$, which we will need in the sequel.

Definition 1.9.1. The *Fourier–Stieltjes transform* $\hat{\mu}$ of a complex measure $\mu \in \mathcal{M}^b(G)$ is defined by

$$\hat{\mu}(\gamma) = \int_G \overline{\gamma(x)}\, \mathrm{d}\mu(x), \quad \gamma \in \hat{G}.$$

If $\mathrm{d}\mu(x) = f(x)\,\mathrm{d}x$ with $f \in \mathcal{L}^1(G)$ where $\mathrm{d}x$ denotes integration with respect to the counting measure on G, then we write $\hat{f}$ for $\hat{\mu}$ and call $\hat{f}$ the *Fourier transform* of f.

If $G = \mathbb{Z}^d$ and we identify $\hat{G}$ with $[0, 2\pi)^d$ (see Remark 1.7.7) then we have

$$\hat{\mu}(t) = \int_{\mathbb{Z}^d} \mathrm{e}^{-\mathrm{i}(t,n)}\, \mathrm{d}\mu(n) = \sum_{n \in \mathbb{Z}^d} \mu(\{n\}) \cdot \mathrm{e}^{-\mathrm{i}(t,n)}, \quad t \in [0, 2\pi)^d$$

for all $\mu \in \mathcal{M}^b(\mathbb{Z}^d)$.

Theorem 1.9.2. *Every complex measure $\mu \in \mathcal{M}^b(G)$ is uniquely determined by its Fourier–Stieltjes transform $\hat{\mu}$ which is a bounded continuous function. For all $\mu, \nu \in \mathcal{M}^b(G)$ we have*

(i) $(\mu + \nu)\hat{} = \hat{\mu} + \hat{\nu}$;

(ii) $(c\mu)\hat{} = c\hat{\mu}$ *for all* $c \in \mathbb{C}$;

(iii) $(\mu * \nu)\hat{} = \hat{\mu} \cdot \hat{\nu}$;

(iv) $(\tilde{\mu})\hat{} = \overline{\hat{\mu}}$.

Proof. The uniqueness follows from Corollary 1.7.13, the remaining properties are simple consequences of the definitions. □

Definition 1.9.3. The *inverse Fourier–Stieltjes transform* $\check{\mu}$ of a complex measure $\mu \in \mathcal{M}^b(\hat{G})$ is defined by

$$\check{\mu}(x) = \int_{\hat{G}} \gamma(x)\, \mathrm{d}\mu(\gamma), \quad x \in G.$$

Note that the function $\gamma \mapsto \gamma(x)$ is bounded and continuous on $\hat{G}$ for all x in G, so the integral exists. If $\mathrm{d}\mu(\gamma) = f(\gamma)\,\mathrm{d}\lambda(\gamma)$ with $f \in \mathcal{L}^1(\hat{G})$ where λ denotes the normalized Haar measure on $\hat{G}$, then we write $\check{f}$ for $\check{\mu}$ and call $\check{f}$ the *inverse Fourier transform* of f.

If $G = \mathbb{Z}^d$ and we identify $\hat{G}$ with $[0, 2\pi)^d$ then we have

$$\check{\mu}(n) = \int_{[0,2\pi)^d} e^{i(t,n)}\, d\mu(t), \quad n \in \mathbb{Z}^d$$

for all $\mu \in \mathcal{M}^b([0, 2\pi)^d)$.

Theorem 1.9.4. *Every complex measure $\mu \in \mathcal{M}^b(\hat{G})$ is uniquely determined by its inverse Fourier–Stieltjes transform $\check{\mu}$ which is a bounded function. For all $\mu, \nu \in \mathcal{M}^b(\hat{G})$ we have*

(i) $(\mu + \nu)\check{} = \check{\mu} + \check{\nu}$;

(ii) $(c\mu)\check{} = c\check{\mu}$ *for all* $c \in \mathbb{C}$;

(iii) $(\tilde{\mu})\check{} = \overline{\check{\mu}}$.

Proof. The uniqueness can be proved by the same arguments used in the proof of Theorem 1.7.8. The remaining properties are simple consequences of the definitions. □

The next theorem is the analogue of Theorem 1.8.15 and can be proved in the same way. We omit the proof.

Theorem 1.9.5. *Assume that $f \in P(G)$ admits the representing measure $\sigma \in \mathcal{M}^b_+(\hat{G})$ and let $\mu \in \mathcal{M}^b(G)$ be arbitrary. Then the function $\mu * \tilde{\mu} * f$ is positive definite and its representing measure σ_μ is given by $d\sigma_\mu(s) = |\hat{\mu}(s)|^2\, d\sigma(s)$.*

The analogue of Theorem 1.8.7 is contained in the next result.

Theorem 1.9.6.

(i) *The relation $f = (\hat{f})\check{}$ holds for all $f \in \mathcal{L}^1(G)$.*

(ii) *If $f \in P(G) \cap \mathcal{L}^1(G)$, then $\hat{f} \geq 0$.*

(iii) *If $f \in \mathcal{L}^1(G)$ and $\hat{f} \geq 0$, then $f \in P(G)$.*

Proof. (i) Applying Fubini's theorem we obtain

$$\begin{aligned}
(\hat{f})\check{}(t) &= \int_{\hat{G}} \left[\int_G f(x)\overline{\gamma(x)}\, dx \right] \gamma(t)\, d\lambda(\gamma) \\
&= \int_G \left[\int_{\hat{G}} \gamma(t - x)\, d\lambda(\gamma) \right] f(x)\, dx \\
&= \int_G f(x) \mathbf{1}_{\{t\}}(x)\, dx = f(t), \quad t \in G.
\end{aligned}$$

(ii) By Theorem 1.7.8 there exists a unique nonnegative measure $\mu \in \mathcal{M}_+^b(\hat{G})$ such that $f = \check{\mu}$. By (i) we also have $f = (\hat{f})\check{}$. Since μ is unique we must have $\mu = \hat{f}\,\mathrm{d}\lambda$. The nonnegativity of μ implies that $\hat{f} \geq 0$. (iii) This statement follows immediately from (i). □

The extended Fourier transform 1.9.7. If $\mu \in \mathcal{M}^f(\mathbb{Z}^d)$ then we can extend the *Fourier–Stieltjes transform* of μ from $\mathbb{T}^d$ to $\mathbb{C}^d \setminus \{0\}$ by setting

$$\hat{\mu}(z) := \int_{\mathbb{Z}^d} z^{-n}\,\mathrm{d}\mu(n), \quad z \in \mathbb{C}^d \setminus \{0\}.$$

In the same way, if f is a complex-valued function on $\mathbb{Z}^d$ having finite support, then we define the *Fourier transform* of f on $\mathbb{C}^d \setminus \{0\}$ by

$$\hat{f}(z) := \sum_{n \in \mathbb{Z}^d} z^{-n} f(n), \quad z \in \mathbb{C}^d \setminus \{0\}.$$

It is easy to see that the relations

$$\left(\tilde{f}\right)\hat{}(z) = \overline{\hat{f}(1/\bar{z})} \tag{1}$$

$$(f * g)\hat{}(z) = \hat{f}(z) \cdot \hat{g}(z) \tag{2}$$

hold for all $z \in \mathbb{C}^d \setminus \{0\}$. In particular,

$$\left(f * \tilde{f}\right)\hat{}(z) = \hat{f}(z) \cdot \overline{\hat{f}(1/\bar{z})}, \quad z \in \mathbb{C}^d \setminus \{0\}.$$

1.10 Basic properties of Gaussian distributions

1.10.1. In Lemma 1.3.4 we already introduced the standard Gaussian density

$$\varphi(x) = \frac{1}{(2\pi)^{d/2}} \cdot \mathrm{e}^{-\frac{1}{2}(x,x)}, \quad x \in \mathbb{R}^d$$

and showed that its characteristic function is

$$g(t) = \mathrm{e}^{-\frac{1}{2}(t,t)}, \quad t \in \mathbb{R}^d.$$

Suppose that a random vector X has characteristic function g and consider the random vector $Y = m + BX$ where $m \in \mathbb{R}^d$ and B is a $d \times d$ real matrix. The covariance matrix of Y is $C = BB^T$ (see page 349). By Theorem 1.1.7, the characteristic function f_Y of Y is given by

$$f_Y(t) = \mathrm{e}^{\mathrm{i}(m,t)} \cdot g(B^T t) = \mathrm{e}^{\mathrm{i}(m,t) - \frac{1}{2}(B^T t, B^T t)} = \mathrm{e}^{\mathrm{i}(m,t) - \frac{1}{2}(Ct,t)}.$$

Since C is a covariance matrix, it is positive semidefinite (cf. Section F.1).

Now let C be an arbitrary positive semidefinite real matrix. By Theorem D.2.11 there exists a symmetric real matrix B such that $B^2 = C$. The above considerations show that $f_Y(t) = \mathrm{e}^{\mathrm{i}(m,t)-\frac{1}{2}(Ct,t)}$ is the characteristic function of $Y = m + BX$ where X is standard Gaussian.

The exponent $\mathrm{i}(m,t) - \frac{1}{2}(Ct,t)$ is a polynomial of t of degree at most two. We will show in Section 3.5 that a function of the form e^P where P is a polynomial of degree greater than two cannot be a characteristic function (cf. Theorem 3.5.1).

Definition 1.10.2. A d-dimensional random vector Y is said to be *Gaussian* (or *normal*) if its characteristic function f has the form

$$f(t) = \mathrm{e}^{\mathrm{i}(m,t)-\frac{1}{2}(Ct,t)}, \quad t \in \mathbb{R}^d$$

with some $m \in \mathbb{R}^d$ and a $d \times d$ positive semidefinite real matrix C.

For such Y we will write $Y \sim N(m, C)$. It follows from Theorem 1.2.1 that $m = \mathbb{E}(Y)$ and $C = \operatorname{cov}(Y)$.

The next lemma follows immediately from the definition of Gaussian random vectors, we omit the proof.

Lemma 1.10.3. *Let X be a d-dimensional Gaussian vector with $X \sim N(m, C)$. For an arbitrary $n \times d$ matrix B, the n-dimensional random vector $Y = BX$ is Gaussian and $Y \sim N(Bm, BCB^T)$.*

In particular, all marginal distributions of X are Gaussian. Moreover, for each $a \in \mathbb{R}^d$ the random variable (a, X) has a univariate Gaussian distribution with mean (a, m) and variance (a, Ca).

The next theorem gives a very useful characterization of Gaussian random vectors.

Theorem 1.10.4. *A d-dimensional random vector X is Gaussian if and only if for each $a \in \mathbb{R}^d$ the random variable (a, X) is Gaussian.*

Proof. The necessity of the condition has already been established in Lemma 1.10.3. To prove the sufficiency suppose that $X_a = (a, X)$ is Gaussian for any $a \in \mathbb{R}^d$. Setting $m = \mathbb{E}(X)$ and $C = \operatorname{cov}(X)$ we have $\mathbb{E}(X_a) = (m, a)$ and $\operatorname{cov}(X_a) = (Ca, a)$. Thus, the characteristic function f_a of X_a is given by

$$f_a(t) = \mathbb{E}\left(\mathrm{e}^{\mathrm{i}t(a,X)}\right) = \mathrm{e}^{\mathrm{i}t(m,a)-\frac{1}{2}t^2(Ca,a)}, \quad t \in \mathbb{R}.$$

Setting here $t = 1$ we obtain the required characteristic function of X, namely

$$\mathbb{E}\left(\mathrm{e}^{\mathrm{i}(a,X)}\right) = \mathrm{e}^{\mathrm{i}(m,a)-\frac{1}{2}(Ca,a)}, \quad a \in \mathbb{R}^d. \qquad \square$$

Theorem 1.10.5. *Let Y be a d-dimensional Gaussian vector with covariance matrix C having rank $k > 0$. Then there exist a k-dimensional standard Gaussian vector X and a $d \times k$ real matrix B of rank k such that $C = BB^T$ and*

$$Y = \mathbb{E}(Y) + BX.$$

Consequently, the distribution of Y is supported by a k-dimensional linear manifold.

Proof. We may assume that $\mathbb{E}(Y) = 0$. Let $Y = (Y_1, \ldots, Y_d)$ and denote by L the linear subspace of $\mathcal{L}_r^2(\Omega, \mathcal{A}, P)$ spanned by the random variables Y_j. The elements of L are Gaussian random variables with zero mean (cf. Theorem 1.10.4). By Theorem D.2.16, the dimension of L is equal to the rank of C. Let $X_1, \ldots, X_k$ be an arbitrary orthonormal basis of L. Then the random vector $X = (X_1, \ldots, X_k)$ is standard Gaussian. The existence of B follows from the fact that each Y_j is a linear combination of the X_i's. Moreover, $\operatorname{rank}(B) = \dim L = k$. □

Using the previous theorem we can now express the density of absolutely continuous Gaussian vectors by means of the standard Gaussian density.

Theorem 1.10.6. *Let Y be a d-dimensional Gaussian vector with $Y \sim N(m, C)$. If* $\det C > 0$ *then Y has the density*

$$p(x) = \frac{1}{(2\pi)^{d/2} \cdot \sqrt{\det C}} \cdot \mathrm{e}^{-\frac{1}{2}(C^{-1}(x-m), x-m)}, \quad x \in \mathbb{R}^d.$$

Proof. We already know that the statement is true for a standard Gaussian vector (cf. 1.10.1). On the other hand, by Theorem 1.10.5, Y can be written as

$$Y = m + BX$$

where X is d-dimensional, standard Gaussian and B is a $d \times d$ matrix. The covariance of $m + BX$ is BB^T and hence $\det C = (\det B)^2$. The theorem follows now immediately from the transformation formula in Corollary B.7.3. □

Theorem 1.10.7. *Let X_n be a d-dimensional Gaussian random vector with mean m_n and covariance matrix $C_n = (c^n_{i,j})$, $n \in \mathbb{N}$. The sequence $\{X_n\}_1^\infty$ converges in law to some random vector X if and only if the sequences $\{m_n\}_1^\infty$ and $\{c^n_{i,j}\}_{n=1}^\infty$ converge for all i and j. In case of convergence, X is Gaussian with mean $m = \lim_n m_n$ and covariance matrix $C = (\lim_n c^n_{i,j})$.*

Proof. The characteristic function of X_n is given by

$$f_n(t) = \mathrm{e}^{\mathrm{i}(m_n,t)-\frac{1}{2}(C_n t,t)}, \quad t \in \mathbb{R}^d.$$

If the sequences $\{m_n\}_1^\infty$ and $\{c_{i,j}^n\}_{n=1}^\infty$ converge, then

$$\lim_{n\to\infty} f_n(t) = \mathrm{e}^{\mathrm{i}(m,t)-\frac{1}{2}(Ct,t)}, \quad t \in \mathbb{R}^d$$

and convergence in law of the sequence $\{X_n\}_1^\infty$ follows from Theorem 1.6.3.

Assume now that $\{X_n\}_1^\infty$ converges in law. By Theorem 1.6.1, the sequence $\{f_n\}_1^\infty$ converges uniformly on every compact set. The same holds for the sequence $\{|f_n|\}_1^\infty$ hence the limit

$$\lim_{n\to\infty} (C_n t, t), \quad t \in \mathbb{R}^d$$

exists. By Lemma D.2.17, the sequence $\{C_n\}_1^\infty$ converges entrywise to some positive semidefinite matrix C. Considering now the sequence $\{f_n/|f_n|\}_1^\infty$ we see that $\mathrm{e}^{\mathrm{i}(m_n,t)}$ converges to some continuous positive definite function. The modulus of this function is 1 and therefore it has the form $\mathrm{e}^{\mathrm{i}(m,t)}$ with some m (cf. Lemma 1.5.1). Thus, the one point measures δ_{m_n} converge weakly to δ_m from which $m_n \to m$ follows. □

Theorem 1.10.8. *If $X = (X_1, \ldots, X_d)$ is Gaussian and the random variables X_j are pairwise uncorrelated, then $X_1, \ldots, X_d$ are (completely) independent.*

Proof. Since the X_j's are uncorrelated the covariance matrix C of X is diagonal: $C = \operatorname{diag}(\sigma_1^2, \ldots, \sigma_d^2)$. Thus,

$$f_X(t) = \mathrm{e}^{\mathrm{i}(\mathbb{E}X,t)-\frac{1}{2}(Ct,t)} = \prod_{j=1}^d \mathrm{e}^{\mathrm{i}t_j \mathbb{E}X_j - \frac{1}{2}\sigma_j^2 t_j^2}, \quad t \in \mathbb{R}^d$$

and hence $X_1, \ldots, X_d$ are independent (cf. Theorem 1.3.10). □

Remark 1.10.9. There exist uncorrelated Gaussian random variables which are not independent. To see an example let

$$h(x) = \sqrt{2} \cdot \mathrm{e}^{-\frac{1}{2}x^2} - \mathrm{e}^{-x^2}, \quad x \in \mathbb{R}$$

and

$$p(x,y) = \frac{1}{2\pi}\left[h(x)\cdot \mathrm{e}^{-y^2} + h(y)\cdot \mathrm{e}^{-x^2}\right], \quad (x,y) \in \mathbb{R}^2.$$

It is easy to check that p is a density and

$$\int_{-\infty}^{\infty} p(x,y)\,\mathrm{d}y = \frac{1}{\sqrt{2\pi}}\cdot \mathrm{e}^{-\frac{1}{2}x^2}, \quad \int_{-\infty}^{\infty} p(x,y)\,\mathrm{d}x = \frac{1}{\sqrt{2\pi}}\cdot \mathrm{e}^{-\frac{1}{2}y^2}.$$

Since p is an even function in x and in y we have

$$\int_{-\infty}^{\infty}\int_{-\infty}^{\infty} xy\,p(x,y)\,\mathrm{d}x\,\mathrm{d}y = 0.$$

Now let $Z = (X, Y)$ be a two-dimensional random vector with density p. Note that Z is not Gaussian. By what we have proved, X and Y are standard Gaussian and uncorrelated. Since p is not the product of the marginal densities, they are not independent.

Theorem 1.10.10. *Let $X = (X_1, \dots, X_d)$ where $X_1, \dots, X_d$ are independent standard Gaussian random variables. Further, let $a, b \in \mathbb{R}^d \setminus \{0\}$ and O be a $d \times d$ real matrix.*[26]

(i) *The random variables*

$$Y_a = (a, X) = a_1 X_1 + \dots + a_d X_d$$

and

$$Y_b = (b, X) = b_1 X_1 + \dots + b_d X_d$$

are independent if and only if $(a, b) = 0$.

(ii) *The coordinates of the random vector OX are independent if and only if O is an orthogonal matrix.*

Proof. The second statement follows immediately from the first one. To prove (i) we consider the 2-dimensional random vector $Y = (Y_a, Y_b)$. It follows from Theorem 1.10.4 that Y is Gaussian. Applying Theorem 1.10.8 we obtain that Y_a and Y_b are independent if and only if

$$0 = \mathbb{E}(Y_a Y_b) = \sum_{j,k=1}^{d} a_j b_k \mathbb{E}(X_j X_k) = (a, b). \qquad \square$$

1.11 Some inequalities

In the present section we mainly consider inequalities which are used in this book.[27]

Theorem 1.11.1. *Let X be a d-dimensional random vector such that $\|X\| \le R$ with some $R > 0$ and $\mathbb{E}(X) = 0$. Then the inequality*

$$|f_X(t)| \le 1 - \frac{1}{8} \cdot (t, \operatorname{cov}(X)t) \tag{1}$$

holds for all $t \in \mathbb{R}^d$ with $\|t\| < \frac{3}{4R}$.

Proof. Assume first that $d = 1$ and denote by μ the distribution of X. Then

$$f(t) = f_X(t) = \int_{-R}^{R} e^{itx} \, d\mu(x), \quad t \in \mathbb{R}.$$

[26] See also Theorem 3.5.7.

[27] Further inequalities can be found in [57].

Using (B.1.6.1), (B.1.6.2), and (B.1.6.3) it is easy to check that for $|s| < \frac{3}{4}$

$$0 \le \cos s \le 1 - \frac{s^2}{4} \quad \text{and} \quad \sin s = s + h(s) \quad \text{where} \quad |h(s)| \le \frac{s^2}{8}.$$

Since $\mathbb{E}(X) = 0$, the inequalities above imply that

$$0 \le \operatorname{Re} f(t) \le 1 - \frac{1}{4}t^2\mathbb{E}(X^2) \quad \text{and} \quad |\operatorname{Im} f(t)| \le \frac{1}{8}t^2\mathbb{E}(X^2), \quad |t| < \frac{3}{4R}$$

and therefore

$$|f(t)| \le 1 - \frac{1}{8}t^2\mathbb{E}(X^2). \tag{2}$$

Now let $d \in \mathbb{N}$ and $t \in \mathbb{R}^d \setminus \{0\}$ be arbitrary. Write $t_0 = t/\|t\|$ and denote by g the characteristic function of the random variable $Y = (t_0, X)$. Then $|Y| \le \|X\| \le R$, $\mathbb{E}(Y) = 0$,

$$f(t) = \mathbb{E}\left(e^{i(t,X)}\right) = \mathbb{E}\left(e^{i\|t\|Y}\right) = g(\|t\|),$$

and $\mathbb{E}(Y^2) = (t_0, \operatorname{cov}(X)t_0) = (t, \operatorname{cov}(X)t)/\|t\|^2$. Application of equation (2) for Y completes the proof. □

Replacing X by $X - \mathbb{E}(X)$ changes neither the modulus of the characteristic function nor the covariance matrix and $\|X - \mathbb{E}(X)\| \le 2R$. Thus we have the following corollary.

Corollary 1.11.2. *If $\|X\| \le R$ then the inequality (1.11.1.1) holds for all $t \in \mathbb{R}^d$ with $\|t\| < \frac{3}{8R}$.*

Lemma 1.11.3. *Let X be a d-dimensional random vector and define the random vector X_R $(R > 0)$ by*

$$X_R(\omega) = \begin{cases} X(\omega) & \text{if} \quad \|X(\omega)\| \le R \\ 0 & \text{if} \quad \|X(\omega)\| > R \end{cases} \qquad \omega \in \Omega.$$

Then the corresponding characteristic functions f and f_R satisfy the inequality

$$|\operatorname{Re} f(t)| \le \operatorname{Re} f_R(t)$$

for all $t \in \mathbb{R}^d$ with $\|t\| \le \frac{\pi}{2R}$.

Proof. Denote by μ and μ_R the corresponding distributions and write

$$B = \{x \in \mathbb{R}^d : \|x\| \le R\}.$$

Note that

$$\mu_R = \mu_B + (1 - \mu(B)) \cdot \delta_0$$

where μ_B is given by $\mu_B(A) = \mu(B \cap A)$, $A \in \mathcal{B}_d$. If $\|t\| \le \frac{\pi}{2R}$ and $\|x\| \le R$, then $\cos((t,x)) \ge 0$. Using this we obtain:

$$\begin{aligned}
|\mathrm{Re}\, f(t)| &= \left| \int_{\mathbb{R}^d} \cos((t,x))\, \mathrm{d}\mu(x) \right| \\
&\le \left| \int_B \cos((t,x))\, \mathrm{d}\mu(x) \right| + \left| \int_{\mathbb{R}^d \setminus B} \cos((t,x))\, \mathrm{d}\mu(x) \right| \\
&= \int_B \cos((t,x))\, \mathrm{d}\mu(x) + \left| \int_{\mathbb{R}^d \setminus B} \cos((t,x))\, \mathrm{d}\mu(x) \right| \\
&\le \int_B \cos((t,x))\, \mathrm{d}\mu(x) + \mu(\mathbb{R}^d \setminus B) \\
&= \int_{\mathbb{R}^d} \cos((t,x))\, \mathrm{d}\mu_B(x) = \mathrm{Re}\, f_R(t). \quad \square
\end{aligned}$$

Theorem 1.11.4. *Let X be a d-dimensional random vector such that the support of its distribution is not contained in some hyperplane of $\mathbb{R}^d$. Then there exist positive numbers δ and ϵ such that the inequality*

$$|f_X(t)| \le 1 - \epsilon \cdot \|t\|^2 \tag{1}$$

holds for all $t \in \mathbb{R}^d$ with $\|t\| < \delta$.[28]

Proof. We may suppose that $f := f_X$ is real-valued. Indeed, if the statement is true for real-valued characteristic functions, then for an arbitrary characteristic function f we have

$$|f(t)|^2 \le 1 - \epsilon \cdot \|t\|^2, \quad \|t\| \le \delta$$

for some positive δ and ϵ, which yields

$$|f(t)| \le 1 - \frac{\epsilon}{2} \cdot \|t\|^2, \quad \|t\| \le \delta.$$

We choose $R > 0$ such that the random variable X_R from Lemma 1.11.3 is not concentrated on a hyperplane.[29] Then the covariance matrix of X_R is strictly positive definite and hence

$$(t, \mathrm{cov}(X_R)t) \ge \lambda \cdot \|t\|^2, \quad t \in \mathbb{R}^d$$

where λ (> 0) is the smallest eigenvalue of $\mathrm{cov}(X_R)$. The theorem now follows from Corollary 1.11.2 and Lemma 1.11.3, using the fact that f and f_R are real valued. $\square$

[28] See also Lemma 1.2.8.

[29] For this it is sufficient to choose R such that $\{x \in \mathrm{supp}\,\mu : \|x\| \le R\}$ has a positive Lebesgue measure.

Theorem 1.11.5. *Let $n \in \mathbb{N}$ and f be a real-valued $2n$-times differentiable characteristic function on $\mathbb{R}$. Then we have*

$$\sum_{k=0}^{2\lfloor\frac{n-1}{2}\rfloor+1} (-1)^k \frac{M_{2k}}{(2k)!} t^{2k} \le f(t) \le \sum_{k=0}^{2\lfloor\frac{n}{2}\rfloor} (-1)^k \frac{M_{2k}}{(2k)!} t^{2k}, \quad t \in \mathbb{R},$$

where M_{2k} is the moment of order $2k$ of the corresponding distribution.

Proof. From Corollary 1.2.3 and inequality (1.1.2.ii) we see that

$$-M_{2k} \le f^{(2k)}(t) \le M_{2k}, \quad t \in \mathbb{R}, \ k = 0, 1, \dots, n.$$

Since f is an even function, all derivatives of odd order are zero at 0. Applying the upper estimate to $f^{(4\lfloor\frac{n}{2}\rfloor)}(\theta t)$ in the Taylor expansion

$$f(t) = \sum_{k=0}^{2\lfloor\frac{n}{2}\rfloor-1} (-1)^k \frac{M_{2k}}{(2k)!} t^{2k} + \frac{t^{4\lfloor\frac{n}{2}\rfloor}}{(4\lfloor\frac{n}{2}\rfloor)!} \cdot f^{(4\lfloor\frac{n}{2}\rfloor)}(\theta t), \quad t \in \mathbb{R}, \ \theta \in [0, 1],$$

we obtain the upper estimate for $f(t)$. The lower estimate can be obtained in the same way. □

As an application of Corollary 1.7.9 we prove two useful inequalities for positive definite functions.

Theorem 1.11.6. *Let f be a positive definite function on a commutative group G such that $f(0) = 1$. Then for all $x \in G$ and $n \in \mathbb{N}$ we have*

(i) $1 - \operatorname{Re} f(nx) \le n[1 - (\operatorname{Re} f(x))^n]$;

(ii) $1 - |f(nx)| \le n[1 - |f(x)|^n]$.

Proof. For each fixed $x \in G$ the function $n \mapsto f(nx)$ is positive definite on $\mathbb{Z}$. We may therefore assume that $G = \mathbb{Z}$ and $x = 1$. Thus, we have to show that

$$1 - \operatorname{Re} f(n) \le n[1 - (\operatorname{Re} f(1))^n] \quad (1)$$

$$1 - |f(n)| \le n[1 - |f(1)|^n]. \quad (2)$$

To prove the first inequality we define the set $K_n \subset \mathbb{R}^2$ by

$$K_n := \{(x, y) : |x| \le 1, |y| \le 1, nx^n + 1 - n \le y\}.$$

Inequality (1) is then equivalent to the relation $(\operatorname{Re} f(1), \operatorname{Re} f(n)) \in K_n$. The set K_n is convex for all $n \in \mathbb{N}$. Using this and Corollary 1.7.9 we see that it is sufficient to show (1) for the positive definite functions $n \mapsto z^n$, $z \in \mathbb{T}$. Thus, it remains to prove that

$$1 - \operatorname{Re}(z^n) \le n[1 - (\operatorname{Re} z)^n], \quad z \in \mathbb{T},$$

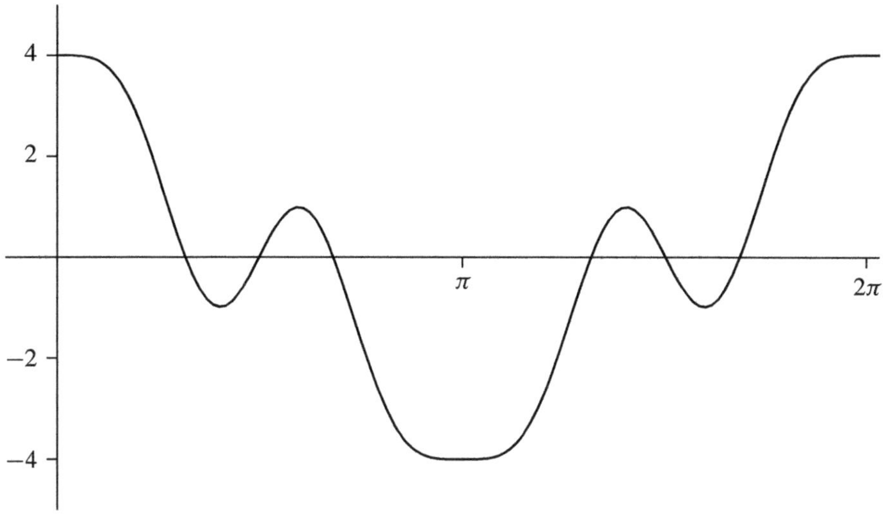

Figure 1.7. The function h_n from the proof of Theorem 1.11.6 with $n = 5$.

or equivalently,

$$h_n(t) := h(t) := n\cos^n t - \cos nt \le n - 1, \quad t \in \mathbb{R} \tag{3}$$

(see Figure 1.7). The function h is periodic and hence it attains its maximum at some $y \in \mathbb{R}$. From

$$h'(y) = -n^2 \cos^{n-1} y \sin y + n \sin ny = 0$$

we see that either $\sin y = 0$ or

$$\sin y \ne 0 \quad \text{and} \quad n\cos^{n-1} y = \frac{\sin ny}{\sin y}. \tag{4}$$

If $\sin y = 0$, then (3) holds. If y satisfies (4), then

$$\begin{aligned} h(y) &= \frac{\cos y \sin ny}{\sin y} - \cos ny = \frac{\cos y \sin ny - \sin y \cos ny}{\sin y} \\ &= \frac{\sin(n-1)y}{\sin y}. \end{aligned}$$

Using induction on n we see that $\sin(n-1)y/\sin y \le n - 1$, $n \in \mathbb{N}$, and therefore $h(y) \le n - 1$.

To prove inequality (2), let $z \in \mathbb{T}$ be such that $\operatorname{Re}(zf(1)) = |f(1)|$. Since the function $n \mapsto z^n f(n)$ is positive definite on $\mathbb{Z}$, inequality (1) shows that

$$n|f(1)|^n + 1 - n = n(\operatorname{Re} zf(1))^n + 1 - n \le \operatorname{Re}(z^n f(n)) \le |f(n)|$$

from which (2) follows. □

Using the inequality

$$1 - r^n \leq n(1 - r), \ -1 \leq r \leq 1,$$

we obtain the following corollary:

Corollary 1.11.7. *With the notation of the previous theorem we have*

(i) $1 - \text{Re}\, f(nx) \leq n^2\, [1 - \text{Re}\, f(x)]$;

(ii) $1 - |f(nx)| \leq n^2\, [1 - |f(x)|]$.

Chapter 2

Correlation functions

In the first section of this chapter we present a few basic results on processes with independent increments which can be obtained easily by using characteristic functions. Then we concentrate on those properties of second order fields which are related to the fact that their correlation functions represent positive semidefinite kernels. In the second part of the chapter we investigate the special case of stationary fields. At the end of the chapter we give an introduction to the harmonic analysis of stationary fields using unitary representations built from the correlation function.

Throughout this chapter the symbol T denotes a nonempty subset of $\mathbb{R}^d$ with some $d \in \mathbb{N}$ and $(\Omega, \mathcal{A}, P)$ is a probability space.

2.1 Random fields

Basic notions 2.1.1. A real (complex) *random field* Z is a mapping from T into the set of all real (complex, respectively) random variables on $(\Omega, \mathcal{A}, P)$. The random variable $Z(t)$, $t \in T$, is also written as Z_t. If $Z(t)$ is square integrable for all $t \in T$, then Z is called a *second order random field*. In the special case $T = \mathbb{R}$ or $T = [0, \infty)$ we also use the terminology *random process* or simply *process* instead of random field. If $T = \mathbb{Z}$ or $T = \mathbb{N}_0$, then Z is usually called a *time series*.

For any finite subset $\{t_1, \ldots, t_n\}$ of T the distribution of the random vector $X = (Z(t_1), \ldots, Z(t_n))$ is called a *finite-dimensional distribution* of Z. If Z is real and all finite-dimensional distributions are Gaussian (see Definition 1.10.2), then Z is said to be a *Gaussian field*. Note that Z is Gaussian if and only if the characteristic function f_X of X is given by

$$f_X(s) = \mathrm{e}^{\mathrm{i}(s,\mathbb{E}(X)) - \frac{1}{2}(Cs,s)}, \quad s = (s_1, \ldots, s_n) \in \mathbb{R}^n$$

where C is the covariance matrix of X. If Z is complex and all random vectors

$$(\operatorname{Re} Z(t_1), \operatorname{Im} Z(t_1), \ldots, \operatorname{Re} Z(t_n), \operatorname{Im} Z(t_n))$$

are Gaussian, then Z is called a *complex Gaussian field.*

A process Z on $T = \mathbb{R}$ or $T = [0, \infty)$ is said to have *independent increments* if the random variables

$$Z(t_2) - Z(t_1), Z(t_3) - Z(t_2), \ldots, Z(t_n) - Z(t_{n-1}),$$

the so-called *increments of* Z, are independent for every $n \geq 3$ and every $t \in \mathbb{R}^n$ such that $t_1 < \cdots < t_n$ and $t_j \in T$.

Next we prove the existence of processes having special properties. The proof uses Kolmogorov's existence theorem formulated in terms of characteristic functions (cf. Remark F.3.4). We need the following simple lemma.

Lemma 2.1.2. *A real process Z on $T = [0, \infty)$ with $Z(0) = 0$ has independent increments if and only if for every $d \geq 1$ and for every $t \in \mathbb{R}^d$ such that $0 \leq t_1 < \cdots < t_d$ the characteristic function f_t of the random vector $(Z(t_1), \ldots, Z(t_d))$ satisfies the equation*

$$f_t(x) = g_1(x_1 + \cdots + x_d) \cdot g_2(x_2 + \cdots + x_d) \cdot \ldots \cdot g_d(x_d), \quad x \in \mathbb{R}^d \tag{1}$$

where g_j, $j \geq 1$, denotes the characteristic function of the random variable $\Delta_j = Z(t_j) - Z(t_{j-1})$, and $t_0 = 0$.

Proof. Denote by g the characteristic function of the random vector $(\Delta_1, \ldots, \Delta_d)$. We have

$$\begin{aligned} f_t(x) &= \mathbb{E}\left(\exp\left[\mathrm{i}x_1 Z(t_1) + \mathrm{i}x_2 Z(t_2) + \cdots + \mathrm{i}x_d Z(t_d)\right]\right) \\ &= \mathbb{E}\left(\exp\left[\mathrm{i}x_1 \Delta_1 + \mathrm{i}x_2(\Delta_1 + \Delta_2) + \cdots + \mathrm{i}x_d(\Delta_1 + \cdots + \Delta_d)\right]\right) \\ &= \mathbb{E}\left(\exp\left[\mathrm{i}(x_1 + \cdots + x_d)\Delta_1 + \mathrm{i}(x_2 + \cdots + x_d)\Delta_2 + \cdots + \mathrm{i}x_d \Delta_d\right]\right) \\ &= g(x_1 + \cdots + x_d, x_2 + \cdots + x_d, \ldots, x_d). \end{aligned}$$

These equations show that (1) holds whenever the random variables Δ_j are independent. Assume now that (1) holds. The equations above show that

$$g(y) = g_1(y_1) \cdot \ldots \cdot g_d(y_d)$$

where $y_j = x_j + \cdots + x_d$, $1 \leq j \leq d$. Since any $y \in \mathbb{R}^d$ can be represented in this way with some $x \in \mathbb{R}^d$ the independence of the Δ_j's follows from Theorem 1.3.10. □

Theorem 2.1.3. *Let h be a complex-valued function on $\mathbb{R}$ such that e^{ph} is a characteristic function for all $p \in [0, \infty)$.*[1] *There exists a process Z on $[0, \infty)$ satisfying the conditions*

(i) $Z(0) = 0$;

(ii) *Z has independent increments;*

(iii) *the characteristic function of $Z(t) - Z(s)$ is $\mathrm{e}^{(t-s)h}$ for all $t \geq s$.*

[1] We will investigate functions with this property in Section 3.11.

Proof. Let $d \geq 1$ and $t_0 = 0$. By assumption, the function

$$f_t(x) = \exp\left[\sum_{j=1}^{d}(t_j - t_{j-1})h(x_j + \cdots + x_d)\right], \quad x \in \mathbb{R}^d$$

is a characteristic function for all $t \in [0,\infty)^d$ such that $t_1 \leq \cdots \leq t_d$. For arbitrary $t \in [0,\infty)^d$ we choose a permutation π such that $t_{\pi(1)} \leq \cdots \leq t_{\pi(d)}$ and define f_t by the consistency condition (F.3.4.i) Then condition (F.3.4.ii) is satisfied as well. Consequently, there exists a process Z on $[0,\infty)$ such that the finite-dimensional distributions of Z have the f_t's as characteristic functions. Obviously, $Z(0) = 0$. The equation above with $d = 2$ and $x_2 = -x_1$ shows that the characteristic function of $Z(t) - Z(s)$ is $\mathrm{e}^{(t-s)h}$ if $t \geq s$ (cf. Theorem 1.1.7). By Lemma 2.1.2, the process Z has independent increments. □

As a corollary we establish the existence of two important processes, the Wiener process and the Poisson process.[2] The next two statements follow immediately from Theorem 2.1.3 by noting that the characteristic function of a Gaussian random variable with mean zero and variance σ^2 is $t \mapsto \mathrm{e}^{-\frac{1}{2}\sigma^2 t^2}$ while the characteristic function of a Poisson distributed random variable with mean $\lambda \geq 0$ is $t \mapsto \mathrm{e}^{\lambda(\mathrm{e}^{\mathrm{i}t}-1)}$ (see Examples 1.1.13 and 1.2.4).

Corollary 2.1.4 (Wiener process). *There exists a process W on $[0,\infty)$ satisfying the conditions*

(i) $W(0) = 0$;

(ii) *W has independent increments;*

(iii) *$W(t) - W(s)$ has a Gaussian distribution with mean zero and variance $t - s$ for all $t \geq s$.*

Corollary 2.1.5 (Poisson process). *For each $\lambda \geq 0$ there exists a process N on $[0,\infty)$ satisfying the conditions*

(i) $N(0) = 0$;

(ii) *N has independent increments;*

(iii) *$N(t) - N(s)$ has a Poisson distribution with mean $\lambda(t - s)$ for all $t \geq s$.*

[2] Usually one also imposes conditions on the *trajectories* $t \mapsto Z_t(\omega)$ of these processes but we do not consider these questions in this book.

2.2 Correlation functions of second order random fields

To give a complete description of a random field it is necessary to specify all of its finite-dimensional distributions. Since the theoretical determination of these distributions is possible only in a few cases, it is natural to restrict oneself to investigating properties of the field which are determined by simple characteristics of the finite-dimensional distributions. The next definition uses only the first and second moments of the field.

Definition 2.2.1. Let Z be a second order field on T. The function $C : T \times T \to \mathbb{C}$ defined by

$$C(s,t) = \mathbb{E}\left(Z(s) \cdot \overline{Z(t)}\right)$$

is called the *correlation function* of Z. The *covariance function* is defined by

$$\sigma(s,t) = \mathbb{E}\left([Z(s) - M(s)] \cdot [\overline{Z(t) - M(t)}]\right)$$

where

$$M(t) = \mathbb{E}(Z(t))$$

is the *mean* of the field.

An easy computation shows that

$$\sigma(s,t) = C(s,t) - M(s) \cdot \overline{M(t)}.$$

If the mean is zero, the correlation function and the covariance function are equal.

Note that all finite-dimensional distributions of a Gaussian field are completely determined if we know its mean and correlation function.

Example 2.2.2. (a) Let $X \in \mathcal{L}^2(\Omega, \mathcal{A}, P)$ and $f : T \to \mathbb{C}$ be arbitrary. Then

$$Z(t) = f(t) \cdot X, \quad t \in T$$

is a second order field with mean $M(t) = f(t) \cdot \mathbb{E}(X)$ and correlation function

$$C(s,t) = f(s)\overline{f(t)} \cdot \mathbb{E}(|X|^2).$$

(b) If $Z_0 = 0$ and Z has independent increments, then for $0 \le s \le t$ we have

$$\begin{aligned} C(s,t) = \mathbb{E}(Z_s Z_t) &= \mathbb{E}(Z_s(Z_t - Z_s)) + \mathbb{E}(Z_s^2) \\ &= \mathbb{E}(Z_s) \cdot \mathbb{E}(Z_t - Z_s) + \mathbb{E}(Z_s^2) \\ &= \mathbb{E}(Z_s) \cdot \mathbb{E}(Z_t) + \mathrm{Var}(Z_s) \end{aligned}$$

and hence

$$\sigma(s,t) = \mathrm{Var}(Z_s), \quad s \le t.$$

In particular, the covariance function of the Wiener process and of the Poisson process with $\lambda = 1$ is given by[3]

$$\sigma(s,t) = \min(s,t), \quad t,s \in [0,\infty).$$

The next theorem gives a characterization of correlation functions.

Theorem 2.2.3. *A complex-valued (real-valued) function K defined on $T \times T$ is the covariance or correlation function of a second order complex (real, respectively) field if and only if for any finite collection $t_1, \dots, t_n$ of elements of T the matrix*

$$\Big(K(t_i,t_j)\Big)_{i,j}^{n}$$

is positive semidefinite (and symmetric, respectively).

Proof. Assume that K is the covariance function of a complex field Z. The case where K is the correlation function can be treated in the same way. We have

$$\begin{aligned}
&\sum_{i,j=1}^{n} K(t_i,t_j)c_i\overline{c_j} \\
&\quad = \sum_{i,j=1}^{n} \mathbb{E}\Big[(Z(t_i) - M(t_i)) \cdot \overline{(Z(t_j) - M(t_j))}\Big] c_i\overline{c_j} \\
&\quad = \mathbb{E}\Big|\sum_{j=1}^{n} c_j(Z(t_j) - M(t_j))\Big|^2 \geq 0
\end{aligned}$$

for all $c_1, \dots, c_n \in \mathbb{C}$, i.e., K is positive semidefinite. If the field is real, then K is symmetric and the relations above hold for all $c_1, \dots, c_n \in \mathbb{R}$. Theorem D.2.4 shows that K is positive semidefinite.

The remaining statements follow from Theorem 2.2.4. □

Theorem 2.2.4. *Let $M : T \to \mathbb{C}$ be arbitrary and let $K : T \times T \to \mathbb{C}$ be such that for any finite collection $t_1, \dots, t_n$ of elements of T the matrix*

$$\Big(K(t_i,t_j)\Big)_{i,j}^{n} \tag{1}$$

is positive semidefinite. Then there exists a Gaussian field Z on T such that

$$\mathbb{E}(Z(t)) = M(t) \tag{2}$$

$$\mathbb{E}(Z(s) \cdot \overline{Z(t)}) - M(s) \cdot \overline{M(t)} = K(s,t) \tag{3}$$

$$\mathbb{E}(Z(s) \cdot Z(t)) = M(s) \cdot M(t). \tag{4}$$

[3] Combining this with Theorem 2.2.3 we obtain an alternative, though not elementary, proof of the positive definiteness of the kernel $(s,t) \mapsto \min(s,t)$ (cf. Theorem D.2.10).

If M and K are real-valued and K is symmetric, then there exists a real Gaussian field Z such that (2) *and* (3) *hold.*[4]

Proof. Assume first that M and K are real-valued and let $\{t_1, \dots, t_n\} \subset T$ be arbitrary. Since K is positive semidefinite there exists a unique Gaussian distribution on $\mathbb{R}^n$ with mean vector $(M(t_1), \dots, M(t_n))$ and covariance matrix (1). The collection of all these distributions satisfies the conditions of Kolmogorov's consistency Theorem F.3.3. Hence there exists a real Gaussian field satisfying (2) and (3).

Now assume that M and K are complex-valued and let $\tilde{T} := T \times \{1, 2\}$. Further, let $\tilde{M}(t, 1) = \operatorname{Re} M(t)$, $\tilde{M}(t, 2) = \operatorname{Im} M(t)$ and define the real kernel $\tilde{K}$ on $\tilde{T} \times \tilde{T}$ by setting

$$\tilde{K}((s,1),(t,1)) = \tilde{K}((s,2),(t,2)) = \frac{1}{2}\operatorname{Re} K(s,t)$$

and

$$\tilde{K}((s,1),(t,2)) = -\tilde{K}((s,2),(t,1)) = -\frac{1}{2}\operatorname{Im} K(s,t), \quad s, t \in T.$$

It follows from Lemma D.2.14 that $\tilde{K}$ is positive semidefinite. By the first part of the proof there exists a real field $\tilde{Z}$ on $\tilde{T}$ such that the analogues of (2) and (3) are satisfied for $\tilde{M}$, $\tilde{K}$ and $\tilde{Z}$. Setting now

$$Z(t) := \tilde{Z}(t,1) + \mathrm{i}\tilde{Z}(t,2), \quad t \in T$$

an easy computation shows that Z satisfies the equations (2)–(4). □

Example 2.2.5. By Lemma D.2.10 and Theorem 2.2.4, there exists a real Gaussian process X on $T = [0, \infty)$ such that $\mathbb{E}(X(t)) = 0$ and

$$\mathbb{E}(X(t)X(s)) = \min(t,s), \quad t, s \in [0, \infty). \tag{1}$$

Let $t_1 \le t_2 \le s_1 \le s_2$. Using (1) we obtain

$$\begin{aligned} \mathbb{E}([X(t_2) - X(t_1)] \cdot [X(s_2) - X(s_1)]) &= C(t_2, s_2) - C(t_2, s_1) \\ &\quad - C(t_1, s_2) + C(t_1, s_1) \\ &= t_2 - t_2 - t_1 + t_1 \\ &= 0 \end{aligned}$$

i.e., the increments $X(t_2) - X(t_1)$ and $X(s_2) - X(s_1)$ are uncorrelated. Since the increments of a Gaussian process are Gaussian, we infer that they are independent (cf. Theorem 1.10.8). Thus, we obtain the Wiener process constructed in Corollary 2.1.4.

Now let $h \in [0, \infty)$ be arbitrary and define the process X^h by

$$X^h(t) = X(t+h) - X(t), \quad t \in [0, \infty).$$

[4] If a real field satisfies (2) and (4), then $E([Z(t) - M(t)]^2) = 0$ and hence $Z(t) = M(t)$.

Denote by C^h the correlation function of this process. The mean of X^h is zero and for $s \le t$ we have

$$\begin{aligned} C^h(s,t) &= \mathbb{E}\left[(X(s+h) - X(s)) \cdot (X(t+h) - X(t))\right] \\ &= C(s+h, t+h) - C(s+h, t) - C(s, t+h) + C(s,t) \\ &= s + h - \min(s+h, t) - s + s \\ &= \max(0, h - t + s). \end{aligned}$$

In the same way we see that $C^h(s,t) = \max(0, h + t - s),\ s \ge t$, and hence

$$C^h(s,t) = \max(0, h - |s - t|).$$

Therefore $C^h(s,t)$ depends only on $s - t$, a property that we will call stationarity (cf. Definition 2.8.1).

Notation 2.2.6. Recall that the set $\mathcal{L}^2(P) := \mathcal{L}^2(\Omega, \mathcal{A}, P)$ of all square integrable complex-valued random variables is a linear space with the usual notion of multiplication by a scalar and addition of random variables. This linear space endowed with the inner product

$$(X, Y) := \int_\Omega X \cdot \overline{Y} \, \mathrm{d}P = \mathbb{E}(X \cdot \overline{Y})$$

is a positive semidefinite inner product space. The inner product space $\mathcal{L}_r^2(P)$ of all square integrable real-valued random variables on $(\Omega, \mathcal{A}, P)$ is introduced in the same way. We denote the corresponding Hilbert spaces by $L^2(P)$ and $L_r^2(P)$, the elements of which are equivalence classes of random variables. In all these cases we will use the notation $\|\cdot\|$ for the norm:

$$\|X\| = \sqrt{(X, X)}.$$

We have by the Cauchy Schwarz inequality

$$|\mathbb{E}(X \cdot Y)| \le \|X\| \cdot \|Y\|.$$

Note that the expectation can be expressed using the inner product:

$$\mathbb{E}(X) = (X, \mathbf{1}_\Omega).$$

Two square integrable random variables with zero mean are orthogonal if and only if they are uncorrelated.

For a field Z of second order on T we define the mapping $\mathcal{Z} : T \to L^2(P)$ by $\mathcal{Z}_t = Z_t + N$ where N is the class of random variables which are zero P-almost everywhere. We call $\mathcal{Z}$ the L^2*-valued random field* corresponding to Z. The notions *mean, covariance function, correlation function* and *Gaussian field* are defined for $\mathcal{Z}$

in the same way as for Z. By $H(Z)$ and $H(\mathcal{Z})$ we denote the closed linear subspaces generated by all Z_t and $\mathcal{Z}_t$, $t \in T$, respectively.

Let $\{e_\alpha\}_{\alpha\in I}$ be an orthonormal basis of $H(\mathcal{Z})$. The index set I may be finite, countable infinite or uncountable.[5] In any case

$$\mathcal{Z}_t = \sum_{\alpha\in I} f_\alpha(t) \cdot e_\alpha < \infty, \quad t \in T \tag{1}$$

where $f_\alpha(t) = (Z_t, e_\alpha)$ (for each t only countable many summands are different from zero). Moreover,

$$\|Z_t\|^2 = \|\mathcal{Z}_t\|^2 = \sum_{\alpha\in I} |f_\alpha(t)|^2. \tag{2}$$

Theorem 2.2.7. *A complex-valued (real-valued) function K on $T \times T$ is the correlation function or covariance function of a complex (real, respectively) field of second order if and only if it can be represented in the form*

$$K(s,t) = \sum_{\alpha\in I} f_\alpha(s)\overline{f_\alpha(t)}, \quad s,t \in T \tag{1}$$

where I is some nonempty set and f_α, $\alpha \in I$, are complex-valued (real-valued, respectively) functions on T such that

$$\sum_{\alpha\in I} |f_\alpha(t)|^2 < \infty, \quad t \in T. \tag{2}$$

Proof. We consider only correlation functions and the complex case, the rest is similar. Assume that (1) and (2) hold. Then K is positive semidefinite:

$$\sum_{j,k=1}^{n} K(t_j, t_k) c_j \overline{c_k} = \sum_{\alpha\in I} \left| \sum_{j=1}^{n} f_\alpha(t_j) c_j \right|^2 \geq 0.$$

Theorem 2.2.3 shows that K is a correlation function.

Now let K be the correlation function of a field $Z : T \to L^2(\Omega, \mathcal{A}, P)$. The relations (1) and (2) follow immediately from (2.2.6.1) and (2.2.6.2) by noting that $(Z_s, Z_t) = (\mathcal{Z}_s, \mathcal{Z}_t)$. □

[5] To give an example where I is uncountable, we take independent random variables Z_t, $t \in \mathbb{R}$, with mean 0 and variance 1. Then $\{\mathcal{Z}_t\}_{t\in\mathbb{R}}$ is an orthonormal basis of $H(\mathcal{Z})$. The fact that such random variables exist follows from Corollary F.3.2.

2.3 Continuity and differentiability

As we will see, continuity and differentiability of a second order field can be completely described in terms of its correlation function.

Definition 2.3.1. A second order field Z is called *strongly continuous* at $t_0 \in T$ if

$$\lim_{t \to t_0} \mathbb{E}\left(|Z(t) - Z(t_0)|^2\right) = 0.$$

The field Z is called *strongly continuous* if it is strongly continuous at each $t_0 \in T$.

Theorem 2.3.2. *A second order field Z on T is strongly continuous at $t \in T$ if and only if its correlation function C is continuous at (t, t). Moreover, if for some $t, s \in T$ the function C is continuous at (t, t) and (s, s), then it is continuous at (t, s) and (s, t) as well.*

Proof. For any $h \in T$ such that $t + h \in T$ we have

$$\|Z(t+h) - Z(t)\|^2 = C(t+h, t+h) - C(t+h, t) - C(t, t+h) + C(t, t).$$

Thus, if C is continuous at (t, t), then Z is strongly continuous at t.

For t and h as above we have

$$\begin{aligned} C(t+h, t+\epsilon) - C(t, t) = \mathbb{E}\Big(& [Z(t+h) - Z(t)] \cdot [\overline{Z(t+\epsilon)} - \overline{Z(t)}] \\ & + [Z(t+h) - Z(t)] \cdot \overline{Z(t)} \\ & + [Z(t+\epsilon) - Z(t)] \cdot \overline{Z(t)}\Big). \end{aligned}$$

Applying the Cauchy–Schwarz inequality we obtain

$$\begin{aligned} |C(t+h, t+\epsilon) - C(t, t)| \le & \|Z(t+h) - Z(t)\| \cdot \|Z(t+\epsilon) - Z(t)\| \\ & + \|Z(t+h) - Z(t)\| \cdot \|Z(t)\| \\ & + \|Z(t+\epsilon) - Z(t)\| \cdot \|Z(t)\|. \end{aligned}$$

Thus, if Z is strongly continuous at t, then C is continuous at (t, t).

To prove the second statement assume that C is continuous at (t, t) and (s, s). As we have seen, this implies that Z is strongly continuous at t and s. The proof is completed by noting that

$$C(t, s) = (Z(t), Z(s)).$$

□

Definition 2.3.3. A process Z of second order on (a, b), $-\infty \leq a < b \leq \infty$, is called *strongly differentiable*, if for all $t \in (a, b)$ the strong limit[6]

$$\lim_{h \to 0} \frac{Z(t+h) - Z(t)}{h}$$

exists. The limit is denoted by $Z'(t)$.

In the same way one can define partial derivatives of a field defined on an open set $T \subset \mathbb{R}^d$.

Next we show that strong differentiability of a process is closely related to the differentiability of its correlation function. For a function f of two variables we write

$$\Delta^2 f(t, s, \epsilon, h) = f(t + \epsilon, s + h) - f(t + \epsilon, s) - f(t, s + h) + f(t, s).$$

Theorem 2.3.4. *Let Z be a process of second order on (a, b) with correlation function C. The process Z is strongly differentiable if and only if the limit*

$$\lim_{\epsilon, h \to 0} \frac{\Delta^2 C(t, t, \epsilon, h)}{\epsilon h}$$

exists for all $t \in (a, b)$.

If Z is strongly differentiable, then the limit

$$D(t, s) := \lim_{\epsilon, h \to 0} \frac{\Delta^2 C(t, s, \epsilon, h)}{\epsilon h}$$

exists for all $t, s \in (a, b)$ and D is the correlation function of Z'. Moreover,

$$\frac{\mathrm{d}}{\mathrm{d}t}(Z(t), v) = (Z'(t), v) \tag{1}$$

for all $v \in L^2(\Omega, \mathcal{A}, P)$. In particular, with $v = \mathbf{1}_\Omega$,

$$\frac{\mathrm{d}}{\mathrm{d}t} \mathbb{E}(Z(t)) = \mathbb{E}(Z'(t)). \tag{2}$$

Proof. By Cauchy's criterion (cf. Lemma D.3.2), the process Z is strongly differentiable at t if and only if the limit

$$\lim_{\epsilon, h \to 0} \left(\frac{Z(t+\epsilon) - Z(t)}{\epsilon}, \frac{Z(t+h) - Z(t)}{h} \right) = \lim_{\epsilon, h \to 0} \frac{\Delta^2 C(t, t, \epsilon, h)}{\epsilon h}$$

exists. This shows the first statement.

---

[6] That is, limit with respect to the norm of $L^2(P)$.

If Z is strongly differentiable, then

$$\begin{aligned}(Z'(t), Z'(s)) &= \left(\lim_{\epsilon\to 0}\frac{Z(t+\epsilon)-Z(t)}{\epsilon}, \lim_{h\to 0}\frac{Z(s+h)-Z(s)}{h}\right)\\ &= \lim_{\epsilon,h\to 0}\left(\frac{Z(t+\epsilon)-Z(t)}{\epsilon}, \frac{Z(s+h)-Z(s)}{h}\right)\\ &= \lim_{\epsilon,h\to 0}\frac{\Delta^2 C(t,s,\epsilon,h)}{\epsilon h} = D(t,s).\end{aligned}$$

Equation (1) follows from the fact that strong convergence implies weak convergence:

$$\left(\lim_{h\to 0}\frac{Z(t+h)-Z(t)}{h}, v\right) = \lim_{h\to 0}\left(\frac{Z(t+h)-Z(t)}{h}, v\right). \qquad \square$$

Applying Lemma B.6.3 we obtain the following corollary:

Corollary 2.3.5. *If C has continuous partial derivatives of second order on $(a,b)\times(a,b)$, then Z is strongly differentiable and the correlation function D of Z' is given by*

$$D(s,t) = \frac{\partial^2 C}{\partial s\partial t}(s,t), \quad s,t\in(a,b).$$

Lemma 2.3.6. *If Z is a strongly differentiable Gaussian process on (a,b), then Z' is Gaussian as well.*

Proof. The random vectors

$$\frac{1}{n}\cdot\big(Z(t_1+1/n)-Z(t_1),\dots,Z(t_k+1/n)-Z(t_k)\big)$$

where $n\in\mathbb{N}$ and $t_j, t_j+1/n\in(a,b)$ are Gaussian and they converge in law to

$$(Z'(t_1),\dots,Z'(t_k))$$

as $n\to\infty$ (see Section F.2). Thus, the statement of the Lemma follows from Theorem 1.10.7. $\square$

2.4 Integration with respect to complex measures

Throughout this section $T\subset\mathbb{R}^d$ denotes a nonempty Borel set and Z is a strongly continuous random field of second order on T. As in Notation 2.2.6 we denote by $\mathcal{Z}$ the corresponding L^2-valued random field. First we define integrals of the form

$$\int_T \mathcal{Z}(t)\,\mathrm{d}\mu(t)$$

where $\mu \in \mathcal{M}^b(T)$ is a complex Radon measure on T. Then we prove basic properties of this integral paying special attention to properties which can be expressed in terms of the correlation function

$$C(s,t) = \mathbb{E}\left(Z(s) \cdot \overline{Z(t)}\right) = \mathbb{E}\left(\mathcal{Z}(s) \cdot \overline{\mathcal{Z}(t)}\right), \quad s,t \in T.$$

Definition 2.4.1. Let $\mu \in \mathcal{M}^b(T)$. The field $\mathcal{Z}$ is called *μ-integrable* if the function $t \mapsto (\mathcal{Z}(t), h)$, $t \in T$, is μ-integrable for all $h \in L^2(P)$ and there exists $I(\mathcal{Z}) \in L^2(P)$ such that

$$\int_T (\mathcal{Z}(t), h)\,\mathrm{d}\mu(t) = (I(\mathcal{Z}), h), \quad h \in L^2(P). \tag{1}$$

Note that $I(\mathcal{Z})$ is uniquely determined by equation (2.4.1.1). We write

$$\int \mathcal{Z}\,\mathrm{d}\mu := \int_T \mathcal{Z}(t)\,\mathrm{d}\mu(t) := I(\mathcal{Z}).$$

For a Lebesgue integrable function f on T let

$$\int_T \mathcal{Z}(t) f(t)\,\mathrm{d}t := \int_T \mathcal{Z}(t)\,\mathrm{d}\mu(t)$$

where $\mathrm{d}\mu(t) = f(t)\,\mathrm{d}t$. If $S \subset T$ is a Borel set, then we write

$$\int_S \mathcal{Z}(t)\,\mathrm{d}\mu(t) = \int_T \mathcal{Z}(t)\mathbf{1}_S(t)\,\mathrm{d}\mu(t).$$

In the special case where $S = [a,b] \subset \mathbb{R}$ we also use the notation $\int_a^b \mathcal{Z}\,\mathrm{d}\mu$ instead of $\int_{[a,b]} \mathcal{Z}\,\mathrm{d}\mu$.

The next lemma follows immediately from the definition of the integral. We omit the proof.

Lemma 2.4.2. *Assuming that all the integrals below exist, we have:*

(i) $$\int_T (\mathcal{Z}(t), h)\,\mathrm{d}\mu(t) = \left(\int_T \mathcal{Z}(t)\,\mathrm{d}\mu(t), h\right), \quad h \in L^2(P)$$

(ii) $$\int_T a_1\mathcal{Z}_1(t) + a_2\mathcal{Z}_2(t)\,\mathrm{d}\mu(t) = a_1\int_T \mathcal{Z}_1(t)\,\mathrm{d}\mu(t) + a_2\int_T \mathcal{Z}_2(t)\,\mathrm{d}\mu(t), \qquad a_1, a_2 \in \mathbb{C}$$

(iii) $$\int_T \mathcal{Z}(t)\,\mathrm{d}(\mu_1 + \mu_2)(t) = \int_T \mathcal{Z}(t)\,\mathrm{d}\mu_1(t) + \int_T \mathcal{Z}(t)\,\mathrm{d}\mu_2(t)$$

(iv) $\displaystyle\int_{S_1 \cup S_2} \mathcal{Z}(t)\,\mathrm{d}\mu(t) = \int_{S_1} \mathcal{Z}(t)\,\mathrm{d}\mu(t) + \int_{S_2} \mathcal{Z}(t)\,\mathrm{d}\mu(t)$

where $S_1,\ S_2 \subset T$ are disjoint Borel sets

(v) *If $\mathcal{Z}(t) = h$ for some $h \in L^2(P)$ and all $t \in T$ then* $\displaystyle\int_T \mathcal{Z}(t)\,\mathrm{d}\mu(t) = \mu(T) \cdot h.$

Example 2.4.3. (a) Let

$$\mu = \sum_{j=1}^{n} c_j \delta_{t_j}, \quad c_j \in \mathbb{C},\ t_j \in T$$

be a measure with finite support. Then

$$\int_T (\mathcal{Z}(t), h)\,\mathrm{d}\mu(t) = \sum_{j=1}^{n} c_j (\mathcal{Z}(t_j), h) = \Big(\sum_{j=1}^{n} c_j \mathcal{Z}(t_j), h \Big), \quad h \in L^2(P).$$

Thus, by Definition 2.4.1,

$$\int \mathcal{Z}\,\mathrm{d}\mu = \sum_{j=1}^{n} c_j \mathcal{Z}(t_j).$$

Moreover, we have

$$\Big\| \int \mathcal{Z}\,\mathrm{d}\mu \Big\|^2 = \Big(\sum_{j=1}^{n} c_j \mathcal{Z}(t_j), \sum_{k=1}^{n} c_k \mathcal{Z}(t_k) \Big) = \sum_{j,k=1}^{n} c_j \overline{c_k} C(t_j, t_k). \qquad (1)$$

(b) Assume that $\mathcal{Z}(t) = f(t)X,\ t \in T = [a,b]$, where f is a continuous function on T and $X \in L^2(P)$. For all $h \in L^2(P)$ we have

$$\begin{aligned} \int_a^b (\mathcal{Z}(t), h)\,\mathrm{d}\mu(t) &= \int_a^b (f(t)X, h)\,\mathrm{d}\mu(t) \\ &= (X, h) \int_a^b f(t)\,\mathrm{d}\mu(t) = \Big(\int_a^b f(t)\,\mathrm{d}\mu(t) \cdot X, h \Big) \end{aligned}$$

and hence

$$\int_a^b \mathcal{Z}(t)\,\mathrm{d}\mu(t) = \int_a^b f(t)\,\mathrm{d}\mu(t) \cdot X.$$

Using this and (2.4.2.ii) we obtain the more general relation

$$\int_a^b \sum_{j=1}^{n} f_j(t) X_j\ \mathrm{d}\mu(t) = \sum_{j=1}^{n} \Big[\int_a^b f_j(t)\,\mathrm{d}\mu(t) \Big] \cdot X_j$$

where $X_j \in L^2(P)$ and f_j is a continuous function on $[a,b]$.

Theorem 2.4.4. *Assume that $t \mapsto (\mathcal{Z}(t), h)$ is μ-integrable for all $h \in L^2(P)$. Then $\mathcal{Z}$ is μ-integrable if and only if there exists a constant $K \in [0, \infty)$ with*

$$\left| \int_T (\mathcal{Z}(t), h) \, \mathrm{d}\mu(t) \right| \le K \cdot \|h\|, \quad h \in L^2(P). \tag{1}$$

Proof. If $\mathcal{Z}$ is μ-integrable, then the inequality (1) holds with $K = \| \int \mathcal{Z} \, \mathrm{d}\mu \|$. This follows from (2.4.1.1) and from the Cauchy–Schwarz inequality

$$|(I(\mathcal{Z}), h)| \le \|I(\mathcal{Z})\| \cdot \|h\|.$$

Assume now that (1) holds with some constant K. Then

$$h \mapsto \int_T (h, \mathcal{Z}(t)) \, \mathrm{d}\overline{\mu}(t)$$

is a bounded linear functional[7] on $L^2(P)$. By a theorem of Riesz (cf. Theorem D.3.4) there exists $I \in L^2(P)$ such that

$$\int_T (h, \mathcal{Z}(t)) \, \mathrm{d}\overline{\mu}(t) = (h, I), \quad h \in L^2(P)$$

i.e.,

$$\int_T (\mathcal{Z}(t), h) \, \mathrm{d}\mu(t) = (I, h), \quad h \in L^2(P).$$

Thus, the field $\mathcal{Z}$ is μ-integrable. □

Corollary 2.4.5. *If*

$$\int_T \|\mathcal{Z}(t)\| \, \mathrm{d}|\mu|(t) < \infty \tag{1}$$

then $\mathcal{Z}$ is μ-integrable and

$$\left\| \int_T \mathcal{Z}(t) \, \mathrm{d}\mu(t) \right\| \le \int_T \|\mathcal{Z}(t)\| \, \mathrm{d}|\mu|(t). \tag{2}$$

Proof. The Cauchy–Schwarz inequality $|(\mathcal{Z}(t), h)| \le \|\mathcal{Z}(t)\| \cdot \|h\|$ and (1) imply that $t \mapsto (\mathcal{Z}(t), h)$ is $|\mu|$-integrable and hence μ-integrable as well. Moreover,

$$\left| \int_T (\mathcal{Z}(t), h) \, \mathrm{d}\mu(t) \right| \le \int_T |(\mathcal{Z}(t), h)| \, \mathrm{d}|\mu|(t) \le \int_T \|\mathcal{Z}(t)\| \, \mathrm{d}|\mu|(t) \cdot \|h\| .$$

By Theorem 2.4.4, the field $\mathcal{Z}$ is μ-integrable. The left-hand side of the inequality above is equal to $\left|\left(\int \mathcal{Z} \, \mathrm{d}\mu, h\right)\right|$. Setting $h = \int \mathcal{Z} \, \mathrm{d}\mu$ and dividing by $\|h\|$ if $\|h\| \neq 0$ we obtain (2). □

---

[7] To obtain a linear functional we use the conjugate of the integral in (1).

Theorem 2.4.6. *If $\mathcal{Z}$ is integrable with respect to μ and ν, then the correlation function C is $\mu \times \overline{\nu}$-integrable and*

$$\Big(\int_T \mathcal{Z}\,\mathrm{d}\mu, \int_T \mathcal{Z}\,\mathrm{d}\nu\Big) = \int_{T\times T} C(t,s)\,\mathrm{d}(\mu\times\overline{\nu})(t,s). \tag{1}$$

In particular,

$$\Big\|\int_T \mathcal{Z}\,\mathrm{d}\mu\Big\|^2 = \int_{T\times T} C(t,s)\,\mathrm{d}(\mu\times\overline{\mu})(t,s). \tag{2}$$

Proof. Equation (2.4.1.1) with $h = \int \mathcal{Z}\,\mathrm{d}\nu$ yields

$$\Big(\int \mathcal{Z}(t)\,\mathrm{d}\mu(t), \int \mathcal{Z}(s)\,\mathrm{d}\nu(s)\Big) = \int \Big(\mathcal{Z}(t), \int \mathcal{Z}(s)\,\mathrm{d}\nu(s)\Big)\,\mathrm{d}\mu(t). \tag{3}$$

Setting now $h = \mathcal{Z}(t)$ in equation (2.4.1.1), replacing the integration variable t by s, and replacing the measure μ by ν we obtain

$$\Big(\mathcal{Z}(t), \int \mathcal{Z}(s)\,\mathrm{d}\nu(s)\Big) = \overline{\int (\mathcal{Z}(s), \mathcal{Z}(t))\,\mathrm{d}\nu(s)} = \int C(t,s)\,\mathrm{d}\overline{\nu}(s).$$

This relation together with equation (3) shows that

$$\Big(\int \mathcal{Z}(t)\,\mathrm{d}\mu(t), \int \mathcal{Z}(s)\,\mathrm{d}\nu(s)\Big) = \int \Big[\int C(t,s)\,\mathrm{d}\overline{\nu}(s)\Big]\,\mathrm{d}\mu(t).$$

By Fubini's theorem C is $\mu \times \overline{\nu}$-integrable and (1) holds. □

Next we characterize integrability of a random field in terms of integrability of its correlation function.

Theorem 2.4.7. *The field $\mathcal{Z}$ is μ-integrable if and only if the correlation function C is $\mu \times \overline{\mu}$-integrable.*

Proof. In view of Theorem 2.4.6 it remains to prove that $\mathcal{Z}$ is μ-integrable whenever C is $\mu \times \overline{\mu}$-integrable. To show this, let $h \in L^2(P)$, $h \neq 0$, be arbitrary and write

$$f_h(t) := (\mathcal{Z}(t), h), \quad \mathcal{Z}_h(t) := \mathcal{Z}(t) - \frac{f_h(t)}{\|h\|^2}\cdot h, \quad t \in T.$$

Then $(\mathcal{Z}_h(t), h) = 0$ and

$$C(t,s) = (\mathcal{Z}(t), \mathcal{Z}(s)) = (\mathcal{Z}_h(t), \mathcal{Z}_h(s)) + \frac{1}{\|h\|^2} f_h(t)\overline{f_h(s)}.$$

Let $S \neq \emptyset$ be an arbitrary compact subset of T. By continuity, the functions $t \mapsto \|\mathcal{Z}(t)\|$ and $t \mapsto \|\mathcal{Z}_h(t)\|$ are bounded on S. Consequently, the restrictions of $\mathcal{Z}$ and

$\mathcal{Z}_h$ to S are integrable with respect to any complex measure (see Corollary 2.4.5). Applying Theorem 2.4.6 we obtain

$$\begin{aligned}\int_S \int_S C(t,s)\,\mathrm{d}\mu(t)\,\mathrm{d}\overline{\mu}(s) &= \int_S \int_S (\mathcal{Z}_h(t), \mathcal{Z}_h(s))\,\mathrm{d}\mu(t)\,\mathrm{d}\overline{\mu}(s) \\ &\quad + \frac{1}{\|h\|^2} \int_S \int_S f_h(t)\overline{f_h(s)}\,\mathrm{d}\mu(t)\,\mathrm{d}\overline{\mu}(s) \\ &\geq \frac{1}{\|h\|^2} \left| \int_S f_h(t)\,\mathrm{d}\mu(t) \right|^2\end{aligned}$$

(note that the first integral after the equation sign is nonnegative by (2.4.6.2)). Consequently,

$$\left| \int_S f_h(t)\,\mathrm{d}\mu(t) \right| \leq \left(\int_{T\times T} |C|\,\mathrm{d}\,|\mu \times \overline{\mu}| \right)^{1/2} \cdot \|h\|.$$

Using Lemma E.1.18 we conclude[8] that f_h is μ-integrable and the inequality above remains valid if we replace S by T. By Theorem 2.4.4 the field $\mathcal{Z}$ is μ-integrable. □

We already know that a correlation function K is positive semidefinite, i.e.,

$$\sum_{i,j=1}^{n} K(t_i, t_j) c_i \overline{c_j} \geq 0$$

for all $n \in \mathbb{N}$, $t_j \in T$ and $c_j \in \mathbb{C}$. The next corollary contains a more general inequality.

Corollary 2.4.8. *Let K be a continuous positive semidefinite kernel on T. Then the inequality*

$$\int_{T\times T} K(t,s)\,\mathrm{d}(\mu \times \overline{\mu})(t,s) \geq 0$$

holds for all $\mu \in \mathcal{M}^b(T)$ such that K is μ-integrable. In particular, the inequality

$$\int_T \int_T K(t,s) h(t) \overline{h(s)}\,\mathrm{d}t\,\mathrm{d}s \geq 0$$

holds for every continuous function $h : T \to \mathbb{C}$ with compact support.

Proof. By Theorems 2.2.3 and 2.3.2 the kernel K is the correlation function of a strongly continuous random field. This field is μ-integrable in view of Theorem 2.4.7. The corollary follows now from (2.4.6.2). □

---

8 To apply this lemma we extend μ to a measure on $\mathbb{R}^d$ by setting $\mu(B) := \mu(B \cap T)$, $B \in \mathcal{B}(\mathbb{R}^d)$. The function f_h is extended to $\mathbb{R}^d$ by setting it equal to zero on $\mathbb{R}^d \setminus T$.

Theorem 2.4.9. *Let $\{\mu_n\}$ be a sequence of complex measures on T converging weakly to some complex measure μ. If $\mathcal{Z}$ is integrable with respect to μ and μ_n then*

$$\int_T \mathcal{Z}\, \mathrm{d}\mu = \lim_{n\to\infty} \int_T \mathcal{Z}\, \mathrm{d}\mu_n. \tag{1}$$

Proof. We may suppose that μ and μ_n are nonnegative. If T is compact, then, by continuity, the function C is bounded and hence integrable with respect to any complex measure on $T \times T$. Using (2.4.6.2) we obtain

$$\begin{aligned}\left\| \int_T \mathcal{Z}\, \mathrm{d}\mu - \int_T \mathcal{Z}\, \mathrm{d}\mu_n \right\|^2 &= \left\| \int_T \mathcal{Z}\, \mathrm{d}(\mu - \mu_n) \right\|^2 \\ &= \int_{T\times T} C\, \mathrm{d}(\mu - \mu_n) \times (\mu - \mu_n)\,.\end{aligned}$$

In view of Theorem E.1.14, the sequence $\{(\mu - \mu_n) \times (\mu - \mu_n)\}$ converges weakly to 0. Since C is bounded, the right-hand side of the relation above tends to zero from which (1) follows.

Now let T be arbitrary. For each $\epsilon > 0$ there exist disjoint Borel sets T_b and T_u such that T_b is compact, $T = T_b \cup T_u$ and

$$\int_{T_u\times T_u} C\, \mathrm{d}(\mu \times \mu) < \epsilon$$

(this integral is nonnegative in view of (2.4.6.2)). Therefore,

$$\left\| \int_T \mathcal{Z}\, \mathrm{d}\mu - \int_{T_b} \mathcal{Z}\, \mathrm{d}\mu \right\|^2 = \left\| \int_{T_u} \mathcal{Z}\, \mathrm{d}\mu \right\|^2 = \int_{T_u\times T_u} C\, \mathrm{d}(\mu \times \mu) < \epsilon.$$

By the first part of the proof, the statement of the theorem is true for the compact set T_b. The inequality above shows that it is true for T as well. □

Using Theorem E.1.8 we obtain the following corollary:

Corollary 2.4.10. *For every complex measure μ there exists a sequence $\{\mu_n\}$ of complex measures with finite support converging weakly to μ and satisfying the relation (2.4.9.1).*

Corollary 2.4.11. *If $\mathcal{Z}$ is a μ_j-integrable Gaussian field ($j = 1, \ldots, d$), then*

$$\left(\int_T \mathcal{Z}\, \mathrm{d}\mu_1, \ldots, \int_T \mathcal{Z}\, \mathrm{d}\mu_d \right)$$

is a Gaussian random vector.

Proof. The statement is obvious if the μ_j's have finite support. The general case follows from Corollary 2.4.10 and from the fact that the limit of weakly convergent Gaussian random vectors is Gaussian (cf. Theorem 1.10.7). □

Theorem 2.4.12 (Newton–Leibniz formula for fields). *Let $\mathcal{Z}$ be a strongly continuous field on (a,b), $-\infty \le a < b \le \infty$, and let $t_0 \in (a,b)$ be arbitrary.*

(i) *The field $t \mapsto \int_{t_0}^t \mathcal{Z}(s)\,\mathrm{d}s$ is strongly differentiable and*

$$\mathcal{Z}(t) = \frac{\mathrm{d}}{\mathrm{d}t}\int_{t_0}^t \mathcal{Z}(s)\,\mathrm{d}s, \quad t \in (a,b).$$

(ii) *If C is twice continuously differentiable, then*

$$\mathcal{Z}(t) = \mathcal{Z}(t_0) + \int_{t_0}^t \mathcal{Z}'(s)\,\mathrm{d}s, \quad t \in (a,b).$$

Proof. (i) We have to show that

$$\frac{1}{h}\left[\int_{t_0}^{t+h} \mathcal{Z}(s)\,\mathrm{d}s - \int_{t_0}^{t} \mathcal{Z}(s)\,\mathrm{d}s\right] = \frac{1}{h}\int_{t}^{t+h} \mathcal{Z}(s)\,\mathrm{d}s$$

converges to $\mathcal{Z}(t)$ as $h \to 0$. This follows from

$$\begin{aligned}\left\|\frac{1}{h}\int_t^{t+h} \mathcal{Z}(s)\,\mathrm{d}s - \mathcal{Z}(t)\right\| &= \frac{1}{|h|}\cdot\left\|\int_t^{t+h} \mathcal{Z}(s) - \mathcal{Z}(t)\,\mathrm{d}s\right\| \\ &\le \frac{1}{|h|}\cdot\int_{[t,t+h]} \|\mathcal{Z}(s) - \mathcal{Z}(t)\|\,\mathrm{d}s \\ &\le \sup_{s\in[t,t+h]} \|\mathcal{Z}(s) - \mathcal{Z}(t)\| \to 0.\end{aligned}$$

(ii) Corollary 2.3.5 shows that $\mathcal{Z}$ is strongly differentiable and the correlation function of $\mathcal{Z}'$ is continuous. Consequently, $\mathcal{Z}'$ is strongly continuous and hence integrable with respect to the Lebesgue measure on $[t_0,t]$. By Theorem 2.3.4, the function $t \mapsto (\mathcal{Z}(t),h)$ is continuously differentiable for all $h \in L^2(P)$. Applying the classical Newton–Leibniz formula to this function we obtain

$$(\mathcal{Z}(t),h) = (\mathcal{Z}(t_0),h) + \int_{t_0}^t \frac{d}{\mathrm{d}s}(\mathcal{Z}(s),h)\,\mathrm{d}s.$$

Equation (2.3.4.1) and the definition of the integral show that

$$(\mathcal{Z}(t),h) = (\mathcal{Z}(t_0),h) + \left(\int_{t_0}^t \mathcal{Z}'(s)\,\mathrm{d}s, h\right), \quad h \in L^2(P).$$

from which the statement follows. □

Lemma 2.4.13 (dominated convergence). *Let $\mathcal{Z}$, $\mathcal{Z}_n$, $n \in \mathbb{N}$, be strongly continuous fields on T such that*

$$\lim_{n\to\infty} \mathcal{Z}_n(t) = \mathcal{Z}(t), \quad t \in T. \tag{1}$$

Assume further that there exists a $|\mu|$-integrable function g on T such that $\|\mathcal{Z}_n(t)\| \leq g(t)$ for all n and t. Then $\mathcal{Z}$ and $\mathcal{Z}_n$ are μ-integrable and

$$\lim_{n\to\infty} \int_T \mathcal{Z}_n(t)\,\mathrm{d}\mu(t) = \int_T \mathcal{Z}(t)\,\mathrm{d}\mu(t).$$

Proof. Relation (1) implies that $\lim_{n\to\infty} \|\mathcal{Z}_n(t)\| = \|\mathcal{Z}(t)\|$ and hence $\|\mathcal{Z}(t)\| \leq g(t)$. By Corollary 2.4.5, the fields $\mathcal{Z}$ and $\mathcal{Z}_n$ are μ-integrable. Using the inequality $\|\mathcal{Z}_n(t) - \mathcal{Z}(t)\| \leq 2g(t)$, the lemma follows from

$$\begin{aligned}\left\| \int_T \mathcal{Z}_n(t)\,\mathrm{d}\mu - \int_T \mathcal{Z}(t)\,\mathrm{d}\mu(t) \right\| &= \left\| \int_T [\mathcal{Z}_n(t) - \mathcal{Z}(t)]\,\mathrm{d}\mu(t) \right\| \\ &\leq \int_T \|\mathcal{Z}_n(t) - \mathcal{Z}(t)\|\,\mathrm{d}|\mu|(t)\end{aligned}$$

and from Lebesgue's theorem on dominated convergence. □

Definition 2.4.14. Assume that $T = \mathbb{R}^d$ or $T = \mathbb{Z}^d$ and let μ be a complex measure on T. Further, let $\mathcal{Z}$ be a continuous field of second order on T such that the field $x \mapsto \mathcal{Z}(t-x)$ is μ-integrable for all $t \in T$. We define the field $\mu * \mathcal{Z}$, the *convolution of μ and $\mathcal{Z}$*, by

$$\mu * \mathcal{Z}(t) = \int_T \mathcal{Z}(t-x)\,\mathrm{d}\mu(x), \quad t \in T.$$

If $f \in \mathcal{L}^1(\lambda)$ or $f \in \mathcal{L}^1(\mathbb{Z}^d)$, then we write

$$f * \mathcal{Z}(t) := \int_{\mathbb{R}^d} \mathcal{Z}(t-x) f(x)\,\mathrm{d}\lambda(x), \quad t \in T$$

and

$$f * \mathcal{Z}(t) := \sum_{n\in\mathbb{Z}^d} \mathcal{Z}(t-n) f(n), \quad t \in T$$

respectively.

In the next theorem we collect basic properties of the convolution.

Theorem 2.4.15. *Assume that all convolutions below exist. Then we have*

$$(\mu + \nu) * \mathcal{Z} = \mu * \mathcal{Z} + \nu * \mathcal{Z} \tag{1}$$

$$(c\mu) * \mathcal{Z} = c \cdot (\mu * \mathcal{Z}), \quad c \in \mathbb{C} \tag{2}$$

$$\mu * \mathcal{Z}(t) = \int \mathcal{Z}\,\mathrm{d}(\delta_t * \mu^{\triangle}) \tag{3}$$

$$(\mu * \nu) * \mathcal{Z} = \mu * (\nu * \mathcal{Z}) \tag{4}$$

for all $t \in T$. If a sequence $\{\mu_n\}$ converges weakly to μ, then

$$\lim_{n\to\infty} \mu_n * \mathcal{Z}(t) = \mu * \mathcal{Z}(t), \quad t \in T. \tag{5}$$

Proof. The first three equations follow immediately from the definition of the convolution, while (5) is a consequence of Theorem 2.4.9.

It is easy to check that (4) holds for measures with finite support. In the general case we approximate μ and ν by measures with finite support (see Corollary 2.4.10) and apply Theorem E.2.5. □

Remark 2.4.16. Let $\mathcal{Z}$ be a not necessarily continuous, second order field on $\mathbb{R}^d$. If the complex measure μ has finite support, then we define the convolution of μ and $\mathcal{Z}$, as in Definition 2.4.14, by

$$\mu * \mathcal{Z}(t) = \int_{\mathbb{R}^d} \mathcal{Z}(t-x)\,\mathrm{d}\mu(x), \quad t \in \mathbb{R}^d.$$

Then equations (1)–(4) in the previous theorem still remain valid.

2.5 The Karhunen–Loève decomposition

In this section we use the notation

$$I := [a,b] := [a_1,b_1] \times \cdots \times [a_d,b_d]$$

where $a, b \in \mathbb{R}^d$ and $-\infty < a_j < b_j < \infty$. We will prove a decomposition theorem for continuous positive semidefinite kernels on $I \times I$. Using this result we derive a decomposition theorem for strongly continuous fields on I.

We denote by $\mathcal{L}^2(a,b)$ the positive semidefinite inner product space of all complex-valued functions on $[a,b]$ which are square integrable with respect to the Lebesgue measure. The corresponding Hilbert space is denoted by $L^2(a,b)$. We write $\int_a^b$ instead of $\int_I$. Finally, $C : I \times I \to \mathbb{C}$ will be a continuous kernel.

Definition 2.5.1. A function $\varphi \in \mathcal{L}^2(a,b)$ is called an *eigenfunction* of C with *eigenvalue* $\lambda \in \mathbb{C}$ if

$$\lambda\varphi(t) = \int_a^b C(t,s)\varphi(s)\,\mathrm{d}s, \quad t \in [a,b].$$

Lemma 2.5.2. *Let $h \in \mathcal{L}^2[a,b]$. The function g defined by*

$$g(t) := \int_a^b C(t,s)h(s)\,\mathrm{d}s, \quad t \in [a,b]$$

is continuous. In particular, if φ is an eigenfunction of C with eigenvalue $\lambda \neq 0$, then φ is continuous.

Proof. For $t, t_0 \in I$ the Cauchy–Schwarz inequality shows that

$$\begin{aligned}|g(t) - g(t_0)|^2 &= \left| \int_a^b (C(t,s) - C(t_0,s))h(s)\,\mathrm{d}s \right|^2 \\ &\le \int_a^b |C(t,s) - C(t_0,s)|^2\,\mathrm{d}s \cdot \int_a^b |h(s)|^2\,\mathrm{d}s.\end{aligned}$$

The kernel C is continuous and hence uniformly continuous on $I \times I$. Using this we see that the right-hand side of the relation above tends to zero as $t \to t_0$. □

Lemma 2.5.3. *The mapping K defined by*[9]

$$Kh(t) := \int_a^b C(t,s)h(s)\,\mathrm{d}s, \quad h \in L^2(a,b),\ t \in [a,b]$$

is a compact linear operator in $L^2(a,b)$.

Proof. Using the fact that C is bounded on $I \times I$ it is easy to see that K is a bounded linear operator in $L^2(a,b)$. We have to show that for every bounded sequence $\{h_n\}$, the sequence $\{Kh_n\}$ has a convergent subsequence. Write $g_n = Kh_n$. To simplify the notation, we also consider g_n as a function defined pointwise by the integral above. Applying the Cauchy–Schwarz inequality as in Lemma 2.5.2, we see that the sequence $\{g_n\}$ is equicontinuous and uniformly bounded. In view of the theorem of Arzelà and Ascoli (cf. Theorem B.2.9), the sequence $\{g_n\}$ has a subsequence converging uniformly on $[a,b]$ to a continuous function. It follows that the operator K is indeed compact. □

Theorem 2.5.4 (Mercer). *An arbitrary continuous positive semidefinite kernel $C : I \times I \to \mathbb{C}$ admits the decomposition*

$$C(t,s) = \sum_{n=1}^{\infty} \lambda_n \varphi_n(t)\overline{\varphi_n(s)}, \quad t,s \in I \tag{1}$$

where

(i) *each φ_n is an eigenfunction of C with eigenvalue $\lambda_n \ge 0$;*

(ii) *the series above is absolutely and uniformly convergent;*

(iii) *$\{\varphi_n + N\}_{n=1}^{\infty}$ is an orthonormal basis of $L^2(a,b)$, where $N = \{\varphi \in \mathcal{L}^2(a,b) : (\varphi,\varphi) = 0\}$.*

Proof. To simplify the notation we identify continuous functions on $[a,b]$ with the corresponding elements of $L^2(a,b)$. We consider the compact linear operator K

[9] To be precise, Kh is the equivalence class represented by the integral on the right-hand side.

defined in Lemma 2.5.3. Using that $C(t,s) = \overline{C(s,t)}$ it is easy to check that $(Kh,g) = (h,Kg)$, i.e., K is self-adjoint. It follows from Corollary 2.4.8 that K is nonnegative and hence all eigenvalues of K are nonnegative. By Remark D.3.13, there exists an orthonormal basis $\{\varphi_n\}_1^\infty$ of $L^2(a,b)$ consisting of eigenvectors of K with eigenvalues $\{\lambda_n\}_1^\infty$. Every $h \in L^2(a,b)$ admits the decomposition

$$h = \sum_{n=1}^{\infty} (h, \varphi_n)\, \varphi_n.$$

For fixed t we apply the above decomposition to the function $h : s \mapsto C(s,t)$:

$$C(t,\cdot) = \overline{C(\cdot,t)} = \sum_{n=1}^{\infty} (\varphi_n, C(\cdot,t))\, \overline{\varphi_n}, \quad t \in [a,b].$$

Since φ_n is an eigenvector of K with eigenvalue λ_n we have

$$(\varphi_n, C(\cdot,t)) = \int_a^b C(t,s)\varphi_n(s)\, \mathrm{d}s = K\varphi_n(t) = \lambda_n \varphi_n(t). \tag{2}$$

Thus,

$$C(t,\cdot) = \sum_{n=1}^{\infty} \lambda_n\, \varphi_n(t)\overline{\varphi_n}.$$

From here, again using (2) and the orthonormality of the vectors φ_n, we obtain

$$\begin{aligned} 0 &= \lim_{n\to\infty} \left\| C(t,\cdot) - \sum_{k=1}^{n} \lambda_k \varphi_k(t)\overline{\varphi_k} \right\|^2 \\ &= \lim_{n\to\infty} \left(\int_a^b |C(t,s)|^2\, \mathrm{d}s - \sum_{k=1}^{n} \lambda_k^2 |\varphi_k(t)|^2 \right), \quad t \in [a,b]. \end{aligned}$$

If $\lambda_k \neq 0$, then the function φ_k is continuous (cf. Lemma 2.5.2). It is easy to see that the function

$$t \mapsto \int_a^b |C(t,s)|^2\, \mathrm{d}s$$

is continuous as well. By Dini's Theorem B.2.5, the series $\sum_{k=1}^{\infty} \lambda_k^2 |\varphi_k(t)|^2$ converges uniformly on $[a,b]$. Using this and the Cauchy–Schwarz inequality

$$\sum_{k=p}^{q} \left| \lambda_k \varphi_k(t)\overline{\varphi_k(s)} \right| \le \left(\sum_{k=p}^{q} \lambda_k |\varphi_k(t)|^2 \right)^{\frac{1}{2}} \cdot \left(\sum_{k=p}^{q} \lambda_k |\varphi_k(s)|^2 \right)^{\frac{1}{2}}$$

we conclude that (ii) holds. To show equation (1) we consider for fixed s the continuous nonnegative function

$$t \mapsto \left| C(t,s) - \sum_{n=1}^{\infty} \lambda_n \varphi_n(t)\overline{\varphi_n(s)} \right|^2.$$

Using (2) we see that its integral over I is equal to zero. Thus, the function above is equal to zero everywhere. □

Theorem 2.5.5 (Karhunen–Loève). *Let $\mathcal{Z} : I \to L^2(\Omega, \mathcal{A}, P)$ be a strongly continuous field with correlation function C. Then $\mathcal{Z}$ admits the expansion*

$$\mathcal{Z}(t) = \sum_{n \in S} \sqrt{\lambda_n}\, \varphi_n(t) X_n, \quad t \in I \tag{1}$$

where

(i) *φ_n and λ_n are as in Mercer's theorem and $S = \{n \in \mathbb{N} : \lambda_n > 0\}$;* [10]

(ii) $X_n = \dfrac{1}{\sqrt{\lambda_n}} \displaystyle\int_a^b \mathcal{Z}(s) \cdot \overline{\varphi_n(s)}\, \mathrm{d}s;$

(iii) *the X_n's form an orthonormal system in $L^2(\Omega, \mathcal{A}, P)$;*

(iv) *the series (1) converges uniformly on I.*

Proof. For $n \in S$ we define X_n by equation (ii) and write $X_n = 0$ if $n \notin S$. Note that the function $s \mapsto \|\mathcal{Z}(s) \cdot \varphi_n(s)\|$ is bounded by continuity, hence the integral in (ii) exists (cf. Theorem 2.4.5). Applying Theorem 2.4.6 we see that

$$\begin{aligned}
(X_n, X_m) &= \frac{1}{\sqrt{\lambda_n \lambda_m}} \left(\int_a^b \mathcal{Z}(t) \cdot \overline{\varphi_n(t)}\, \mathrm{d}t, \int_a^b \mathcal{Z}(s) \cdot \overline{\varphi_m(s)}\, \mathrm{d}s \right) \\
&= \frac{1}{\sqrt{\lambda_n \lambda_m}} \int_a^b \int_a^b C(t,s) \overline{\varphi_n(t)} \varphi_m(s)\, \mathrm{d}s\, \mathrm{d}t \\
&= \sqrt{\frac{\lambda_m}{\lambda_n}} \int_a^b \overline{\varphi_n(t)} \varphi_m(t)\, \mathrm{d}t \\
&= \delta_{n,m}, \quad n, m \in S
\end{aligned}$$

[10] The correlation function C is positive semidefinite, hence we may apply Mercer's Theorem 2.5.4 to it. If $S = \emptyset$, then the sum (1) is zero by definition.

i.e., (iii) holds. Using property (2.4.2.i) of the integral and the definition of X_n, we obtain

$$\begin{aligned}(\mathcal{Z}(t), X_n) &= \frac{1}{\sqrt{\lambda_n}}\left(\mathcal{Z}(t), \int_a^b \mathcal{Z}(s)\overline{\varphi_n(s)}\,\mathrm{d}s\right)\\ &= \frac{1}{\sqrt{\lambda_n}}\int_a^b C(t,s)\varphi_n(s)\,\mathrm{d}s\\ &= \sqrt{\lambda_n}\varphi_n(t), \quad n \in \mathbb{N}.\end{aligned}$$

Applying this we see that

$$\begin{aligned}\left\|\mathcal{Z}(t) - \sum_{n=1}^{k}\sqrt{\lambda_n}\varphi_n(t)X_n\right\|^2 &= C(t,t) - 2\mathrm{Re}\sum_{n=1}^{k}\sqrt{\lambda_n}\cdot\overline{\varphi_n(t)}(\mathcal{Z}(t), X_n)\\ &\quad + \sum_{n=1}^{k}\lambda_n|\varphi_n(t)|^2\\ &= C(t,t) - \sum_{n=1}^{k}\lambda_n|\varphi_n(t)|^2, \quad k \in \mathbb{N}.\end{aligned}$$

In view of Theorem 2.5.4, the right-hand side converges to zero uniformly on I as $k \to \infty$. This completes the proof. □

Remark 2.5.6. If $\mathcal{Z}$ is Gaussian, then in view of Corollary 2.4.11 the random variables X_n in Theorem 2.5.5 are Gaussian. If the mean of $\mathcal{Z}$ is zero, then $\mathbb{E}(X_n) = 0$. This follows from (2.4.2.i) and from $\mathbb{E}(X_n) = (X_n, \mathbf{1})$. By orthogonality, the random variables X_n are then uncorrelated and hence independent (cf. Theorem 1.10.8).

The next theorem is the converse of the previous one and it is easier to prove.

Theorem 2.5.7. *Let $\emptyset \neq S \subset \mathbb{N}$, for each $n \in S$ let λ_n be a positive number and $\{X_n\}_{n\in S}$ be an orthonormal system in $L^2(\Omega, \mathcal{A}, P)$. Further, let $\{\varphi_n\}_{n\in S}$ be an orthonormal system of continuous functions in $\mathcal{L}^2(a,b)$ such that the series*

$$\mathcal{Z}(t) = \sum_{n\in M}\sqrt{\lambda_n}\,\varphi_n(t)X_n, \quad t \in I$$

is uniformly convergent on I. Then this series defines a strongly continuous second order field $\mathcal{Z}$ and the function φ_n is an eigenfunction with eigenvalue λ_n of the correlation function of $\mathcal{Z}$.

Proof. To simplify the notation we assume that $S = \mathbb{N}$, the general case can be treated in the same way. It is easy to check that $\mathcal{Z}$ is a field of second order with correlation function

$$C(t,s) = (\mathcal{Z}(t), \mathcal{Z}(s)) = \sum_{n=1}^{\infty} \lambda_n \varphi_n(t)\overline{\varphi_n(s)}, \quad t, s \in I.$$

where the series is uniformly convergent on I. From this we conclude that C and hence $\mathcal{Z}$ is continuous. Since $\{\varphi_n\}_{n\in\mathbb{N}}$ is an orthonormal system, we have

$$\begin{aligned}\int_a^b C(t,s)\varphi_k(s)\,\mathrm{d}s &= \int_a^b \sum_{n=1}^{\infty} \lambda_n \varphi_n(t)\overline{\varphi_n(s)}\varphi_k(s)\,\mathrm{d}s \\ &= \sum_{n=1}^{\infty} \lambda_n \varphi_n(t) \int_a^b \overline{\varphi_n(s)}\varphi_k(s)\,\mathrm{d}s \\ &= \lambda_k \varphi_k(t), \quad t \in I,\ k \in \mathbb{N}. \quad \square\end{aligned}$$

Example 2.5.8. We consider the Wiener process W from Corollary 2.1.4 on a finite interval $I = [0, T]$, $T > 0$. The correlation function of W is given by $C(t,s) = \min(t,s)$. First we compute the eigenfunctions of C with positive eigenvalues. The equation

$$\int_0^T C(t,s)\varphi(s)\,\mathrm{d}s = \lambda\varphi(t), \quad 0 \le t \le T,\ \lambda > 0$$

can be rewritten as

$$\int_0^t s\varphi(s)\,\mathrm{d}s + t\int_t^T \varphi(s)\,\mathrm{d}s = \lambda\varphi(t).$$

From here we see that $\varphi(0) = 0$. The left-hand side, hence also the right-hand side, is differentiable with respect to t.[11] We have,

$$\int_t^T \varphi(s)\,\mathrm{d}s = \lambda\varphi'(t), \quad t \in [0, T].$$

This equation implies that $\varphi'(T) = 0$. Differentiating again we obtain

$$-\varphi(t) = \lambda\varphi''(t).$$

The solutions of this equation satisfying the condition $\varphi(0) = 0$ are given by

$$\varphi(t) = A\sin\frac{t}{\sqrt{\lambda}}, \quad t \in [0, T].$$

Taking into account the condition $\varphi'(T) = 0$ it is easy to check that

$$\lambda_n = \frac{T^2}{(n+\frac{1}{2})^2\pi^2}, \quad n \in \mathbb{N}_0$$

[11] At $t = 0$ and $t = T$ we have only one-sided derivatives.

are the only possible values of λ. We denote by φ_n the corresponding function and set $A := \sqrt{\frac{2}{T}}$, i.e.,

$$\varphi_n(t) = \sqrt{\frac{2}{T}} \cdot \sin\left[\left(n + \frac{1}{2}\right)\frac{\pi t}{T}\right].$$

The L^2-norm of φ_n is equal to 1. From Theorem 2.5.5 we obtain the representation

$$W(t) = \sqrt{2T} \sum_{n=1}^{\infty} \frac{\sin\left[\left(n + \frac{1}{2}\right)\frac{\pi t}{T}\right]}{\left(n + \frac{1}{2}\right)\pi} X_n^T, \quad 0 \le t \le T.$$

Remark 2.5.6 shows that the random variables X_n^T are independent and standard Gaussian.

Example 2.5.9. Let $Y_1, \dots, Y_n \in L^2(\Omega, \mathcal{A}, P)$ be uncorrelated with mean zero and standard deviation $\sigma_j > 0$. Further, let $k_1, \dots, k_n \in \mathbb{N}$. The correlation function of the process

$$\mathcal{Z}(t) = \sum_{j=1}^{n} \mathrm{e}^{\mathrm{i}tk_j} Y_j, \quad t \in [0, 2\pi]$$

is given by

$$C(t,s) = \sum_{j=1}^{n} \sigma_j^2 \mathrm{e}^{\mathrm{i}tk_j} \cdot \overline{\mathrm{e}^{\mathrm{i}sk_j}} = \sum_{j=1}^{n} \sigma_j^2 \mathrm{e}^{\mathrm{i}(t-s)k_j}, \quad t \in [0, 2\pi].$$

The functions $\varphi_j(t) = \frac{1}{\sqrt{2\pi}} \cdot \mathrm{e}^{\mathrm{i}tk_j}$, $t \in [0, 2\pi]$, build an orthonormal system in $\mathcal{L}^2(0, 2\pi)$. Write $X_j := Y_j/\sigma_j$. Then $X_1, \dots, X_n$ is an orthonormal system in $L^2(\Omega, \mathcal{A}, P)$,

$$\mathcal{Z}(t) = \sqrt{2\pi} \sum_{j=1}^{n} \sigma_j \varphi_j(t) X_j, \quad t \in [0, 2\pi]$$

and φ_j is an eigenfunction of C with eigenvalue $\sqrt{2\pi} \cdot \sigma_j^2$ (cf. Theorem 2.5.7).

2.6 Integration with respect to orthogonal random measures

The results and methods presented in this section will be used to prove integral representations for random fields. Throughout the present section $(W, \mathcal{B})$ denotes a measurable space.

Definition 2.6.1. A mapping $\zeta : \mathcal{B} \to L^2(\Omega, \mathcal{A}, P)$ is called a *random orthogonal measure* on $\mathcal{B}$ (or on $(W, \mathcal{B})$), if

(i) $(\zeta(B), \zeta(D)) = 0$ for all $B, D \in \mathcal{B}$ with $B \cap D = \emptyset$;

(ii) $\zeta(\cup_{j=1}^{\infty} B_j) = \sum_{j=1}^{\infty} \zeta(B_j)$ for all mutually disjoint sets $B_j \in \mathcal{B}$.

If ζ is a random orthogonal measure, then the mapping $\sigma : \mathcal{B} \to [0, \infty)$, defined by

$$\sigma(B) := (\zeta(B), \zeta(B)) = \|\zeta(B)\|^2, \quad B \in \mathcal{B}$$

is called the *structure measure* of ζ.[12]

By $H(\zeta)$ we denote the smallest subspace of $L^2(\Omega, \mathcal{A}, P)$ containing the set $\{\zeta(B) : B \in \mathcal{B}\}$.

Lemma 2.6.2. *The structure measure σ of a random orthogonal measure ζ is a finite nonnegative measure on $\mathcal{B}$.*

Proof. Equation (2.6.1.i) with $B = D = \emptyset$ shows that $\sigma(\emptyset) = 0$. If the sets $B_j \in \mathcal{B}$ are mutually disjoint, then by (2.6.1.i), the $\zeta(B_j)$'s are mutually orthogonal. From here and from (2.6.1.ii) we obtain

$$\sigma\left(\bigcup_{j=1}^{\infty} B_j\right) = \left\|\zeta\left(\bigcup_{j=1}^{\infty} B_j\right)\right\|^2 = \left\|\sum_{j=1}^{\infty} \zeta(B_j)\right\|^2$$

$$= \sum_{j=1}^{\infty} \|\zeta(B_j)\|^2 = \sum_{j=1}^{\infty} \sigma(B_j). \quad \square$$

Theorem 2.6.3. *Let ζ be a random orthogonal measure on $\mathcal{B}$ with structure measure σ. Then there exists an isometric linear operator I_ζ from $L^2(\sigma)$ onto $H(\zeta)$ such that*

$$I_\zeta\left(\sum_{j=1}^{n} c_j \mathbf{1}_{B_j}\right) = \sum_{j=1}^{n} c_j \zeta(B_j)$$

holds for all $n \in \mathbb{N}$ and all $c_j \in \mathbb{C}$, $B_j \in \mathcal{B}$.

Proof. Let L denote the linear manifold of all functions f of the form $f = \sum_{j=1}^{n} c_j \mathbf{1}_{B_j}$. This manifold is dense in $L^2(\sigma)$. For $f \in L$ we define $I_\zeta(f)$ by the equation above. It is easy to check that this definition is correct, i.e., the right-hand side does not depend on the special representation of f. It is clear that I_ζ is a linear operator from L into $H(\zeta)$. By the definition of $H(\zeta)$, the range of I_ζ is dense in $H(\zeta)$.

---

12 It is possible to define random orthogonal measures where the structure measure is not finite (see for example [19]). However, we do not need this more general notion in the present book.

An arbitrary step function f can be written in the form above with mutually disjoint sets B_j. By orthogonality, we obtain

$$(I_\zeta(f), I_\zeta(f)) = \sum_{j=1}^{n} |c_j|^2 \sigma(B_j) = (f, f)$$

(we denote the inner product in both spaces by $(\cdot,\cdot)$). Thus, the mapping I_ζ is isometric. By Theorem D.3.6, this mapping can be uniquely extended to an isometric linear operator from $L^2(\sigma)$ onto $H(\zeta)$. □

In the sequel we will use the notation

$$\int_W f(t)\,\mathrm{d}\zeta(t) := \int f(t)\,\mathrm{d}\zeta(t) := \int f\,\mathrm{d}\zeta := I_\zeta(f), \quad f \in L^2(\sigma).$$

For a set $A \in \mathcal{B}$ we write

$$\int_A f(t)\,\mathrm{d}\zeta(t) := \int \mathbf{1}_A(t) f(t)\,\mathrm{d}\zeta(t), \quad f \in L^2(\sigma).$$

In the next theorem we list the basic properties of this integral which follow immediately from Theorem 2.6.3.

Theorem 2.6.4. *Let ζ be a random orthogonal measure. For all $f, g, f_n \in L^2(\sigma)$ and $a, b \in \mathbb{C}$ we have*

(i) $\displaystyle\int [af(t) + bg(t)]\,\mathrm{d}\zeta(t) = a\int f(t)\,\mathrm{d}\zeta(t) + b\int g(t)\,\mathrm{d}\zeta(t)$

(ii) $\displaystyle\Big(\int f(t)\,\mathrm{d}\zeta(t), \int g(t)\,\mathrm{d}\zeta(t)\Big) = \int f(t)\overline{g(t)}\,\mathrm{d}\sigma(t)$

(iii) $\displaystyle\lim_{n\to\infty} \int f_n(t)\,\mathrm{d}\zeta(t) = \int \lim_{n\to\infty} f_n(t)\,\mathrm{d}\zeta(t)$

whenever the sequence $\{f_n\}$ is convergent.

Example 2.6.5. Let $t_j \in W$ and $X_j \in L^2(\Omega, \mathcal{A}, P)$, $j \in \mathbb{N}$, where the t_j's are mutually distinct and the X_j's are mutually orthogonal such that

$$\sum_{j=1}^{\infty} \|X_j\|^2 < \infty.$$

For an arbitrary subset $B \subset W$ we define

$$\zeta(B) := \sum_{j:\, t_j \in B} X_j.$$

Thus,

$$\zeta = \sum_{j=1}^{\infty} X_j \delta_{t_j}. \tag{1}$$

Then ζ is a random orthogonal measure with a discrete structure measure

$$\sigma = \sum_{j=1}^{\infty} \|X_j\|^2 \delta_{t_j}.$$

Choosing $B_j = \{t_j\}$ in Theorem 2.6.3 and taking the limit $n \to \infty$ we obtain

$$\int f(t)\,\mathrm{d}\zeta(t) = \sum_{j=1}^{\infty} f(t_j) X_j$$

where f is an arbitrary function on W.[13]

It is easy to see that the following converse statement is true as well: If σ is discrete, then ζ can be represented in the form (1).

Definition 2.6.6. A second order process $\mathcal{Z}$ on $T = [0, \infty)$ is called a *process with orthogonal increments*, if

$$(\mathcal{Z}(t_2) - \mathcal{Z}(t_1), \mathcal{Z}(t_4) - \mathcal{Z}(t_3)) = 0, \quad 0 \le t_1 \le t_2 \le t_3 \le t_4,\ t_j \in T.$$

The Wiener process and the Poisson process on $[0, \infty)$ have orthogonal increments (cf. Theorem 2.1.4 and Theorem 2.1.5). The next theorem shows that for each process with orthogonal increments there is an associated random orthogonal measure.

Theorem 2.6.7. *Let $\mathcal{Z}$ be a bounded, continuous*[14] *process with orthogonal increments on $T = [0, \infty)$. Then there exists a unique orthogonal random measure ζ on $\mathcal{B}(T)$ such that $H(\zeta) \subset H(\mathcal{Z})$ and*

$$\zeta([t, s)) = \mathcal{Z}(s) - \mathcal{Z}(t), \quad 0 \le t < s < \infty. \tag{1}$$

Proof. We define the function F by

$$F(t) = \|\mathcal{Z}(t) - \mathcal{Z}(0)\|^2, \quad t \in [0, \infty).$$

This function is continuous and bounded. Since the increments are orthogonal we have

$$\begin{aligned} F(t) &= \|\mathcal{Z}(t) - \mathcal{Z}(s) + \mathcal{Z}(s) - \mathcal{Z}(0)\|^2 \\ &= \|\mathcal{Z}(t) - \mathcal{Z}(s)\|^2 + F(s) \end{aligned}$$

[13] If $\sigma(\{t_j\}) > 0$ it is meaningful to write $f(t_j)$ for $f \in L^2$. Otherwise $X_j = 0$ and hence it is not necessary to evaluate f at t_j.

[14] Actually, it is sufficient to assume continuity from the left.

where $0 \le s \le t$. Thus, the function F is increasing and

$$F(t) - F(s) = \|Z(t) - Z(s)\|^2, \quad 0 \le s \le t.$$

By basic results of measure theory, there exists a finite nonnegative measure σ on $\mathcal{B}(T)$ such that

$$\sigma([s,t)) = F(t) - F(s), \quad s \le t.$$

For a set $B \subset T$ of the form $B = [t,s)$ we define $\zeta(B)$ by (1). In the same way as in the proof of Theorem 2.6.3 we define $I : L^2(\sigma) \to H(\mathcal{Z})$ for step functions from $L^2(\sigma)$, we use however only Borel sets of this special form. These functions span a dense linear manifold and hence I can be extended to an isometric operator from $L^2(\sigma)$ into $H(\mathcal{Z})$ (see Theorem D.3.6). Now, let $B \subset T$ be an arbitrary Borel set and write $\zeta(B) := I(\mathbf{1}_B)$. The fact that I is isometric implies that ζ is an orthogonal random measure. The uniqueness follows from the density of the linear manifold used above. □

Example 2.6.8. Let $\mathcal{Z}$ be the Wiener process or the Poisson process on $[0,\infty)$. Denote by ζ and σ the corresponding random orthogonal measure and the structure measure, in the sense of the previous theorem, respectively. For $B = [0,t) \subset [0,\infty)$ we have

$$\sigma(B) = (\zeta(B), \zeta(B)) = (\mathcal{Z}(t), \mathcal{Z}(t)) = t.$$

From this we conclude that $\sigma = \lambda|_{[0,\infty)}$.

Definition 2.6.9. Let $\mathcal{Z}$ and ζ be as in Theorem 2.6.7. We define the integral $\int_a^b f(t)\,\mathrm{d}\mathcal{Z}(t), 0 \le a \le b$, by

$$\int_a^b f(t)\,\mathrm{d}\mathcal{Z}(t) := \int_a^b f(t)\,\mathrm{d}\zeta(t), \quad f \in L^2(\sigma)$$

where σ is the structure measure of ζ.

Theorem 2.6.10 (integration by parts). *Let f be an absolutely continuous function on $[0,t]$, $t > 0$, and let $\mathcal{Z}$ be a continuous process on $[0,\infty)$ with orthogonal increments. Then we have*

$$\int_0^t f(s)\,\mathrm{d}\mathcal{Z}(s) = f(t)\mathcal{Z}(t) - f(0)\mathcal{Z}(0) - \int_0^t \mathcal{Z}(s) f'(s)\,\mathrm{d}s.$$

Proof. Denote by σ the structure measure of the random orthogonal measure corresponding to $\mathcal{Z}$ (cf. Theorem 2.6.7). We write

$$t_{i,n} := \frac{t \cdot i}{n}, \quad 0 \le i \le n$$

and define the function f_n on $[0,t]$ by $f_n(s) = f(t_{i,n})$ if $s \in [t_{i,n}, t_{i+1,n})$ and $f_n(t) = f(t)$. The sequence $\{f_n\}$ converges uniformly, hence also in $L^2(\sigma)$, to f. Using Theorem 2.6.4 we obtain

$$\int_0^t f(s)\,\mathrm{d}\mathcal{Z}(s) = \int_0^t \lim_{n\to\infty} f_n(s)\,\mathrm{d}\mathcal{Z}(s) = \lim_{n\to\infty} \sum_{i=0}^{n-1} f_n(t_{i,n})[\mathcal{Z}(t_{i+1,n}) - \mathcal{Z}(t_{i,n})].$$

Using summing by parts[15] the last sum can be written as

$$f(t)\mathcal{Z}(t) - f(0)\mathcal{Z}(0) - \sum_{i=0}^{n-1} \mathcal{Z}(t_{i,n})[f(t_{i+1,n}) - f(t_{i,n})].$$

The sum above is equal to

$$\int_0^t \mathcal{Z}(g_n(s)) f'(s)\,\mathrm{d}s \tag{1}$$

where $g_n(s) = t_{i,n}$ whenever $s \in [t_{i,n}, t_{i+1,n})$. Denote by C the correlation function of $\mathcal{Z}$. Using the inequality $|g_n(s) - s| < \frac{1}{n}$, the equation

$$\|\mathcal{Z}(s) - \mathcal{Z}(g_n(s))\|^2 = C(s,s) - C(s, g_n(s)) - C(g_n(s), s) + C(g_n(s), g_n(s))$$

and the uniform continuity of C on $[0,t]\times[0,t]$ we see that for any $\epsilon > 0$ the inequality

$$\|\mathcal{Z}(s) - \mathcal{Z}(g_n(s))\|^2 < \epsilon$$

holds if n is sufficiently large. Applying this and inequality (2.4.5.2) we conclude that the integral (1) converges to

$$\int_0^t \mathcal{Z}(s) f'(s)\,\mathrm{d}s$$

as $n \to \infty$. The proof is complete. □

We end this section with a lemma which will be useful in Section 5.4.

Lemma 2.6.11. *Let ζ be a random orthogonal measure on $(W, \mathcal{B})$ with structure measure σ and let $g \in L^2(\sigma)$. Then*

$$\zeta_g(A) := \int_W \mathbf{1}_A g\,\mathrm{d}\zeta = \int_A g\,\mathrm{d}\zeta, \quad A \in \mathcal{B}.$$

defines a random orthogonal measure ζ_g on $(W, \mathcal{B})$ with structure measure σ_g given

---

15 $\sum_{k=m}^{n} a_k(b_{k+1} - b_k) = [a_{n+1}b_{n+1} - a_m b_m] - \sum_{k=m}^{n} b_k(a_{k+1} - a_k)$

by $d\sigma_g = |g|^2\, d\sigma$. *Moreover, we have*

(i)
$$\int f\, d\zeta_g = \int fg\, d\zeta, \quad f \in L^2(\sigma_g);$$

(ii) *the* σ_g*-measure of the set of zeros of* g *is equal to* 0, *the function* $1/g$ *is in* $L^2(\sigma_g)$ *and*

$$\zeta(A) = \int_A \frac{1}{g}\, d\zeta_g, \quad A \in \mathcal{B}.$$

Proof. By equation (2.6.4.ii),

$$(\zeta_g(A), \zeta_g(B)) = \int_W \mathbf{1}_A \mathbf{1}_B |g|^2\, d\zeta = \int_{A\cap B} |g|^2\, d\sigma$$

and hence (2.6.1.i) holds. Let $B_j \in \mathcal{B}$, $j \in \mathbb{N}$, be mutually disjoint. Lebesgue's dominated convergence theorem shows that

$$\lim_{N\to\infty} \mathbf{1}_{\cup_1^N B_j} = \mathbf{1}_{\cup_1^\infty B_j} \quad \text{and} \quad \lim_{N\to\infty} g\mathbf{1}_{\cup_1^N B_j} = g\mathbf{1}_{\cup_1^\infty B_j}$$

the convergence being in $L^2(\sigma)$. From (2.6.4.iii) we see that (2.6.1.ii) holds, i.e., ζ_g is a random orthogonal measure.

The relation (i) is true for step functions by the definition of ζ_g. Let $\{f_n\}$ be a sequence of step functions converging in $L^2(\sigma_g)$ to f. Since $d\sigma_g = |g|^2\, d\sigma$, the sequence $\{f_n g\}$ converges in $L^2(\sigma)$ to fg, from which (i) follows.

The first two statements in (ii) follow from $d\sigma_g = |g|^2\, d\sigma$, while the third one follows from (i) replacing f by $\mathbf{1}_A \cdot \frac{1}{g}$. □

2.7 The theorem of Karhunen

Let ζ be a random orthogonal measure on the measurable space $(W, \mathcal{B})$ with structure measure σ. Further, let $g : T \times W \to \mathbb{C}$ be such that $g(t, \cdot) \in L^2(\sigma)$ for all $t \in T$. We define the second order field $\mathcal{Z}$ by

$$\mathcal{Z}(t) = \int_W g(t, x)\, d\zeta(x), \quad t \in T.$$

By virtue of (2.6.4.ii) the correlation function C of $\mathcal{Z}$ is given by

$$C(t, s) = \int_W g(t, x)\overline{g(s, x)}\, d\sigma(x), \quad t, s \in T.$$

In the next theorem we consider the converse direction. We assume that the correlation function admits an integral representation of this kind and deduce from it an integral representation for the field.

Theorem 2.7.1 (Karhunen). *Let $\mathcal{Z} : T \to L^2(\Omega, \mathcal{A}, P)$ be a field with correlation function C. Assume that C admits the representation*

$$C(t,s) = \int_W g(t,x)\overline{g(s,x)}\, \mathrm{d}\sigma(x), \quad t, s \in T \tag{1}$$

where

(i) *σ is a finite nonnegative measure on a measurable space $(W, \mathcal{B})$;*

(ii) *g is a complex-valued function on $T \times W$ with $g(t,\cdot) \in L^2(\sigma)$ for all $t \in T$;*

(iii) *the linear manifold $L_g :=$ span $\{g(t,\cdot),\ t \in T\}$ is dense in $L^2(\sigma)$.*

Then there exists a uniquely determined random orthogonal measure ζ on $(W, \mathcal{B})$ with values in $L^2(\Omega, \mathcal{A}, P)$ and with structure measure σ such that

$$\mathcal{Z}(t) = \int_W g(t,x)\, \mathrm{d}\zeta(x), \quad t \in T \tag{2}$$

and $H(\mathcal{Z}) = H(\zeta)$.

Proof. Let

$$f(x) = \sum_{k=1}^{n} c_k g(t_k, x), \quad t_k \in T,\ x \in W,\ c_k \in \mathbb{C}$$

be an arbitrary function in L_g. We define the linear mapping $I : L_g \to H(\mathcal{Z})$ by

$$I(f) = \sum_{k=1}^{n} c_k \mathcal{Z}(t_k).$$

We have to show that this definition is correct, i.e., the sum above depends only on f but not on the special choice of the c_k's and t_k's. Using equation (1) we obtain

$$\begin{aligned}\left(\sum_{k=1}^{n} c_k \mathcal{Z}(t_k), \mathcal{Z}(s)\right) &= \sum_{k=1}^{n} c_k C(t_k, s)\\ &= \sum_{k=1}^{n} c_k \int_W g(t_k,x)\overline{g(s,x)}\, \mathrm{d}\sigma(x)\\ &= \int_W f(x)\overline{g(s,x)}\, \mathrm{d}\sigma(x), \quad s \in T.\end{aligned}$$

Using this and the fact that the linear span of the set $\{\mathcal{Z}(s) : s \in T\}$ is dense in $H(\mathcal{Z})$ we conclude that $\sum_{k=1}^{n} c_k \mathcal{Z}(t_k)$ depends only on f.

It is clear that I is linear. A similar computation as above shows that

$$(I(f), I(f)) = \sum_{j,k=1}^{n} c_j \overline{c_k} C(t_j, t_k) = \int_W f(x)\overline{f(x)}\, \mathrm{d}\sigma(x) = (f, f)$$

i.e., I is isometric. Consequently, I can be extended to an isometric linear operator from $\overline{L_g} = L^2(\sigma)$ into $H(\mathcal{Z})$ (cf. Theorem D.3.6). We denote also the extension by I and define ζ by

$$\zeta(B) = I(\mathbf{1}_B), \quad B \in \mathcal{B}. \tag{3}$$

Using the fact that I is isometric we see that ζ is a random orthogonal measure with structure measure σ. Write

$$X(t) := \int g(t,x)\, \mathrm{d}\zeta(x), \quad t \in T.$$

Next we show that $\mathcal{Z} = X$. We have

$$(\mathcal{Z}(t), \zeta(B)) = (I(g(t,\cdot)), I(\mathbf{1}_B)) = \int g(t,x)\mathbf{1}_B(x)\, \mathrm{d}\sigma(x), \quad B \in \mathcal{B}. \tag{4}$$

For fixed t, let $g_n(s) = \sum_k c_k^n \mathbf{1}_{B_k^n}$, $n \in \mathbb{N}$, be step functions converging to $g(t,\cdot)$ in $L^2(\sigma)$ as $n \to \infty$. Then, by (2.6.4.iii),

$$\lim_{n\to\infty} \sum_k c_k^n \zeta(B_k^n) = \lim_{n\to\infty} \int g_n(t,x)\, \mathrm{d}\zeta(x) = X(t).$$

Combining this with (4) we conclude that

$$(\mathcal{Z}(t), X(s)) = \int g(t,x)\overline{g(s,x)}\, \mathrm{d}\sigma(x)$$

and hence

$$(\mathcal{Z}(t), X(s)) = (\mathcal{Z}(t), \mathcal{Z}(s)), \quad t, s \in T.$$

Since span $\{\mathcal{Z}(t) : t \in T\}$ is dense in $H(\mathcal{Z})$ we must have $X(s) = \mathcal{Z}(s)$.

The representation (2) shows that $H(Z) \subset H(\zeta)$. By the definition of ζ we also have $H(\zeta) \subset H(Z)$, i.e., $H(\zeta) = H(Z)$.

Finally, let η be a random orthogonal measure on $(W, \mathcal{B})$ with structure measure σ such that

$$\int_W g(t,x)\, \mathrm{d}\eta(x) = \int_W g(t,x)\, \mathrm{d}\zeta(x), \quad t \in T.$$

The assumption (iii) implies that

$$\int_W h\, \mathrm{d}\eta = \int_W h\, \mathrm{d}\zeta, \quad h \in L^2(\sigma).$$

Setting $h = \mathbf{1}_B$, $B \in \mathcal{B}$, we obtain $\eta = \zeta$. □

It is possible to prove a similar result (see [34]) without using the density condition (2.7.1.iii). In this case, besides losing the equality of the spaces $H(\mathcal{Z})$ and $H(\zeta)$, the random orthogonal measure has values in a possibly larger space. This is the motivation for the next theorem where we use the same notation as in Theorem 2.7.1.

Theorem 2.7.2. *Assume that* C *admits the integral representation (2.7.1.1) satisfying the conditions (2.7.1.i) and (2.7.1.ii). Then the integral representation (2.7.1.2) with a random orthogonal measure* $\zeta : \mathcal{B} \to L^2(\Omega, \mathcal{A}, P)$ *having structure measure* σ *is possible if and only if* $\dim L_g^{\perp} \leq \dim H(\mathcal{Z})^{\perp}$.

Proof. Assume first that $\dim L_g^{\perp} \leq \dim H(Z)^{\perp}$ and consider the orthogonal decomposition

$$L^2(W, \mathcal{B}, \sigma) = L_g \oplus L_g^{\perp}.$$

Choose an orthonormal basis $\{e_\iota : \iota \in I\}$ of $L_g^{\perp}$, where I is a nonempty index set such that $T \cap I = \emptyset$. By assumption there exists an orthonormal system $\{Y_\iota : \iota \in I\}$ in $H(Z)^{\perp}$. Now we consider the random field

$$\widetilde{X}(t) := \begin{cases} \mathcal{Z}(t), & \text{if } t \in T \\ Y_t, & \text{if } t \in I, \end{cases}$$

for all $t \in T \cup I$. We write further for $t \in T \cup I$

$$\widetilde{g}(t, \cdot) := \begin{cases} g(t, \cdot), & \text{if } t \in T \\ e_t, & \text{if } t \in I. \end{cases}$$

Then $\widetilde{X}(t) \in L^2(\Omega, \mathcal{A}, P)$ for all $t \in T \cup I$ and

$$\widetilde{C}(s,t) = (\widetilde{X}(s), \widetilde{X}(t)) = \int_W \widetilde{g}(s,x)\overline{\widetilde{g}(t,x)}\, \mathrm{d}\sigma(x).$$

Since $L_{\widetilde{g}}$ is dense in $L^2(W, \mathcal{B}, \sigma)$, Theorem 2.7.1 yields the existence of a uniquely determined random orthogonal measure ζ with structure measure σ and values in $L^2(\Omega, \mathcal{A}, P)$ such that

$$\widetilde{X}(t) = \int_W \widetilde{g}(t,x)\, \mathrm{d}\zeta(x).$$

In particular,

$$\mathcal{Z}(t) = \int_W g(t,x)\, \mathrm{d}\zeta(x)$$

for all $t \in T$.

For the converse direction assume that

$$\dim H(\mathcal{Z})^{\perp} < \dim L_g^{\perp}.$$

As before, choose an orthonormal basis $\{e_\iota : \iota \in I\}$ of $L_g^{\perp}$, where I is nonempty and $T \cap I = \emptyset$. Assume $\zeta : \mathcal{B} \to L^2(\Omega, \mathcal{A}, P)$ to be a random orthogonal measure satisfying (2.7.1.2) and having structure measure σ. Set for $\iota \in I$

$$Y_\iota := \int_W e_\iota(x)\, \mathrm{d}\zeta(x) \in L^2(\Omega, \mathcal{A}, P).$$

We then have

$$
\begin{aligned}
(\mathcal{Z}(s), Y_\iota) &= \left(\int_W g(s,x)\, \mathrm{d}\zeta(x), \int_W e_\iota(x)\, \mathrm{d}\zeta(x) \right) \\
&= \int_W g(s,x)\overline{e_\iota(x)}\, \mathrm{d}\sigma(x) = 0, \quad s \in T,\ \iota \in I
\end{aligned}
$$

and

$$
\begin{aligned}
(Y_\iota, Y_\jmath) &= \left(\int_W e_\iota(x)\, \mathrm{d}\zeta(x), \int_W e_\jmath(x)\, \mathrm{d}\zeta(x) \right) \\
&= \int_W e_\iota(x)\overline{e_\jmath(x)}\, \mathrm{d}\sigma(x) = \delta_{\iota,\jmath}, \quad \iota, \jmath \in I.
\end{aligned}
$$

Thus, $\{Y_\iota : \iota \in I\}$ is an orthonormal system in $H(\mathcal{Z})^\perp$. Hence, this system can be extended to an orthonormal basis of $H(\mathcal{Z})^\perp$. Since $\dim L_g^\perp$ is equal to the cardinality of I, we obtain

$$\dim L_g^\perp \le \dim H(\mathcal{Z})^\perp$$

a contradiction to the assumption. The proof is complete. □

Remark 2.7.3. For fixed functions C and g the representing measure σ in (2.7.1.1) is in general not uniquely determined. To see classical examples consider the case of $W = \mathbb{R}$ equipped with the usual Borel σ-algebra $\mathcal{B}$.

Let first

$$g(t,x) = x^t, \quad n \in T := \mathbb{N}_0,\ x \in \mathbb{R}.$$

A bounded positive measure σ on the real line is in general not uniquely determined by the values of the integrals

$$\int_{-\infty}^{\infty} g(t,x)\overline{g(s,x)}\, \mathrm{d}\sigma(x) = \int_{-\infty}^{\infty} x^{t+s}\, \mathrm{d}\sigma(x), \quad t, s \in \mathbb{N}_0$$

(see also Remark 3.3.7). The description of all measures with given moments is the *power moment problem*. For more information on the power moment problem we refer the reader to the monographs [1] and [55].

Another example is given by the *trigonometric moment problem*. In this case $T = [-a, a],\ a > 0$, and

$$g(t,x) = \mathrm{e}^{\mathrm{i}tx}, \quad t \in [-a,a],\ x \in \mathbb{R}.$$

We will see in Chapter 4 that σ is in general not uniquely determined by the values of the integrals

$$\int_{-\infty}^{\infty} g(t,x)\overline{g(s,x)}\, \mathrm{d}\sigma(x) = \int_{-\infty}^{\infty} \mathrm{e}^{\mathrm{i}(t-s)x}\, \mathrm{d}\sigma(x), \quad t, s \in [-a,a].$$

The paper [5] contains more details on the connection between Karhunen's theorem, moment problems and quadrature formulae.

2.8 Stationary fields

Recall that throughout this chapter T denotes a nonempty subset of $\mathbb{R}^d$.

Definition 2.8.1. A second order field Z on T is said to be *stationary*,[16] if the expectation $M := \mathbb{E}(Z(t))$ does not depend on t and $\mathbb{E}(Z(t) \cdot \overline{Z(s)})$ is a function of $t - s$:

$$\mathbb{E}\left(Z(t) \cdot \overline{Z(s)}\right) = C(t-s), \quad t, s, t-s \in T$$

where C is a complex-valued function defined on the set

$$T - T = \{t - s : t, s \in T\}.$$

With a little abuse of terminology we call the function C, like in Definition 2.1.1, the *correlation function* of Z. If Z is stationary, then the covariance

$$\mathbb{E}\left((Z(t) - M) \cdot (\overline{Z(s) - M})\right)$$

is a function of $t - s$ as well. This function is called the *covariance function* of Z. Denoting it by K we have

$$K(h) = C(h) - |M|^2, \quad h \in T - T.$$

The notions above are defined for $\mathcal{Z}$ (cf. Notation 2.2.6) in the same way as for Z.

Definition 2.8.2. A field Z on T is said to be *strongly stationary* or *stationary in the strong sense* when all its finite-dimensional distributions remain the same when shifted. That is, for all $n \in \mathbb{N}$ and all $t_1, \dots t_n, h \in \mathbb{R}^d$ the random vectors

$$(Z(t_1), \dots, Z(t_n)) \quad \text{and} \quad (Z(t_1 + h), \dots, Z(t_n + h))$$

have the same distribution whenever $t_1 + h, \dots, t_n + h \in T$.

If Z is a strongly stationary field of second order then it is stationary. Indeed, the random variables $Z(t)$ and $Z(t+h)$ have the same distribution and hence the same expectation. Moreover, the random vectors

$$(Z(t), Z(s)) \quad \text{and} \quad (Z(t+h), Z(s+h))$$

have the same distribution and hence the same correlation.[17]

Next we show that for Gaussian fields the notions stationarity and strong stationarity are the same.

---

[16] Stationary fields are also called *stationary in the wide sense*, *widely stationary* or *second order stationary*. These are the main fields studied in this book, hence we call them simply stationary for brevity.

[17] Actually, we used only the cases $n = 1$ and $n = 2$ of the definition of strong stationarity.

Theorem 2.8.3. *Every stationary Gaussian field Z is strictly stationary.*

Proof. Since Z is Gaussian the random vectors

$$(Z(t_1), \dots, Z(t_n)) \quad \text{and} \quad (Z(t_1 + h), \dots, Z(t_n + h))$$

$t_j, t_j + h \in T$, $h \in \mathbb{R}^d$, are Gaussian. Since Z is stationary, they have the same expectation and the same covariance. Consequently, they have the same distribution. □

Lemma 2.8.4. *Let Z be stationary on $T = \mathbb{R}^d$ or $T = \mathbb{Z}^d$ with a continuous correlation function C. Then we have*

$$(\mu * Z(t), \nu * Z(s)) = (\mu * \tilde{\nu} * Z(t), Z(s)) = \mu * \tilde{\nu} * C(t - s) \tag{1}$$

for all complex measures μ, ν on T and for all $s, t \in T$.

Proof. It is easy to check that the lemma holds for measures with finite support. In the general case we approximate μ and ν by measures with finite support (see Corollary 2.4.10) and apply Theorem E.2.5. □

Definition 2.8.5. A second order field Z is called *white noise* with expectation μ and variance σ^2, $\sigma \geq 0$, if the random variables $X(t)$, $t \in T$, are uncorrelated and $\mathbb{E}(Z(t)) = \mu$ and $\mathrm{Var}(Z(t)) = \sigma^2$ for all $t \in T$.

We will use the notation $Z \sim WN(\mu, \sigma^2)$ to indicate that Z is white noise with expectation μ and variance σ^2. White noise is stationary with covariance function

$$K(h) = \begin{cases} \sigma^2, & h = 0 \\ 0, & h \neq 0 . \end{cases}$$

If the random variables $Z(t)$, $t \in T$, are not identically distributed, then Z is not strongly stationary.

Example 2.8.6.

(1) Let $X \in \mathcal{L}^2(\Omega, \mathcal{A}, P)$, $T = \mathbb{R}^d$ and $x \in \mathbb{R}^d$. The field

$$Z(t) = \mathrm{e}^{\mathrm{i}(t,x)} X, \quad t \in \mathbb{R}^d \tag{1}$$

is stationary since

$$(Z(t + h), Z(t)) = \mathrm{e}^{\mathrm{i}(t+h,x)} \cdot \mathrm{e}^{-\mathrm{i}(t,x)} (X, X) = \mathrm{e}^{\mathrm{i}(h,x)} (X, X).$$

The question arises of under which conditions on a complex-valued function f on $\mathbb{R}^d$ is the field $Z : t \mapsto f(t)X$ stationary? Without loss of generality assume that $(X, X) = 1$ and $f(0) = 1$. The field Z is stationary if and only if the relation

$$(f(t+h)X, f(t)X) = f(t+h)\overline{f(t)} = g(h), \quad t, h \in \mathbb{R}^d$$

holds with some function g. Setting $t = 0$ we see that $f = g$ while setting $h = 0$ we obtain $|f| = 1$. Thus, f is a character of the group $\mathbb{R}^d$ (cf. Definition 1.4.5). If f is continuous then, by Theorem C.9.2, the function f has the form $f(h) = \mathrm{e}^{\mathrm{i}(h,x)}$ with some $x \in \mathbb{R}^d$.

(2) The example given by equation (1) can be generalized as follows. Let $X_1, \ldots, X_n \in \mathcal{L}^2(\Omega, \mathcal{A}, P)$ be an orthonormal system such that $\mathbb{E}(X_j) = m$ and let $x_1, \ldots, x_n \in \mathbb{R}^d$. Then the field Z defined by

$$Z(t) = \sum_{j=1}^{n} \mathrm{e}^{\mathrm{i}(t,x_j)} X_j, \quad t \in \mathbb{R}^d$$

is stationary with correlation function

$$C(h) = \sum_{j=1}^{n} (X_j, X_j) \cdot \mathrm{e}^{\mathrm{i}(h,x_j)}, \quad h \in \mathbb{R}^d.$$

(3) To give an example of a real stationary field let $A, B \in \mathcal{L}_r^2(\Omega, \mathcal{A}, P)$ be such that $(A, A) = (B, B) = 1$ and $(A, B) = (A, \mathbf{1}) = (B, \mathbf{1}) = 0$. Further, let $\theta \in \mathbb{R}^d$ and put

$$Z(t) = A \cos \theta t + B \sin \theta t, \quad t \in \mathbb{R}^d$$

where we write θt instead of $((\theta, t))$, for brevity. Then $\mathbb{E}(Z(t)) = 0$. Using trigonometric identities we obtain

$$\begin{aligned}
(Z(t+h), Z(t)) &= (A \cos \theta(t+h) + B \sin \theta(t+h), A \cos \theta t + B \sin \theta t) \\
&= \cos \theta(t+h) \cdot \cos \theta t + \sin \theta(t+h) \cdot \sin \theta t \\
&= \cos \theta h.
\end{aligned}$$

Thus, Z is stationary with correlation function $h \mapsto \cos((\theta, h))$.

Example 2.8.7. (1) In this example we generalize the field defined in equation (2.8.6.1) by making the exponent random. Let X be a real random variable and Y be a d-dimensional real random vector. Assume that X and Y are independent, $\mathbb{E}(X) = 0$ and $\mathrm{Var}(X) = \sigma^2 < \infty$. We define the field Z by

$$Z(t) = \mathrm{e}^{\mathrm{i}(t,Y)} \cdot X, \quad t \in \mathbb{R}^d.$$

This field is stationary since

$$\mathbb{E}(Z(t)) = \mathbb{E}(X) \cdot \mathbb{E}\big(e^{i(t,Y)}\big) = 0$$

and

$$\begin{aligned}(Z(t+h), Z(t)) &= \big(e^{i(t+h,Y)} \cdot X, e^{i(t,Y)} \cdot X\big) \\ &= \mathbb{E}\big(e^{i(h,Y)} \cdot X^2\big) \\ &= \sigma^2 \cdot f_Y(h), \quad h \in \mathbb{R}^d\end{aligned}$$

where f_Y denotes the characteristic function of Y.

(2) Now we construct a real stationary field in a similar way. Let X and Y be as above and choose a real random variable Ψ such that $X,\ Y,\ \Psi$ are independent and[18]

$$\mathbb{E}(\cos\Psi) = \mathbb{E}(\sin\Psi) = 0 \quad \text{and} \quad \mathbb{E}(\cos^2\Psi) = \mathbb{E}(\sin^2\Psi) = \frac{1}{2}. \tag{1}$$

We define the field Z by

$$Z(t) = X \cdot \cos((t, Y) + \Psi), \quad t \in \mathbb{R}^d.$$

Then $\mathbb{E}(Z(t)) = 0$ and

$$(Z(t+h), Z(t)) = \sigma^2 \cdot \mathbb{E}\big(\cos((t+h, Y) + \Psi) \cdot \cos((t, Y) + \Psi)\big).$$

Applying trigonometric identities and using (1) we obtain

$$\begin{aligned}(Z(t+h), Z(t)) &= \frac{\sigma^2}{2} \cdot \mathbb{E}\left(\cos((t+h)Y)\cos(tY) + \sin((t+h)Y)\sin(tY)\right) \\ &= \frac{\sigma^2}{2} \cdot \mathbb{E}\cos((h, Y)) \\ &= \frac{\sigma^2}{2} \cdot \operatorname{Re} f_Y(h)\end{aligned}$$

where f_Y is the characteristic function of Y.

Example 2.8.8. (1) Let $Z_j,\ j \in \mathbb{Z}$, be independent and identically distributed such that $\mathbb{E}(Z_j) = 0$ and $\operatorname{Var}(Z_j) = \sigma^2$. For $\theta \in \mathbb{R}$ we define the time series X by[19]

$$X_j = Z_j + \theta Z_{j-1}, \quad j \in \mathbb{Z}.$$

[18] These equations hold, e.g., if Ψ is uniformly distributed in $[0, 2\pi]$.

[19] A time series of this form is called a *moving average sequence* of order 1.

This time series is stationary:

$$(X_{j+h}, X_j) = (Z_{j+h} + \theta Z_{j-1+h}, Z_j + \theta Z_{j-1})$$
$$= \begin{cases} (1+\theta^2)\sigma^2, & h = 0, \\ \theta\sigma^2, & h = \pm 1, \\ 0, & |h| > 0. \end{cases}$$

It is even strongly stationary. To prove this we show that the characteristic function f_Y of the random vector

$$Y = (X_{1+h}, X_{2+h}, \dots, X_{n+h}), \quad h \in \mathbb{Z},\ n \in \mathbb{N}$$

does not depend on h. Denote by f the characteristic function of Z_j. For $t = (t_1, \dots, t_n) \in \mathbb{R}^n$ we have

$$(Y,t) = \sum_{j=1}^{n} (Z_{j+h} + \theta Z_{j-1+h}) t_j = \theta t_1 Z_h + t_n Z_{n+h} + \sum_{j=1}^{n-1} (t_j + \theta t_{j+1}) Z_{j+h}$$

and therefore

$$f_Y(t) = \mathbb{E}\left(e^{i(Y,t)}\right) = f(\theta t_1) \cdot f(t_n) \cdot \prod_{j=1}^{n-1} f(t_j + \theta t_{j+1}).$$

Thus, f_Y does not depend on h.

(2) Let $T = \mathbb{Z}$, $Z \sim WN(0, \sigma^2)$ and $a \in \mathbb{C}$ such that $|a| < 1$. We are looking for a stationary time series X satisfying the equation[20]

$$X_n = Z_n + a X_{n-1}, \quad n \in \mathbb{Z}. \tag{1}$$

Applying this equation iteratively we obtain

$$\begin{aligned} X_n &= Z_n + a Z_{n-1} + a^2 X_{n-2} \\ &= \cdots \\ &= Z_n + a Z_{n-1} + \cdots + a^k Z_{n-k} + a^{k+1} X_{n-k-1}. \end{aligned}$$

If X is stationary, then $\|X_n\|$ is constant, so

$$\left\| X_n - \sum_{j=0}^{k} a^j Z_{n-j} \right\|^2 = |a|^{2k+2} \|X_{n-k-1}\|^2 \to 0, \quad k \to \infty.$$

Thus,

$$X_n = \sum_{j=0}^{\infty} a^j Z_{n-j}, \quad n \in \mathbb{Z}. \tag{2}$$

[20] A time series satisfying this equation is called an *autoregressive sequence* of order 1.

By Theorem F.2.3 this series converges with probability one. Now we show that the time series $X = \{X_n\}_{n\in\mathbb{N}}$ is stationary. Indeed,

$$(X_n, \mathbf{1}) = \sum_{j=0}^{\infty} a^j (Z_{n-j}, \mathbf{1}) = 0$$

for all n and for $h \geq 0$ we have

$$\begin{aligned}
(X_{n+h}, X_n) &= \left(\sum_{j=0}^{\infty} a^j Z_{n+h-j}, \sum_{k=0}^{\infty} a^k Z_{n-k} \right) \\
&= \left(\sum_{j=h}^{\infty} a^j Z_{n+h-j}, \sum_{k=0}^{\infty} a^k Z_{n-k} \right) \\
&= \left(\sum_{j=0}^{\infty} a^{j+h} Z_{n-j}, \sum_{k=0}^{\infty} a^k Z_{n-k} \right) \\
&= \sigma^2 a^h \sum_{j=0}^{\infty} |a|^{2j} \\
&= \frac{\sigma^2}{1-|a|^2} \cdot a^h .
\end{aligned}$$

We find that X is stationary and the correlation function C is given by

$$C(h) = \frac{\sigma^2}{1-|a|^2} \cdot (\overline{a})^{|h|}, \quad h < 0.$$

We have thus shown that equation (1) has exactly one stationary solution which is given by (2).

If $a \neq 0$ then equation (1) has a unique solution for each fixed X_0 which can be determined iteratively: $X_1 = Z_1 + aX_0$, $X_{-1} = (X_0 - Z_0)/a$, and so on. If X_0 is not given by (2), i.e.,

$$X_0 \neq \sum_{j=0}^{\infty} a^j Z_{-j}$$

then the solution is not stationary.

Example 2.8.9. Let Z be the Wiener process or the Poisson process constructed in Corollary 2.1.4 and Corollary 2.1.5, respectively. Further, let $h \in [0, \infty)$ be arbitrary and define the process X^h by

$$X^h(t) = Z(t+h) - Z(t), \quad t \in [0, \infty).$$

The computations in Example 2.2.5 show that X^h is stationary and its correlation function C^h is given by

$$C^h(t) = \max(0, h-t), \quad t \in [0, \infty).$$

2.9 Spectral representation of stationary fields

First we generalize the notion of positive definiteness introduced in Definition 1.4.1.[21]

Definition 2.9.1. A function $f : T - T \to \mathbb{C}$ is said to be *T-positive definite* if the inequality

$$\sum_{i,j=1}^{n} f(t_i - t_j) c_i \overline{c_j} \geq 0$$

holds for every positive integer n, for all $t_1, \dots, t_n \in T$ and for all $c_1, \dots, c_n \in \mathbb{C}$.

Now we reformulate some of our previous results in terms of stationary fields. The next theorem follows from Theorem 2.2.3.

Theorem 2.9.2. *The correlation function of a real or complex stationary field on T is T-positive definite. Conversely, let C be a real- or complex-valued function defined on $T - T$. If C is T-positive definite, then there exists a real or complex, respectively, stationary Gaussian field on T with correlation function C. Analogous statements hold for the covariance function.*

In the remaining part of this section we consider only the correlation function. Combining the previous theorem with Bochner's Theorem 1.7.3 and Herglotz's Theorem 1.7.5 we obtain the following characterizations of correlation functions of stationary fields on $\mathbb{R}^d$ and $\mathbb{Z}^d$.

Theorem 2.9.3. *A continuous complex-valued function C on $\mathbb{R}^d$ is the correlation function of a continuous stationary field on $\mathbb{R}^d$ if and only if it can be represented in the form*

$$C(t) = \int_{\mathbb{R}^d} e^{i(t,x)} \, d\sigma(x), \quad t \in \mathbb{R}^d$$

with some nonnegative finite Borel measure σ on $\mathbb{R}^d$

Theorem 2.9.4. *A complex-valued function C on $\mathbb{Z}^d$ is the correlation function of a stationary field on $\mathbb{Z}^d$ if and only if it can be represented in the form*

$$C(n) = \int_{[0,2\pi)^d} e^{i(n,t)} \, d\sigma(t), \quad n \in \mathbb{Z}^d$$

with some nonnegative finite Borel measure σ on $[0, 2\pi)^d$.

The measure σ in the previous theorems is called the *spectral measure* of the field. If σ is absolutely continuous with respect to the Lebesgue measure, then the corresponding density is called the *spectral density*.

[21] See also Definition 4.1.1.

Example 2.9.5. Let $\mathcal{Z} \sim WN(0, \sigma^2)$ be a white noise on $\mathbb{Z}^d$. Since

$$\int_{[0,2\pi)^d} \mathrm{e}^{\mathrm{i}(n,x)}\, \mathrm{d}\lambda_d(x) = 0, \quad n \in \mathbb{Z}^d \setminus \{0\}$$

the correlation function C can be represented as

$$C(n) = \frac{\sigma^2}{(2\pi)^d} \int_{[0,2\pi)^d} \mathrm{e}^{\mathrm{i}(n,x)}\, \mathrm{d}\lambda_d(x), \quad n \in \mathbb{Z}^d.$$

Thus, the spectral density is constant. Conversely, if a field $\mathcal{Z}$ on $\mathbb{Z}^d$ has a constant spectral density, then $\mathcal{Z}$ is a white noise.

Using Theorem 1.7.8 we can characterize the correlation functions of not necessarily continuous fields on $\mathbb{R}^d$. By Γ_d we denote the set of all characters of the group $\mathbb{R}^d$. Recall that Γ_d consists of all complex-valued functions $\gamma \neq 0$ on $\mathbb{R}^d$ such that $\gamma(x+y) = \gamma(x)\gamma(y)$ and $\gamma(-x) = \overline{\gamma(x)}$ for all x and y. Endowing Γ_d with the topology of pointwise convergence, Γ_d becomes a compact Hausdorff space (cf. Remark 1.7.6).

Theorem 2.9.6. *A complex-valued function C on $\mathbb{R}^d$ is the correlation function of a stationary field on $\mathbb{R}^d$ if and only if it can be represented in the form*

$$C(t) = \int_{\Gamma_d} \gamma(t)\, \mathrm{d}\sigma(\gamma), \quad t \in \mathbb{R}^d$$

with some nonnegative finite Borel measure σ on Γ_d.

Example 2.9.7. Let $\mathcal{Z} \sim WN(0, 1)$ be a white noise on $\mathbb{R}^d$. Then $C = \mathbf{1}_{\{0\}}$ and the measure σ in the previous theorem is equal to the normalized Haar measure on Γ_d (see Corollary 1.7.11 and its proof).

We are now able to prove integral representations for stationary fields using Karhunen's Theorem 2.7.1.

Theorem 2.9.8. *Let $\mathcal{Z}$ be a continuous stationary field on $\mathbb{R}^d$ with correlation function C and spectral measure σ. Then there exists a uniquely determined random orthogonal measure ζ on $(\mathbb{R}^d, \mathcal{B}_d)$ with structure measure σ such that*

$$\mathcal{Z}(t) = \int_{\mathbb{R}^d} \mathrm{e}^{\mathrm{i}(t,x)}\, \mathrm{d}\zeta(x), \quad t \in \mathbb{R}^d$$

and $H(\mathcal{Z}) = H(\zeta)$.

Conversely, let ζ be a random orthogonal measure on $(\mathbb{R}^d, \mathcal{B}_d)$ with structure measure σ. Then the equation above defines a continuous stationary field $\mathcal{Z}$ with spectral measure σ.

Proof. We have

$$C(t-s) = \int_{\mathbb{R}^d} e^{i(t-s,x)}\, d\sigma(x) = \int_{\mathbb{R}^d} e^{i(t,x)}\overline{e^{i(s,x)}}\, d\sigma(x), \quad t, s \in \mathbb{R}^d.$$

The linear span L of the functions $e^{i(t,\cdot)}$, $t \in \mathbb{R}^d$, is dense in $L^2(\sigma)$. To see this assume that $g \in L^{\perp}$. Then

$$\int_{\mathbb{R}^d} e^{i(t,x)}\overline{g(x)}\, d\sigma(x) = 0, \quad t \in \mathbb{R}^d.$$

Since $g \in L^2(\sigma) \subset L^1(\sigma)$, the measure $\mu := \bar{g}\sigma$ is finite. The equation above shows that $\check{\mu} = 0$ and hence $g = 0$. Thus, L is dense in $L^2(\sigma)$. The first statement of the theorem follows immediately from Karhunen's Theorem 2.7.1, while the second one is a consequence of basic properties of the integral with respect to ζ (cf. Theorem 2.6.4). □

Example 2.9.9. Let $\mathcal{Z}$ be a continuous stationary field on $\mathbb{R}^d$. If the spectral measure σ of $\mathcal{Z}$ is discrete, i.e.,

$$\sigma = \sum_{j=1}^{\infty} p_j \delta_{x_j}, \quad x_j \in \mathbb{R}^d,\ p_j \geq 0,\ \sum_{j=1}^{\infty} p_j < \infty$$

then

$$C(t) = \sum_{j=1}^{\infty} p_j e^{i(t,x_j)}, \quad t \in \mathbb{R}^d$$

and

$$\mathcal{Z}(t) = \sum_{j=1}^{\infty} e^{i(t,x_j)} \cdot X_j, \quad t \in \mathbb{R}^d$$

where the X_j's are orthogonal and $(X_j, X_j) = p_j$ (see Example 2.6.5). If $\mathcal{Z}$ is a Gaussian field then the X_j's are even independent.

The next two theorems can be proved in the same way as Theorem 2.9.8. We omit the proofs.

Theorem 2.9.10. *Let $\mathcal{Z}$ be a stationary field on $\mathbb{Z}^d$ with correlation function C and spectral measure σ. Then there exists a uniquely determined random orthogonal measure ζ on $\left([0, 2\pi)^d, \mathcal{B}\left([0, 2\pi)^d\right)\right)$ with structure measure σ such that*

$$\mathcal{Z}(n) = \int_{[0,2\pi)^d} e^{i(n,x)}\, d\zeta(x), \quad n \in \mathbb{Z}^d$$

and $H(\mathcal{Z}) = H(\zeta)$.

Conversely, let ζ be a random orthogonal measure on $([0,2\pi)^d, \mathcal{B}([0,2\pi)^d))$ with structure measure σ. Then the equation above defines a stationary field $\mathcal{Z}$ with spectral measure σ.

Theorem 2.9.11. *Let $\mathcal{Z}$ be a stationary field on $\mathbb{R}^d$ with correlation function C and spectral measure σ. Then there exists a uniquely determined random orthogonal measure ζ on $(\Gamma_d, \mathcal{B}(\Gamma_d))$ with structure measure σ such that*

$$\mathcal{Z}(t) = \int_{\Gamma_d} \gamma(t)\,\mathrm{d}\zeta(\gamma), \quad t \in \mathbb{R}^d$$

and $H(\mathcal{Z}) = H(\zeta)$. Conversely, if ζ is a random orthogonal measure on $(\Gamma_d, \mathcal{B}(\Gamma_d))$, then the equation above defines a stationary field $\mathcal{Z}$.

We will call the random orthogonal measure ζ occurring in Theorems 2.9.8, 2.9.10 and 2.9.11, the *representing measure* of the field $\mathcal{Z}$.

Remark 2.9.12. Let $\mathcal{Z}$ be a continuous stationary field on $\mathbb{R}^d$ (analogous considerations can be carried out for fields on $\mathbb{Z}^d$ or for not necessarily continuous fields on $\mathbb{R}^d$). We have seen in the proof of Karhunen's Theorem 2.7.1 that the mapping $\mathrm{e}^{\mathrm{i}(t,\cdot)} \mapsto \mathcal{Z}(t)$, $t \in \mathbb{R}^d$, can be uniquely extended to an isometric linear operator $I_{\mathcal{Z}}$ from $L^2(\sigma)$ onto $H(\mathcal{Z}) = H(\zeta)$. Using the inverse of $I_{\mathcal{Z}}$ we can reformulate problems that are formulated in the space $H(\mathcal{Z})$ to problems formulated in $L^2(\sigma)$.

To see examples, let X be a stationary time series on $\mathbb{Z}$ with spectral measure σ. By isometry we have

$$\left\| X_{n+1} - \sum_{k=0}^{n} c_k X_k \right\|^2 = \int_0^{2\pi} \left| \mathrm{e}^{\mathrm{i}(n+1)x} - \sum_{k=0}^{n} c_k \mathrm{e}^{\mathrm{i}kx} \right|^2 \mathrm{d}\sigma(x).$$

This equation, for example, is important for the prediction of stationary processes.[22] In this case the problem is to find those c_j's which minimize the left-hand side.

Another problem is to determine $\zeta([a,b))$, $a < b$, where ζ is the random orthogonal measure in the spectral representation of X. For this we choose a sequence

$$p_n(x) = \sum_j c_{j,n} \mathrm{e}^{\mathrm{i}k_{j,n}x}, \quad n \in \mathbb{N},\ k_{j,n} \in \mathbb{Z},\ x \in [0,2\pi)$$

of trigonometric polynomials converging in $L^2(\sigma)$ to $\mathbf{1}_{[a,b)}$. Since $I_{\mathcal{Z}}$ is isometric we have $\lim_{n\to\infty} I_{\mathcal{Z}}(p_n) = I_{\mathcal{Z}}(\mathbf{1}_{[a,b)})$. Using the definition of ζ (see equation (2.7.1.3)) we obtain

$$\zeta([a,b)) = \lim_{n\to\infty} \sum_j c_{j,n} X_{k_{j,n}}.$$

[22] We will generalize the equation above in Theorem 2.9.13.

We will use this method in Theorems 2.9.16 and 2.9.17 to prove inversion formulae. In Section 2.10 we will investigate the interconnection between stationary fields and unitary representations which will provide another point of view for the operator $I_{\mathcal{Z}}$.

Theorem 2.9.13. *Let $\mathcal{Z}$ be a continuous stationary field on $\mathbb{R}^d$ with spectral measure σ, correlation function C, and let μ and ν be complex measures on $\mathbb{R}^d$. Then*

$$\begin{aligned}\left(\int_{\mathbb{R}^d} \mathcal{Z}\,\mathrm{d}\mu, \int_{\mathbb{R}^d} \mathcal{Z}\,\mathrm{d}\nu\right) &= \int_{\mathbb{R}^d} \check{\mu}\cdot\overline{\check{\nu}}\,\mathrm{d}\sigma \\ &= \int_{\mathbb{R}^d}\int_{\mathbb{R}^d} C(t-s)\,\mathrm{d}\mu(t)\,\mathrm{d}\overline{\nu}(s) \\ &= \int_{\mathbb{R}^d} C\,\mathrm{d}(\mu * \tilde{\nu})\end{aligned}$$

and

$$\int_{\mathbb{R}^d} \mathcal{Z}\,\mathrm{d}\mu = I_{\mathcal{Z}}(\check{\mu}) \tag{1}$$

where $I_{\mathcal{Z}}$ is the isometric operator from $L^2(\sigma)$ onto $H(Z)$ introduced in Remark 2.9.12.

Proof. The integrability of $\mathcal{Z}$ is a consequence of Theorem 2.4.4, since $\|\mathcal{Z}(t)\|$ is constant. The equations

$$\left(\int_{\mathbb{R}^d} \mathcal{Z}\,\mathrm{d}\mu, \int_{\mathbb{R}^d} \mathcal{Z}\,\mathrm{d}\nu\right) = \int_{\mathbb{R}^d}\int_{\mathbb{R}^d} C(t-s)\,\mathrm{d}\mu(t)\,\mathrm{d}\overline{\nu}(s) = \int_{\mathbb{R}^d} C\,\mathrm{d}(\mu * \tilde{\nu})$$

follow from Theorem 2.4.6 and from the definition of the convolution (cf. Section E.2).

To prove the remaining two equations it suffices to consider nonnegative (finite) measures. Using the definition of $I_{\mathcal{Z}}$ and the fact that it is isometric we see that the equations hold for measures with finite support. By Corollary 2.4.10 there exist sequences $\{\mu_n\}$ and $\{\nu_n\}$ of nonnegative measures with finite support converging weakly to μ and ν, respectively, such that

$$\int_{\mathbb{R}^d} \mathcal{Z}\,\mathrm{d}\mu = \lim_{n\to\infty}\int_{\mathbb{R}^d} \mathcal{Z}\,\mathrm{d}\mu_n \quad \text{and} \quad \int_{\mathbb{R}^d} \mathcal{Z}\,\mathrm{d}\nu = \lim_{n\to\infty}\int_{\mathbb{R}^d} \mathcal{Z}\,\mathrm{d}\nu_n.$$

Theorem 1.6.1 shows that the sequences $\{\check{\mu}_n\}$ and $\{\check{\nu}_n\}$ converge uniformly on compact sets (and hence also in $L^2(\sigma)$) to $\check{\mu}$ and $\check{\nu}$, respectively. Now the equations in question follow by taking the limit $n \to \infty$. □

The next theorem can be proved in the same way as the previous one. We omit the proof.

Theorem 2.9.14. *If $\mathcal{Z}$ is a stationary field on $\mathbb{Z}^d$ with spectral measure σ and μ and ν are complex measures on $\mathbb{Z}^d$, then*

$$\left(\int_{\mathbb{Z}^d} \mathcal{Z}\,\mathrm{d}\mu, \int_{\mathbb{Z}^d} \mathcal{Z}\,\mathrm{d}\nu\right) = \int_{[0,2\pi)^d} \check{\mu}\cdot\overline{\check{\nu}}\,\mathrm{d}\sigma$$

and

$$\int_{\mathbb{Z}^d} \mathcal{Z}\,\mathrm{d}\mu = I_{\mathcal{Z}}(\check{\mu}). \tag{1}$$

As an application of the operator $I_{\mathcal{Z}}$ we present the so-called *sampling theorem* attributed to E. T. Whittaker, H. Nyquist, C. Shannon, V. Kotelnikov, and É. Borel. It implies, in terms of signal processing, that a bandlimited signal can be recovered from uniformly spaced discrete samples as long as the sampling rate is not smaller than twice the bandwidth.

Theorem 2.9.15 (sampling theorem). *Let X be a continuous stationary process on $\mathbb{R}$ such that the support of its spectral measure σ is contained in a finite interval $[-a,a]$, $a>0$, and $\sigma(\{-a\}) = \sigma(\{a\}) = 0$. Then we have*

$$X(t) = \sum_{n=-\infty}^{\infty} \frac{\sin(at-\pi n)}{at-\pi n}\cdot \mathcal{Z}\left(\frac{\pi n}{a}\right), \quad t\in\mathbb{R}.^{23}$$

Proof. The correlation function C of X admits the representation

$$C(t) = \int_{-a}^{a} \mathrm{e}^{\mathrm{i}tx}\,\mathrm{d}\sigma(x), \quad t\in\mathbb{R}.$$

Let $t\in\mathbb{R}$ be fixed and consider the function $f_t : x\mapsto \mathrm{e}^{\mathrm{i}tx}$ on the interval $[-a,a]$. Its Fourier series is given by

$$\sum_{n=-\infty}^{\infty} \frac{\sin(at-\pi n)}{at-\pi n}\cdot \mathrm{e}^{\mathrm{i}x\frac{\pi n}{a}}, \quad x\in[-a,a].$$

Since f_t is continuously differentiable, the series converges in $L^2(\sigma)$ to f_t (see Corollary C.3.2). Applying the isometry $I_{\mathcal{Z}}$ we obtain the statement of the theorem. □

Next we present two inversion theorems.

Theorem 2.9.16. *Let $\mathcal{Z}$ be a continuous stationary field on $\mathbb{R}^d$ with representing measure ζ and spectral measure σ. If the σ-measure of the boundary of the set*

$$J = (a_1,b_1)\times\cdots\times(a_d,b_d)\subset\mathbb{R}^d, \quad a_j<b_j$$

[23] We use the convention $\frac{\sin 0}{0} := 1$.

is equal to 0, *then*[24]

$$\zeta(J) = \lim_{T\to\infty} \frac{1}{(2\pi)^d} \int_{-T}^{T} \cdots \int_{-T}^{T} \prod_{j=1}^{d} \frac{e^{-ia_j t_j} - e^{-ib_j t_j}}{it_j} \cdot X(t)\, dt.$$

Proof. We have

$$\begin{aligned} K_T(x) &:= \frac{1}{(2\pi)^d} \int_{-T}^{T} \cdots \int_{-T}^{T} \prod_{j=1}^{d} \frac{e^{-ia_j t_j} - e^{-ib_j t_j}}{it_j} \cdot e^{i(x,t)}\, dt \\ &= \frac{1}{(2\pi)^d} \int_{-T}^{T} \cdots \int_{-T}^{T} \prod_{j=1}^{d} \frac{e^{i(x_j - a_j)t_j} - e^{i(x_j - b_j)t_j}}{it_j}\, dt \\ &= \prod_{j=1}^{d} \left(\frac{1}{2\pi} \int_{-T}^{T} \left[\frac{\sin(x_j - a_j)t_j}{t_j} - \frac{\sin(x_j - b_j)t_j}{t_j} \right] dt_j \right) \\ &=: \prod_{j=1}^{d} K_j(T, x). \end{aligned}$$

By Corollary B.1.8,

$$\lim_{T\to\infty} K_j(T, x) = \begin{cases} 1, & x_j \in (a_j, b_j) \\ \frac{1}{2}, & x_j = a_j \text{ or } x_j = b_j \\ 0, & x_j \notin [a_j, b_j] \end{cases}.$$

and hence

$$\lim_{T\to\infty} K_T = \mathbf{1}_J, \qquad \sigma\text{-almost everywhere.}$$

Theorem B.1.9 shows that $|K_j(T, x)| < 2$ and therefore we also have convergence in $L^2(\sigma)$. The isometric operator $I_{\mathcal{Z}}$ being continuous we conclude that $\lim_{T\to\infty} I_{\mathcal{Z}}(K_T) = \zeta(J)$. Now the theorem follows immediately from equation (2.9.13.1). □

The next theorem can be proved in the same way by using Corollary C.3.2.

Theorem 2.9.17. *Let $\mathcal{Z}$ be a stationary field on $\mathbb{Z}^d$ with representing measure ζ and spectral measure measure σ. If the σ-measure of the boundary of the set*

$$J = (a_1, b_1) \times \cdots \times (a_d, b_d), \quad 0 \le a_j < b_j < 2\pi$$

is equal to 0, *then*

$$\zeta(J) = \lim_{n\to\infty} \frac{1}{(2\pi)^d} \sum_{|k|\le n} \prod_{j=1}^{d} \frac{e^{-ia_j k_j} - e^{-ib_j k_j}}{ik_j} \cdot \mathcal{Z}(k).$$

[24] In case of $t_j = 0$ we use the same convention as in Theorem 1.3.1.

Theorem 2.9.18. *Let $\mathcal{Z}$ be a continuous stationary field on $\mathbb{R}^d$ or $\mathbb{Z}^d$ with representing measure ζ and spectral measure σ. The following properties are equivalent:*

(i) *$H(\mathcal{Z})$ is finite-dimensional;*

(ii) *the support of σ is finite;*

(iii) $\mathcal{Z}(t) = \sum_{j=1}^{n} e^{i(t,x_j)} X_j$
with some $n \in \mathbb{N}$, mutually orthogonal $X_j \in H(\mathcal{Z})$, and $x_j \in \mathbb{R}^d$ or $x_j \in [0, 2\pi)^d$, respectively.

Proof. Noting that $L^2(\sigma)$ is finite dimensional if and only if $\operatorname{supp} \sigma$ is finite, the equivalence of (i) and (ii) follows from the fact that $H(\mathcal{Z})$ and $L^2(\sigma)$ are isomorphic.

Assume that $\sigma = \sum_{j=1}^{n} p_j \delta_{x_j}$, where the x_j's are mutually distinct. Then the functions $\mathbf{1}_{\{x_j\}}$ build an orthogonal basis of $L^2(\sigma)$. For all functions f we have

$$f = \sum_{j=1}^{n} f(x_j)\mathbf{1}_{\{x_j\}}$$

equality being in $L^2(\sigma)$, and hence

$$\int f(x)\, d\zeta(x) = \sum_{j=1}^{n} f(x_j) X_j,$$

where the random variables $X_j := \int \mathbf{1}_{\{x_j\}}\, d\zeta$ are orthogonal. Relation (iii) follows from the equation above by setting $f(x) := f_t(x) := e^{i(t,x)}$ and from the spectral representation of $\mathcal{Z}$.

If (iii) holds, then $H(\mathcal{Z}) = \operatorname{span}\{X_1, \dots, X_n\}$, from which (i) follows. □

Theorem 2.9.19. *Let $\mathcal{Z}$ be a continuous stationary field on $T = \mathbb{R}^d$ or on $T = \mathbb{Z}^d$ with correlation function C and spectral measure σ. Then $\mu * \mathcal{Z}$ is stationary with correlation function $\mu * \tilde{\mu} * C$ and spectral measure σ_μ given by*

$$d\sigma_\mu(x) = |\hat{\mu}(x)|^2\, d\sigma(x). \tag{1}$$

Moreover, $\mu * \mathcal{Z}$ admits the representation

$$\mu * \mathcal{Z}(t) = \int e^{i(t,x)} \hat{\mu}(x)\, d\zeta(x), \quad t \in T,$$

where ζ is the representing measure of $\mathcal{Z}$ and integration is over $\mathbb{R}^d$ or $[0, 2\pi)^d$, respectively.

Proof. The statement on stationarity and on the correlation function follows from Lemma 2.8.4 with $\nu = \mu$ while equation (1) follows from Theorem 1.8.15 and Theorem 1.9.5, respectively.

The statement on the representation of $\mathcal{Z}$ can be easily checked when the support of μ is finite. In the general case we approximate μ by measures with finite support using (2.4.15.5)[25] and (2.6.4.iii). □

2.10 Unitary representations

This section contains an introduction to unitary representations and their connection to stationary fields. This connection is useful for example in the study of ergodic properties of stationary fields (see Section 5.3).

Throughout the present section H is a complex Hilbert space and Z denotes a stationary field on T, where $T = \mathbb{R}^d$ or $T = \mathbb{Z}^d$.

Unitary Operators 2.10.1. Recall that a linear operator U from H onto H is called unitary if $(Uh, Uh) = (h, h)$ for all $h \in H$. If U is unitary then

$$(Uh, Ug) = (h, g) \quad \text{and} \quad (Uh, g) = (h, U^{-1}g), \quad h, g \in H.$$

To see examples, let $H = L^2(\lambda_d)$. Since the Lebesgue measure is invariant under translations, the equation[26]

$$U_t h(x) := h(x - t), \quad h \in H,\ x \in \mathbb{R}^d$$

defines a unitary operator U_t in H for all $t \in \mathbb{R}^d$. Another example is given by

$$Uh(x) := g(x)h(x), \quad h \in H,\ x \in \mathbb{R}^d$$

where g is a Lebesgue measurable function on $\mathbb{R}^d$ such that $|g| = 1$.

Definition 2.10.2. A *unitary representation* (U_t) of T in H is a mapping $t \mapsto U_t$ from T into the set of all unitary operators in H such that

$$U_{t+s} = U_t U_s, \quad t, s \in T.$$

The representation (U_t) is called *continuous* if the mapping $t \mapsto U_t h$ is continuous for all $h \in H$ with respect to the norm topology on H.
If $T = \mathbb{R}^d$ and the function $t \mapsto (g, U_t h)$ is Lebesgue measurable for all $g, h \in H$, then (U_t) is said to be *weakly measurable*.

[25] See also Corollary 2.4.10.

[26] Strictly speaking this definition is not correct since h is an equivalence class and not a function. It can easily be made correct by using the representants of h.

Lemma 2.10.3. *Let (U_t) be a unitary representation of T in H. Then $U_0 = I$ (the identity operator), $U_{-t} = U_t^{-1}$ and*

$$(U_t v, w) = (v, U_{-t} w), \quad t \in T, \ v, w \in H.$$

Moreover, (U_t) is continuous if and only if the complex-valued functions

$$t \mapsto (U_t v, w)$$

are continuous for all $v, w \in L$, where L is a dense subset of H.

Proof. Since $U_0 = U_{0+0} = U_0 U_0$ we see that $U_0 = I$. From $I = U_{t-t} = U_t U_{-t}$ it follows that $U_{-t} = U_t^{-1}$ and hence

$$(U_t v, w) = (v, U_t^* w) = (v, U_t^{-1} w) = (v, U_{-t} w).$$

If (U_t) is continuous, then the functions $t \mapsto (U_t v, w)$ are obviously continuous for all $v, w \in H$. Assume that these functions are continuous for all $v, w \in L$. Let $h \in H$ and $w \in L$ be arbitrary and choose a sequence $\{v_n\}$ in L converging to h. Using the Cauchy–Schwarz inequality we get

$$|(U_t v_n, w) - (U_t h, w)| = |(U_t(v_n - h), w))| \leq \|v_n - h\| \cdot \|w\|.$$

Thus, the function $t \mapsto (U_t h, w)$ is continuous since it is the uniform limit of the sequence $t \mapsto (U_t v_n, w)$ of continuous functions. Applying the same argument once more, we see that $t \mapsto (U_t h, g)$ is continuous for all $h, g \in H$. The continuity of (U_t) follows from the identity

$$\|U_t h - U_s h\|^2 = 2\|h\|^2 - (U_t h, U_s h) - (U_s h, U_t h).$$

□

Example 2.10.4. Continuous unitary representations of $\mathbb{R}^d$ in $H = L^2(\lambda_d)$ are given by

$$U_t h(x) := h(x - t), \quad h \in H, \ x, t \in \mathbb{R}^d$$

and

$$U_t h(x) := \mathrm{e}^{\mathrm{i}(t,x)} h(x), \quad h \in H, \ t, x \in \mathbb{R}^d.$$

The continuity of the first example follows from Lemma 2.10.3 and from the fact that the function

$$t \mapsto \int h(x - t) g(x) \, \mathrm{d}\lambda(x), \quad t \in \mathbb{R}^d$$

is continuous for all $h, g \in \mathcal{L}^2(\mathbb{R}^d)$ (cf. Theorem E.2.4).

Replacing $\mathbb{R}^d$ by $\mathbb{Z}^d$ and λ_d by the counting measure on $\mathbb{Z}^d$ we obtain unitary representations of $\mathbb{Z}^d$.

Definition 2.10.5. Let (U_t) be a unitary representation of T in H. A subspace L of H is said to be *(U_t)-invariant* if

$$U_t L \subset L, \quad t \in T.$$

A vector $h \in H$, $h \neq 0$, is called an *eigenvector* of (U_t) if there exists a complex-valued function γ on T such that

$$U_t h = \gamma(t) h, \quad t \in T.$$

Let γ be as above. It is easy to check that $\gamma(0) = 1$ and

$$\gamma(t + s) = \gamma(t)\gamma(s), \quad t, s \in T.$$

Thus, γ is a character of T. If the representation is continuous, then γ is continuous as well.

Definition 2.10.6. Let (U_t) be a unitary representation of T in H. A vector $v \in H$ is called a *cyclic vector* if the linear space

$$\operatorname{span}\{U_t : t \in T\}$$

is dense in H. Unitary representations having a cyclic vector are called *cyclic*.

Example 2.10.7. The constant function $\mathbf{1} \in H = L^2([0, 2\pi))$ is a cyclic vector for the unitary representation

$$U_t h(x) := \mathrm{e}^{\mathrm{i}tx} h(x), \quad h \in H, \ x \in [0, 2\pi), \ t \in \mathbb{Z}$$

of $\mathbb{Z}$ since the trigonometric polynomials build a dense linear space in H. This follows immediately from the Stone–Weierstraß Theorem B.2.4.

Example 2.10.8. Let H be a subspace of $L^2(\Omega, \mathcal{A}, P)$ and let (U_t) be a unitary representation of T in H. We choose an arbitrary random variable $X \in H$ and define the second order field Z by

$$Z(t) := U_t X, \quad t \in T.$$

Then Z is stationary. Indeed,

$$(Z(t), Z(s)) = (U_t X, U_s X) = (U_{-s} U_t X, X) = (U_{t-s} X, X), \quad t, s \in T.$$

It is easy to check that $H(Z)$ is (U_t)-invariant. Moreover, the restriction of (U_t) to $H(Z)$ is a unitary representation with cyclic vector X.

Lemma 2.10.9. *Let* (U_t) *be a unitary representation of* T *in* H*. The function*

$$f(t) = (h, U_t h), \quad t \in T$$

is positive definite for all $h \in H$*.*

Proof. Let $t_j \in T$ and $c_j \in \mathbb{C}$, $j = 1, \dots, n$, be arbitrary. Using that $U_{t_j - t_i} = U_{t_j} U_{t_i}^{-1}$ we obtain

$$\begin{aligned} \sum_{i,j=1}^{n} f(t_j - t_i) c_i \overline{c_j} &= \sum_{i,j=1}^{n} (c_i h, c_j U_{t_j - t_i} h) \\ &= \sum_{i,j=1}^{n} (c_i U_{t_i} h, c_j U_{t_j} h) \\ &= \left\| \sum_{i=1}^{n} c_i U_{t_i} h \right\|^2 \geq 0, \end{aligned}$$

i.e., f is positive definite. □

Definition 2.10.10. Two unitary representations (U_t) and (V_t) of T in H_1 and H_2, respectively, are called *equivalent* if there exists an isometric linear operator M from H_1 onto H_2 such that

$$M U_t = V_t M, \quad t \in T.$$

Lemma 2.10.11. *Let* (U_t) *and* (V_t) *be equivalent unitary representations of* $\mathbb{R}^d$ *in* H_1 *and* H_2*, respectively. Then* (U_t) *is continuous if and only if* (V_t) *is continuous.*

Proof. Let M be as in Definition 2.10.10. For all $v, w \in H_1$ we have

$$(M U_t v, w) = (U_t v, M^{-1} w) = (V_t M v, w), \quad t \in \mathbb{R}^d.$$

Since M maps H_1 onto H_2, the statement follows from Lemma 2.10.3. □

Lemma 2.10.12. *Let* (U_t) *and* (V_t) *be cyclic unitary representations of* T *in* H_1 *and* H_2 *with cyclic vectors* h_1 *and* h_2*, respectively. If*

$$(h_1, U_t h_1) = (h_2, V_t h_2), \quad t \in T,$$

then these representations are equivalent.[27]

[27] We use the same notation for the inner products in H_1 and H_2.

Proof. We define the mapping M from $L := \operatorname{span}\{U_t h_1 : t \in T\}$ into H by

$$M\Big(\sum_{i=1}^{n} c_i U_{t_i} h_1\Big) := \sum_{i=1}^{n} c_i V_{t_i} h_2.$$

That this definition is correct, i.e., the left-hand side does not depend on the special choice of c_i and t_i, follows from the relation

$$\begin{aligned}\Big(\sum_{i=1}^{n} c_i U_{t_i} h_1, \sum_{j=1}^{m} d_j U_{s_j} h_1\Big) &= \sum_{i=1}^{n}\sum_{j=1}^{m} c_i \overline{d_j}\,(h_1, U_{s_j - t_i} h_1) \\ &= \Big(\sum_{i=1}^{n} c_i V_{t_i} h_2, \sum_{j=1}^{m} d_j V_{s_j} h_2\Big)\end{aligned}$$

and from the fact that h_1 and h_2 are cyclic vectors. This mapping is linear and

$$\begin{aligned} M h_1 &= h_2, \\ (g, h) &= (Mg, Mh), \\ M U_t g &= V_t M g, \quad t \in T,\ g \in L. \end{aligned}$$

By Theorem D.3.6, M can be extended to an isometric linear operator $\tilde{M}$ from H_1 onto H_2. It is clear that $\tilde{M} U_t = V_t \tilde{M}$ for all t, i.e., (U_t) and (V_t) are equivalent. □

Lemma 2.10.13. *Let μ and ν be complex measures with finite support on T. If the equation $\mu * Z(s) = \nu * Z(s)$ holds for some $s \in T$, then*

$$\mu * Z(t) = \nu * Z(t)$$

for all $t \in T$.

Proof. Indeed, applying Lemma 2.8.4 we see that

$$\|(\mu - \nu) * Z(t)\|^2 = (\mu - \nu) * (\mu - \nu)\tilde{} * C(0).$$

Thus, $\|(\mu - \nu) * Z(t)\|$ does not depend on t. Since it is equal to zero if $t = s$, it is zero for all t. □

The next theorem establishes a close relation between unitary representations and stationary fields.

Theorem 2.10.14. *For every (continuous) stationary field Z on T there exists a (continuous) cyclic unitary representation (U_t) of T in the Hilbert space $H(Z)$ such that*

$$U_t Z(s) = Z(s - t), \quad s, t \in T. \tag{1}$$

We call this representation the canonical unitary representation of Z in $H(Z)$. *Conversely, if (V_t) is a (continuous) cyclic unitary representation of T in a Hilbert space H, then there exists a (continuous) stationary field Z on T such that the canonical unitary representation of Z in $H(Z)$ is equivalent to (V_t).*

Proof. Denote by $L(Z)$ the linear space of all vectors $v \in H(Z)$ of the form

$$v = \mu * Z(0) = \sum_j c_j Z(-x_j)$$

where $\mu = \sum_j c_j \delta_{x_j}$ is a complex measure with finite support. For v as above we define $U_t v$ by

$$U_t v := \delta_t * \mu * Z(0) = \mu * Z(-t), \quad t \in T.$$

It follows from Lemma 2.10.13 that this definition is correct, i.e., $U_t v$ does not depend on the special representation of v. It is clear that U_t is a linear operator from $T(Z)$ onto $T(Z)$. Moreover,

$$U_t U_s v = U_{t+s} v, \quad t, s \in T,$$

in particular, $U_t U_{-t} v = v$. It follows from Lemma 2.8.4 that U_t is isometric:

$$(U_t v, U_t v) = (v, v), \quad v \in L(Z).$$

We conclude that U_t can be extended to a unitary operator on $H(Z) = \overline{L(Z)}$, we denote this operator also by U_t. It is easy to check that (U_t) is a unitary representation of T with cyclic vector $Z(0)$. If Z is continuous then, by the definition of U_t, the function $t \mapsto U_t v$ is continuous for all $v \in L$. In view of Lemma 2.10.3, the representation (U_t) is continuous.

To prove the second part of the theorem, assume that (V_t) is a cyclic unitary representation of T in some Hilbert space H and let v_0 be a cyclic vector. The function

$$f(t) = (v_0, V_t v_0), \quad t \in T$$

is positive definite in view of Lemma 2.10.9. By Theorem 1.7.8 there exists a unique measure $\mu \in \mathcal{M}^+(\Gamma)$ such that

$$f(t) = \int_\Gamma \gamma(t)\, \mathrm{d}\mu(\gamma), \quad t \in T,$$

where Γ denotes the character group of T. For each $t \in T$ the function

$$g_t(\gamma) = \gamma(t), \quad \gamma \in \Gamma$$

is bounded and continuous on Γ, hence $g_t \in L^2(\mu)$. We define the second order random field $Z : T \to L^2(\mu)$ by $Z(t) = g_t$. Since

$$(Z(t), Z(s)) = \int_{\Gamma^d} \gamma(t)\overline{\gamma(s)}\, \mathrm{d}\mu(\gamma) = \int_{\Gamma^d} \gamma(t-s)\, \mathrm{d}\mu(\gamma) = f(t-s)$$

this field is stationary. Denoting by (U_t) the canonical unitary representation in $H(Z)$ we have

$$U_t Z(s) = Z(s-t) = g_{s-t} = g_{-t} \cdot g_s = g_{-t} \cdot Z(s).$$

From this we see that the operator U_t acts as multiplication with the function g_{-t}. By the uniqueness of μ, the constant function $\mathbf{1}$ is a cyclic vector for (U_t). Moreover,

$$(\mathbf{1}, U_t \mathbf{1}) = \int_\Gamma \gamma(t)\, \mathrm{d}\mu(\gamma) = f(t) = (v_0, V_t v_0).$$

Lemma 2.10.12 shows that the representations (U_t) and (V_t) are equivalent. If (V_t) is continuous, then (U_t) is continuous as well (cf. Lemma 2.10.11). The continuity of Z follows from $U_t Z(0) = Z(-t)$. □

Remark 2.10.15. Let Z be a continuous stationary field on $\mathbb{R}^d$. The proof of the second part of the previous theorem can be carried out by using Bochner's Theorem 1.7.3 instead of Theorem 1.7.8, as follows.

Let σ be the spectral measure of Z and consider the isometric mapping $I_Z : L^2(\sigma) \to H(Z)$ with $I_Z\left(\mathrm{e}^{\mathrm{i}(t,\cdot)}\right) = Z(t)$ (cf. Remark 2.9.12). We define the unitary representation (V_t) by

$$V_t g(x) := \mathrm{e}^{-\mathrm{i}(t,x)} g(x), \quad g \in L^2(\sigma).$$

The function $\mathbf{1}$ is a cyclic vector of this representation.[28] Moreover, (V_t) is equivalent with the unitary representation (U_t) of Z in $H(Z)$. This follows from the relation

$$(\mathbf{1}, V_t \mathbf{1}) = \left(\mathbf{1}, \mathrm{e}^{\mathrm{i}(t,\cdot)}\right) = \left(I_Z(\mathbf{1}), I_Z\left(\mathrm{e}^{\mathrm{i}(t,\cdot)}\right)\right) = (Z(0), Z(-t)) = (Z(0), U_t Z(0))$$

and from Lemma 2.10.12.[29] We call (V_t) the *canonical unitary representation of Z in $L^2(\sigma)$*. In Section 2.11 we will encounter another unitary representation for Z which is constructed using the correlation function of Z.

Lemma 2.10.16. *Let (U_t) be a unitary representation of T in H, let H_0 be a subspace of H and denote by P_0 the orthogonal projection from H onto H_0. The following statements are equivalent:*

(i) *H_0 is (U_t)-invariant;*

(ii) $U_t P_0 = P_0 U_t, \quad t \in T$;

(iii) *$H_0^\perp$ is (U_t)-invariant.*

[28] This follows from the fact that σ is uniquely determined by its inverse Fourier–Stieltjes transform.

[29] Note that $(Z(0), Z(-t)) = (Z(s), Z(s-t))$ by stationarity, so that $Z(0)$ can be replaced by $Z(s)$ where $s \in \mathbb{R}^d$ is arbitrary.

Proof. The equivalence of (i) and (iii) follows from the relations

$$\begin{aligned} H_0 \text{ is } (U_t) - \text{invariant} &\iff U_t h \in H_0, \quad t \in T,\ h \in H_0 \\ &\iff U_t h \perp g, \quad h \in H_0,\ g \in H_0^\perp \\ &\iff h \perp U_{-t} g, \quad t,\ h \in H_0,\ g \in H_0^\perp \\ &\iff U_{-t} g \in H_0^\perp, \quad t,\ g \in H_0^\perp \\ &\iff H_0^\perp \text{ is } (U_t) - \text{invariant.} \end{aligned}$$

Assume that (i) holds. An arbitrary element $h \in H$ can be written as

$$h = h_0 + h_0^\perp, \quad h_0 \in H_0,\ h_0^\perp \in H_0^\perp.$$

From this we see that $U_t P_0 h = U_t h_0$. Since (iii) holds we obtain

$$P_0 U_t h = P_0 U_t h_0 + P_0 U_t h_0^\perp = U_t h_0.$$

Thus the relation (i) implies (ii).

Now assume that (ii) holds. For all $h_0 \in H_0$ we have $P_0 U_t h_0 = U_t P_0 h_0 = U_t h_0$ and hence $U_t h_0 \in H_0$. Thus, H_0 is (U_t)-invariant, completing the proof. □

Lemma 2.10.17. *Let Z be a stationary field on T with canonical unitary representation (U_t) in $H(Z)$. If $H(Z)$ admits the decomposition*

$$H(Z) = H_1 \oplus H_2$$

where H_1 and H_2 are (U_t)-invariant subspaces, then Z can be decomposed as

$$Z = Z_1 + Z_2$$

where Z_1 and Z_2 are orthogonal stationary fields with $H(Z_1) = H_1$ and $H(Z_2) = H_2$.

Proof. Denote by P_i the orthogonal projection onto H_i. Setting

$$Z_i(t) := P_i Z(t) = P_i U_{-t} Z(0), \quad t \in V,\ i = 1, 2,$$

we obviously have $Z = Z_1 + Z_2$ and $Z_1 \perp Z_2$. By Lemma 2.10.16 we have $Z_i(t) = U_{-t} P_i Z(0)$ and hence Z_i is stationary (cf. Example 2.10.8) and $H(Z_i) \subset H_i$. From the relation

$$H_1 \oplus H_2 = H(Z) = H(Z_1 + Z_2) \subset H(Z_1) \oplus H(Z_2) \subset H_1 \oplus H_2$$

we conclude that $H(Z_1) \oplus H(Z_2) = H_1 \oplus H_2$ and therefore $H(Z_i) = H_i$. □

Corollary 2.10.18. *If $H(Z)$ is n-dimensional where $n \in \mathbb{N}$, then there exist one-dimensional (U_t)-invariant subspaces H_j, $j = 1, \dots, n$, such that*[30]

$$H(Z) = H_1 \oplus \cdots \oplus H_n.$$

Moreover,

$$Z(t) = \gamma_1(t)X_1 + \cdots + \gamma_n(t)X_n, \quad t \in T,$$

where $X_j \in H_j$ and γ_j is a character of T.

Proof. Since the unitary operators U_t commute, they have a common eigenvector. Denoting by H_1 the subspace spanned by this eigenvector X_1 we have

$$H(Z) = H_1 \oplus H_1^{\perp}.$$

By Lemma 2.10.16, the subspace $H_1^{\perp}$ is (U_t)-invariant. Using this, the lemma follows easily by induction on n (see also the discussion in the first part of Example 2.8.6). □

2.11 Unitary representations and positive definite functions

In Theorem 2.10.14 we constructed a unitary representation for a given stationary field by using the integral representation of the correlation function. In the present section we follow another method where we construct a unitary representation without using the integral representation. We also present some applications of this method.

Throughout this section we assume that $T = \mathbb{R}^d$ or $T = \mathbb{Z}^d$. In the linear space of all complex-valued functions g on T we introduce the *translation operator* E_t by

$$E_t g(x) := g(x - t), \quad t, x \in T.$$

We have $E_t g = \delta_t * g$ and $E_{s+t} = E_s E_t$. For a complex-valued function g on T, we denote by $\mathcal{T}(g)$ the linear space spanned by all translates of g. Note that

$$\mathcal{T}(g) = \{\mu * g : \mu \in \mathcal{M}^f(T)\}.$$

Theorem 2.11.1. *To every positive definite function f on T there corresponds a Hilbert space $H(f)$ with inner product $(\cdot,\cdot) = (\cdot,\cdot)_f$ such that*

(i) *elements of $H(f)$ are bounded complex-valued functions on T and $H(f)$ is invariant under translations;*

(ii) *$\mathcal{T}(f) \subset H(f)$ and $\mathcal{T}(f)$ is dense in $H(f)$;*

(iii) *setting $U_t g := E_t g$, $g \in H(f)$, $t \in T$, we obtain a unitary representation (U_t) of T in $H(f)$ with cyclic vector f;*

---

[30] See also Theorem 2.9.18.

(iv) $g(t) = (g, U_t f)$ *for all* $g \in H(f)$ *and* $t \in T$*; in particular,* $f(t) = (f, U_t f)$*;*

(v) *convergence of a sequence* $\{g_n\}$ *in* $H(f)$ *implies uniform convergence on* T*;*

(vi) *weak convergence of a net* $\{g_\alpha\}$ *in* $H(f)$ *implies pointwise convergence;*

(vii) *if* $f \in P^c(\mathbb{R}^d)$*, then the representation* (U_t) *and each* $g \in H(f)$ *are continuous;*

(viii) *if* $f \in P^m(\mathbb{R}^d)$*, then each* $g \in H(f)$ *is measurable and the representation* (U_t) *is weakly measurable.*

Proof. Let $g, h \in \mathcal{T}(f)$ and choose finitely supported measures $\mu = \sum_{i=1}^n a_i \delta_{x_i}$ and $\nu = \sum_{j=1}^m b_j \delta_{y_j}$ such that $g = \mu * f$ and $h = \nu * f$. We write

$$(g, h) := \tilde{\nu} * \mu * f(0) = \sum_{i=1}^{n} \sum_{j=1}^{m} f(y_j - x_i) a_i \overline{b_j}\,.$$

Since

$$(g, h) = \sum_{j=1}^{m} g(y_j)\overline{b_j} = \sum_{i=1}^{n} \overline{h(x_i)} a_i,$$

the definition of (g, h) does not depend on the particular choice of μ and ν. Using Lemma 1.4.3 and Lemma 1.4.8 we see that $(\cdot, \cdot)$ is a positive semidefinite inner product on $\mathcal{T}(f)$. It follows from the definition of $(\cdot, \cdot)$ that

$$g(t) = (g, U_t f), \quad t \in T,\ g \in \mathcal{T}(f). \tag{1}$$

Moreover, the translation operators E_t are isometric on $\mathcal{T}(f)$:

$$(E_t g, E_t g) = (g, g), \quad g \in \mathcal{T}(f).$$

Using equation (1), together with the Cauchy–Schwarz inequality, we obtain

$$|g(t)|^2 = |(g, E_t f)|^2 \le (g, g)(E_t f, E_t f) = (g, g)(f, f). \tag{2}$$

From this we conclude that $(g, g) = 0$ if and only if $g = 0$, so $\mathcal{T}(f)$ is a pre-Hilbert space. Moreover, every function $g \in \mathcal{T}(f)$ is bounded.

Let $\{g_n\}$ be a Cauchy sequence in $\mathcal{T}(f)$. It follows from (2) that $\{g_n\}$, as a sequence of functions on T, is a Cauchy sequence in the metric of uniform convergence on T. Consequently, there exists a complex-valued function g on T with

$$\lim_{n\to\infty} g_n(t) = g(t), \quad t \in T \tag{3}$$

where the convergence is uniform. Let $H(f)$ be the completion of $\mathcal{T}(f)$ constructed by means of functions on T. These functions are obviously bounded. It follows easily from (3) and from the fact that E_t is isometric on $\mathcal{T}(f)$ that $H(f)$ is invariant under E_t and $(E_t g, E_t g) = (g, g)$ holds for all $g \in H(f)$. Denote by U_t the restriction of E_t to $H(f)$. Since $U_{-t}U_t = U_t U_{-t} = I$ (the identity operator), the operator U_t is invertible and hence unitary. It is easy to check that (U_t) is a unitary representation of T in $H(f)$. The linear space spanned by the set $\{U_t f : t \in T\}$ is equal to $\mathcal{T}(f)$ and hence it is dense in $H(f)$.

Since $g \mapsto (g, U_t f)$ is continuous on $H(f)$ and $\mathcal{T}(f)$ is dense in $H(f)$, equation (1) holds for all $g \in H(f)$. Using the Cauchy–Schwarz inequality again, we obtain

$$|g_n(t) - g(t)|^2 = |(g_n - g, U_t f)|^2 \le \|g_n - g\|^2 \|f\|^2.$$

Thus, $g_n \to g$ uniformly on T whenever $\{g_n\}$ converges to g in $H(f)$.

Property (vi) follows immediately from the equation $g_\alpha(t) = (g_\alpha, U_t f)$.

To prove (vii), let $g, h \in H(f)$ be arbitrary and choose $\mu_n, \nu_n \in \mathcal{M}^f(T)$ such that $\lim_n \mu_n f = g$ and $\lim_n \nu_n f = h$, the convergence being in $H(f)$. Then the function $t \mapsto (g, U_t h)$ is the uniform limit of the functions

$$t \mapsto (\mu_n f, U_t \nu_n f) = \tilde{\nu}_n * \mu_n * f(t)$$

which are continuous. Thus, the representation (U_t) is continuous. The continuity of the functions in $H(f)$ now follows from (iv). The last statement can be proved using the same arguments. □

Throughout the rest of this section (U_t) denotes the unitary representation of T in $H(f)$, constructed in the previous theorem. If $f = C$ is the correlation function of a stationary process Z, then (U_t) is called the *canonical unitary representation of Z in $H(C)$*.

Theorem 2.11.2. *Let $f \in P(T)$ and let H_1 and H_2 be (U_t)-invariant subspaces of $H(f)$ such that $H(f) = H_1 \oplus H_2$. Then the orthogonal projection f_i of f onto H_i $(i = 1, 2)$ is a positive definite function. Moreover, the equation*

$$g_i(t) = (g_i, U_t f_i)$$

holds for all $t \in T$ and $g_i \in H_i$ $(i = 1, 2)$.

Proof. Using that $H_1 \perp H_2$ and the fact that H_i is (U_t)-invariant, we obtain

$$g_i(t) = (g_i, U_t f) = (g_i, U_t(f_1 + f_2)) = (g_i, U_t f_i), \quad g_i \in H_i.$$

In particular, $f_i(t) = (f_i, U_t f_i)$ and hence f_i is positive definite in view of Lemma 2.10.9. □

Theorem 2.11.3. *If $f = f_1 + f_2$, where $f, f_1, f_2 \in P(T)$, then $f_1, f_2 \in H(f)$ and there exists a nonnegative operator Q on $H(f)$ such that $\|Q\| \leq 1$ and*

$$QU_t = U_t Q \tag{1}$$
$$f_1(t) = (Qf, U_t f) \tag{2}$$
$$f_2(t) = ((I - Q)f, U_t f) \tag{3}$$

for all $t \in T$. Conversely, if $f \in P(T)$ and an operator Q on $H(f)$ has the properties above, then the functions f_1 and f_2 defined by (2) and (3) are positive definite and $f = f_1 + f_2$.

Proof. We define in $\mathcal{T}(f)$ a new positive semidefinite inner product $(\cdot,\cdot)_1$ by setting

$$(\mu * f, \nu * f)_1 := \tilde{\nu} * \mu * f_1(0), \quad \mu, \nu \in \mathcal{M}^f(T). \tag{4}$$

Since f_1 and $f - f_1$ are positive definite, we have

$$0 \leq (g, g)_1 \leq (g, g), \quad g \in \mathcal{T}(f).$$

From Theorem D.3.5 we conclude that the inner product $(\cdot,\cdot)_1$ can be extended to a positive semidefinite inner product on $H(f)$, which we also denote by $(\cdot,\cdot)_1$. By the second part of Theorem D.3.5 there exists a nonnegative linear operator Q on $H(f)$ such that $\|Q\| \leq 1$ and

$$(g, h)_1 = (Qg, h), \quad g, h \in H(f).$$

From equation (4) we see that $(U_t g, h)_1 = (g, U_{-t} h)_1$ holds for all $g, h \in H(f)$ and $t \in T$. Using this, we obtain

$$(U_t Qg, h) = (Qg, U_{-t} h) = (g, U_{-t} h)_1 = (U_t g, h)_1 = (QU_t g, h).$$

This being true for all $g, h \in H(f)$, we must have $U_t Q = QU_t$, $t \in T$. Taking into account the definition of $(\cdot,\cdot)_1$, we obtain

$$f_1(t) = (f, U_t f)_1 = (Qf, U_t f) = (Qf)(t)$$

and

$$f_2(t) = f(t) - f_1(t) = ((I - Q)f, U_t f) = ((I - Q)f)(t).$$

Therefore, $f_1 = Qf \in H(f)$ and $f_2 = (I - Q)f \in H(f)$.

Suppose now that Q is a nonnegative operator on $H(f)$ with $\|Q\| \leq 1$ and $QU_t = U_t Q$, $t \in T$. Then the operator $I - Q$ is nonnegative as well. Setting $Q_1 := \sqrt{Q}$ and $Q_2 := \sqrt{I - Q}$, the operators Q_1 and Q_2 commute with U_t for every $t \in T$ (cf. Theorem D.3.7). Thus, we have

$$f_1(t) = (Qf, U_t f) = (Q_1 f, U_t Q_1 f)$$

and

$$f_2(t) = ((I - Q)f, U_t f) = (Q_2 f, U_t Q_2 f).$$

It is obvious that $f = f_1 + f_2$. It follows from Lemma 2.10.9 that f_1 and f_2 are positive definite. □

Corollary 2.11.4. *Let f_1, f_2 be positive definite functions on $\mathbb{R}^d$.*
If $f_1 + f_2$ is continuous, then f_1 and f_2 are also continuous.
If $f_1 + f_2$ is measurable (vanishes λ-almost everywhere), then f_1 and f_2 are measurable (vanish λ-almost everywhere, respectively).

Proof. By the previous theorem $f_1, f_2 \in H(f)$ and therefore the corollary follows from the last two statements of Theorem 2.11.1. □

Theorem 2.11.5. *Let $f \in P^m(\mathbb{R}^d)$, $h \in L^1(\mathbb{R}^d)$ and define the mapping A_h by $A_h g := h * g$, $g \in H(f)$.*[31] *Then*

(i) *A_h is a bounded linear operator from $H(f)$ into $H(f)$;*

(ii) *if H_0 is a closed (U_t)-invariant subspace of $H(f)$, then H_0 is invariant under A_h;*

(iii) $(A_h)^* = A_{\tilde{h}}$.

Proof. By Theorem 2.11.1, the function $t \mapsto (U_t g_1, g_2)$ is measurable for all $g_1, g_2 \in H(f)$. We define the sesquilinear functional B_h by

$$B_h(g_1, g_2) := \int_{\mathbb{R}^d} (U_t g_1, g_2) h(t)\, \mathrm{d}t, \quad g_1, g_2 \in H(f).$$

Since $|(U_t g_1, g_2)| \le \|g_1\|\, \|g_2\|$, we have

$$|B_h(g_1, g_2)| \le \|g_1\|\, \|g_2\| \int_{\mathbb{R}^d} |h(t)|\, \mathrm{d}t.$$

Consequently, by the second part of Theorem D.3.5, there exists a bounded linear operator $C_h : H(f) \longrightarrow H(f)$ for which $B_h(g_1, g_2) = (C_h g_1, g_2)$ for all $g_1, g_2 \in H(f)$. Using the definition of B_h we obtain

$$(C_h g)(t) = (C_h g, U_t f) = \int_G (U_s g, U_t f) h(s)\, \mathrm{d}s = \int_G g(t-s) h(s)\, \mathrm{d}s = h * g(t)$$

and so $A_h g = C_h g \in H(f)$ if $g \in H(f)$.

[31] Note that $h * g$ is a continuous function because g is bounded and measurable by Theorem 2.11.1.

Let H_0 be a closed (U_x)-invariant subspace of $H(f)$. If $g_1 \in H_0$, then

$$(A_h g_1, g_2) = \int_G (U_t g_1, g_2) h(t)\,\mathrm{d}t = 0$$

holds for every $g_2 \in H_0^{\perp}$. Thus, $A_h g_1 \in H_0^{\perp\perp} = H_0$, i.e., H_0 is invariant under A_h.

We now compute the adjoint A_h^* of A_h:

$$\begin{aligned}(A_h^* g)(x) &= (A_h^* g, U_x f) = (g, A_h U_x f) = \overline{(A_h U_x f, g)} \\ &= \overline{\int_G (U_{x+y} f, g) h(y)\,\mathrm{d}y} \\ &= \overline{\int_G \overline{g(x+y)} h(y)\,\mathrm{d}y} \\ &= (\tilde{h} * g)(x) = (A_{\tilde{h}} g)(x).\end{aligned}$$

Thus, $A_h^* = A_{\tilde{h}}$. □

Corollary 2.11.6. *For all $f \in P^m(\mathbb{R}^d)$ we have:*[32]

(i) $(h * g_1, g_2) = (g_1, \tilde{h} * g_2), \quad g_1, g_2 \in H(f),\ h \in L^1(G)$;

(ii) $\displaystyle\int_{\mathbb{R}^d}\int_{\mathbb{R}^d} f(y-x) h(x) \overline{h(y)}\,\mathrm{d}x\,\mathrm{d}y \geq 0, \quad h \in \mathcal{L}^1(\mathbb{R}^d).$

Proof. Property (i) is a consequence of (2.11.5.iii) while (ii) is proved as follows:

$$\begin{aligned}\int_{\mathbb{R}^d}\int_{\mathbb{R}^d} f(y-x) h(x)\overline{h(y)}\,\mathrm{d}x\,\mathrm{d}y &= \tilde{h} * h * f(0) \\ &= (\tilde{h} * h * f, f) = (h * f, h * f) \geq 0.\end{aligned}$$ □

We are now able to prove a decomposition theorem for measurable positive definite functions.

Theorem 2.11.7. *Let $f \in P^m(\mathbb{R}^d)$. Then*

$$f = f_c + f_0$$

where $f_c \in P^c(\mathbb{R}^d)$, $f_0 \in P^m(\mathbb{R}^d)$ and $f_0 = 0$ λ-almost everywhere.

Proof. Denote by H_c the set of all continuous functions in $H(f)$ and by H_0 the orthogonal complement of H_c. Then H_c and H_0 are closed (U_t)-invariant subspaces of $H(f)$ and

$$H(f) = H_c \oplus H_0.$$

[32] See also Lemma 1.5.8 and Corollary 2.4.8.

By Theorem 2.11.2 the function f can be decomposed as

$$f = f_c + f_0, \quad f_c \in P^c(\mathbb{R}^d) \cap H_c, \ f_0 \in P(\mathbb{R}^d) \cap H_0.$$

It remains to prove that $f_0 = 0$, λ-almost everywhere. Theorem 2.11.5 shows that $h * f_0 \in H_0$ for all $h \in \mathcal{L}^1(\mathbb{R}^d)$. But $h * f_0$ is continuous, i.e., $h * f_0 \in H_c$, and therefore we must have $h * f_0 = 0$. This being true for all $h \in \mathcal{L}^1(\mathbb{R}^d)$, we obtain $f_0 = 0$ λ-almost everywhere. □

Theorem 2.11.8. *If $f \in P(\mathbb{R}^d)$ is Lebesgue measurable on a neighborhood V of 0, then f is Lebesgue measurable on $\mathbb{R}^d$.*

Proof. Let B be an open ball with center 0 such that $B+B \subset V$ and denote by H_B the subspace of $H(f)$ spanned by the functions $U_t f$, $t \in B$. As f is measurable on V and $B + B \subset V$, the function $U_t f$ is measurable on B for all $t \in B$. Since convergence in $H(f)$ implies uniform convergence (cf. Theorem 2.11.1) every function in H_B is measurable on B. If $g^\perp \in H_B^\perp$, then $g^\perp(t) = (g^\perp, U_t f) = 0$, $t \in B$. Applying the decomposition

$$H(f) = H_B \oplus H_B^\perp$$

we see that every function $h \in H(f)$ can be written as

$$h = g + g^\perp$$

where g is measurable on B and $g^\perp$ is identically zero on B. Thus, h is measurable on B. In particular, $U_t f$ is measurable on B for all $t \in \mathbb{R}^d$. From this it follows that f is measurable on $\mathbb{R}^d$. □

Chapter 3

Special properties

3.1 Strict positive definiteness

In several applications, for example in interpolation problems, it is more advantageous to use strictly positive definite functions.[1] The aim of the present section is to collect basic properties of these functions and to give necessary and sufficient conditions for a function to be strictly positive definite.[2]

Definition 3.1.1. A function $f : \mathbb{R}^d \to \mathbb{C}$ is said to be *strictly positive definite* if the matrix

$$A = \big(f(t_i - t_j)\big)_{i,j=1}^{n}$$

is positive definite for every positive integer n and for all pairwise distinct $t_1, \dots, t_n \in \mathbb{R}^d$.

We denote by $SP(\mathbb{R}^d)$ the set of strictly positive definite functions on $\mathbb{R}^d$ while $SP^c(\mathbb{R}^d)$ is the set of all continuous functions in $SP(\mathbb{R}^d)$. It is easy to check that $SP(\mathbb{R}^d) \subset P(\mathbb{R}^d)$.

The next lemma follows immediately from the definition. We omit the proof.

Lemma 3.1.2. *Let f be a complex-valued function on $\mathbb{R}^d$. Then $f \in SP(\mathbb{R}^d)$ if and only if the inequality*

$$\sum_{i,j=1}^{n} f(t_i - t_j) c_i \overline{c_j} > 0$$

holds for every positive integer n, for all pairwise distinct $t_1, \dots, t_n \in \mathbb{R}^d$, and for all $c \in \mathbb{C}^n \setminus \{0\}$.

Theorem 3.1.3. *Let $f \in SP(\mathbb{R}^d)$ and $g \in P(\mathbb{R}^d)$. We have*

(i) $f(0) > 0$ *and* $|f(t)| < f(0)$ *for all* $t \neq 0$;

(ii) $\overline{f},\ f + g \in SP(\mathbb{R}^d)$. *In particular,* $\operatorname{Re} f \in SP(\mathbb{R}^d)$;

(iii) *If* $r \in \mathbb{R} \setminus \{0\}$, *then* $f(r\cdot) \in SP(\mathbb{R}^d)$;

(iv) *If* $g \neq 0$, *then* $fg \in SP(\mathbb{R}^d)$. *In particular,* $|f|^2 \in SP(\mathbb{R}^d)$.

---

1 We refer to [58] for an introduction to multivariate scattered data approximation.

2 More specific results and references can be found in [23].

Proof. Setting $n = 1$ and $c = 1$ in Lemma 3.1.2 shows that $f(0) > 0$ while setting $n = 2,\ x_1 = x,\ x_2 = 0,\ c_1 = 1$, and $c_2 = -1$ yields

$$f(0)^2 - f(x)f(-x) = f(0)^2 - |f(x)|^2 > 0$$

whenever $x \neq 0$.

(iv) is a consequence of Schur's Theorem D.2.12 while the remaining statements follow immediately from the definition of strictly positive definiteness. □

Recall that (complex) *trigonometric polynomials* on $\mathbb{R}^d$ are functions P of the form

$$P(t) = \sum_{j=1}^{n} c_j \mathrm{e}^{\mathrm{i}(t,x_j)}, \quad t \in \mathbb{R}^d$$

where $n \in \mathbb{N},\ c \in \mathbb{C}^n$ and $x \in \mathbb{R}^d$.

Theorem 3.1.4. *Let $f \in P^c(\mathbb{R}^d)$ and let μ be the corresponding spectral measure. Then $f \in SP^c(\mathbb{R}^d)$ if and only if there is no trigonometric polynomial $P \neq 0$ on $\mathbb{R}^d$ which vanishes on the support of μ.*

Proof. Let $P \neq 0$ be a trigonometric polynomial on $\mathbb{R}^d$. We write P in the form

$$P(t) = \sum_{j=1}^{n} c_j \mathrm{e}^{\mathrm{i}(t,x_j)}, \quad t \in \mathbb{R}^d,$$

where $c \in \mathbb{C}^n \setminus \{0\}$ and the x_j's are pairwise distinct. The proof of inequality (1.1.2.v) shows that

$$\sum_{i,j=1}^{n} f(t_i - t_j) c_i \overline{c_j} = \int_{\mathbb{R}^d} |P(t)|^2 \, \mathrm{d}\mu(t).$$

Since $|P|^2$ is continuous and nonnegative, the integral on the right-hand side is zero if and only if P vanishes on the support of μ. The theorem follows immediately from this observation. □

The next two theorems provide simple sufficient conditions to ensure strict positive definiteness on $\mathbb{R}^d$.

Theorem 3.1.5. *If $f \in P^c(\mathbb{R}^d)$, $d \geq 2$, is radial and not constant, then f is strictly positive definite.*

Proof. By Lemma 3.6.3 the spectral measure μ of f is radial. Since f is not constant the support of μ contains a sphere of positive radius centered at the origin. Therefore, Theorem C.10.3 shows that there is no trigonometric polynomial $P \neq 0$ vanishing on the support of μ. From Theorem 3.1.4 we infer that f is strictly positive definite. □

Theorem 3.1.6. *Let $f \in P^c(\mathbb{R}^d)$ and let μ be the corresponding spectral measure. If the support of μ contains a nonempty open set then f is strictly positive definite.*

Proof. The statement follows by the same argument as in the proof of Theorem 3.1.5. Instead of Theorem C.10.3 we apply Theorem C.10.2. □

We now turn to the special case $d = 1$.[3]

Proposition 3.1.7. *Let $f \in P^c(\mathbb{R})$ and let μ be the corresponding spectral measure. If the support of μ has accumulation points, then f is strictly positive definite.*

Proof. If a trigonometric polynomial vanishes on the support of μ then, being analytic, it vanishes on $\mathbb{R}$. Thus, the statement follows from Theorem 3.1.4. □

Corollary 3.1.8. *If the support of μ is not countable, then f is strictly positive definite. In particular, if μ is not purely discrete, then $f \in SP^c(\mathbb{R})$.*

Corollary 3.1.9. *If $f \in P^c(\mathbb{R})$ and*

$$\limsup_{x\to\infty} |f(x)| < f(0),$$

then f is strictly positive definite.

Proof. Theorem 1.8.5 with $s = 0$ shows that the spectral measure of f cannot be purely discrete so we can apply Corollary 3.1.8. □

3.2 Infinitely differentiable and rapidly decreasing functions

In this section we construct positive definite functions which are rapidly decreasing (cf. Definition 3.2.6). These functions can be useful for example as technical tools in proofs. We also consider the restriction of the Fourier transformation to the Schwartz space $\mathcal{S}(\mathbb{R}^d)$ of all rapidly decreasing functions (see Theorem 3.2.8).

3.2.1. Recall that $C^\infty(\mathbb{R}^d)$ denotes the set of infinitely differentiable functions on $\mathbb{R}^d$ while $C^\infty_{00}(\mathbb{R}^d)$ stands for the set of functions in $C^\infty(\mathbb{R}^d)$ which have compact support. It is easy to check that the function

$$g(t) := \begin{cases} e^{-1/t} & \text{if } t > 0, \\ 0 & \text{if } t \le 0 \end{cases} \tag{1}$$

(see Figure 3.1) is infinitely differentiable on $\mathbb{R}$. Since $x \mapsto \|x\|^2$ is infinitely differ-

[3] For special results in the one-dimensional case we refer to [12].

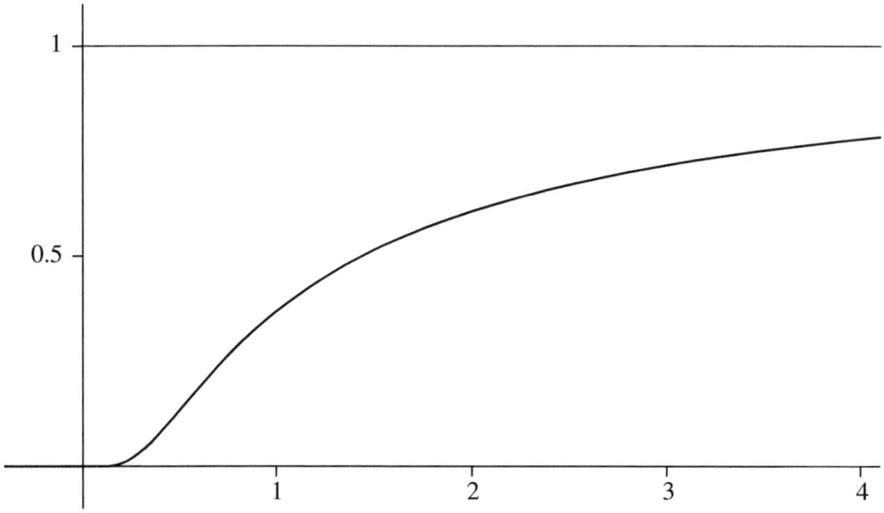

Figure 3.1. The function g from equation (3.2.1.1).

entiable on $\mathbb{R}^d$ we see that the function

$$h(x) := g(1 - \|x\|^2), \quad x \in \mathbb{R}^d \tag{2}$$

(see Figure 3.2) belongs to $C_{00}^{\infty}(\mathbb{R}^d)$. Moreover, h is radial and its support is equal to the closed unit ball $B^c(1)$.

The support of

$$f := h * \tilde{h} \tag{3}$$

is the ball $B^c(2)$. In addition, f is positive definite, infinitely differentiable and radial (cf. Lemma 3.6.5).

Theorem 3.2.2. *A set $S \subset \mathbb{R}^d$ has the form $S = \{x \in \mathbb{R}^d : f(x) \neq 0\}$ for some infinitely differentiable positive definite function $f \neq 0$ if and only if*

(i) $0 \in S$;

(ii) *S is open;*

(iii) $S = -S$.

Proof. If $f \in P^c(\mathbb{R}^d)$ and $f \neq 0$, then the set

$$S := \{x \in \mathbb{R}^d : f(x) \neq 0\}$$

obviously has the properties (i)–(iii).

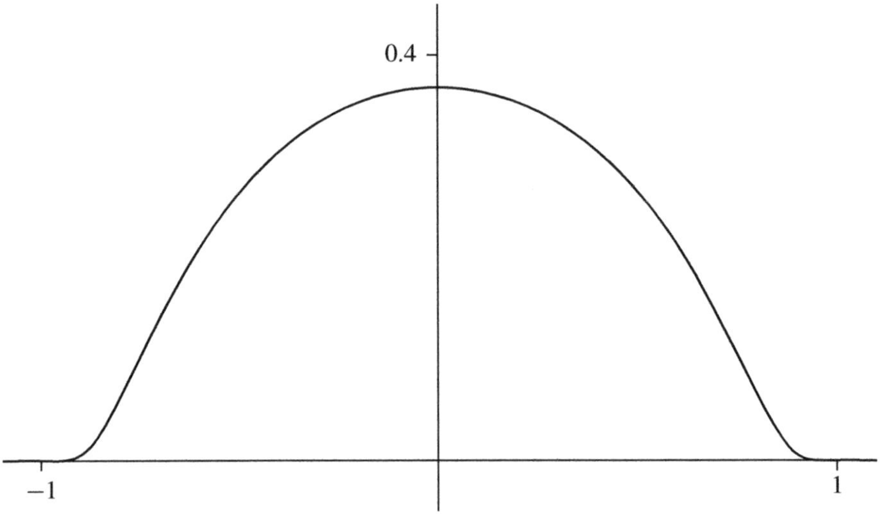

Figure 3.2. The function h from equation (3.2.1.2) with $d = 1$.

Assume that $S \subset \mathbb{R}^d$ is such that (i)–(iii) hold and let $s_1, s_2, \ldots$ be all points of S having rational coordinates. For each $j \in \mathbb{N}$ choose $r_j > 0$ such that

$$B^c(r_j) \cup [B^c(r_j) + s_j] \cup [B^c(r_j) - s_j] \subset S. \tag{1}$$

Let $h \in P^c(\mathbb{R}^d) \cap \mathrm{C}^\infty(\mathbb{R}^d)$ be such that $h \geq 0$ and $\operatorname{supp} h = B^c(1)$. We define the function $g_j,\ j \in \mathbb{N}$, by

$$g_j(x) = 2h(x/r_j) + h((x + s_j)/r_j) + h((x - s_j)/r_j), \quad x \in \mathbb{R}^d. \tag{2}$$

By Lemma 1.4.18, the function g_j is positive definite. In addition, g_j is nonnegative, $g_j \in \mathrm{C}^\infty(\mathbb{R}^d)$ and

$$\operatorname{supp} g_j = B^c(r_j) \cup [B^c(r_j) + s_j] \cup [B^c(r_j) - s_j] \subset S.$$

Using equation (2) and Theorem 1.2.1 we see that

$$|D^\alpha g_j| \leq \frac{4A_\alpha}{r_j^{|\alpha|}}, \quad \alpha \in \mathbb{N}_0^d$$

where A_α denotes the absolute moment corresponding to h. Using this, Lemma B.1.16 shows the existence of a sequence $\{p_k\}_1^\infty$ of positive real numbers such that the series

$$\sum_{j=1}^{\infty} p_j |D^\alpha g_j|$$

converges for all $\alpha \in \mathbb{N}_0^d$. Setting

$$f := \sum_{j=1}^{\infty} p_j g_j$$

we have $f \in P^c(\mathbb{R}^d)$. Theorem B.2.10 shows that $f \in C^\infty(\mathbb{R}^d)$. Noting that $f(s_j) \geq g_j(s_j) > 0$, the relation

$$\{x \in \mathbb{R}^d : f(x) \neq 0\} = S$$

follows from equation (1) and from the fact that for each $s \in S$ there exists a subsequence $\{s_{k_j}\}$ converging to s. □

The next three lemmata are often useful as technical tools in proofs.

Lemma 3.2.3. *There exists a sequence $\{g_n\}_1^\infty$ of radial positive definite functions $g_n \in C_{00}^\infty(\mathbb{R}^d)$ of the form $g_n = h_n * \tilde{h}_n$ where $h_n \in C_{00}^\infty(\mathbb{R}^d)$ such that $g_n(0) = 1$ and $\{g_n\}_1^\infty$ tends to 1 uniformly on compact sets.*

Proof. This lemma can be proved in the same way as Lemma 1.5.6 by using the function f from (3.2.1.3). □

Lemma 3.2.4. *There exists a sequence $\{f_n\}$ of radial positive definite functions $f_n \in C_{00}^\infty(\mathbb{R}^d)$ such that*

(i) $\hat{f}_n \to 1$ *uniformly on compact sets;*

(ii) $\int_{\mathbb{R}^d} f_n h \, d\lambda \to h(0)$ *for every continuous function h.*

Proof. Consider the positive definite function $f := h * \tilde{h}$ where h is the function defined in (3.2.1.3). It is clear that f is radial and $f \in C_{00}^\infty(\mathbb{R}^d)$. Scaling and normalizing this function we obtain a sequence $\{f_n\}$ of radial positive definite functions such that f_n is a density and the support of f_n is contained in $B^c(1/n)$. It is a routine exercise to show that (ii) holds. Replacing h in (ii) by the characters of $\mathbb{R}^d$, we see that $\hat{f}_n \to 1$ pointwise. The uniform convergence follows from Theorem 1.5.11. □

Lemma 3.2.5. *For all $\epsilon, \delta \in (0, \infty)$ there exists a radial positive definite function $p \in C_{00}^\infty(\mathbb{R}^d)$ such that p is a density and*

$$0 \leq \hat{p}(t) < \delta, \quad t \in \mathbb{R}^d \setminus B^o(\epsilon).$$

Proof. Let $f \in C_{00}^\infty(\mathbb{R}^d)$ be an arbitrary radial positive definite function which is a density (cf. the proof of Lemma 3.2.4). Then $0 \leq \hat{f}(t) \leq 1 = f(0)$ for all t. By the Riemann–Lebesgue Lemma 1.8.4, the function $\hat{f}$ is in $C_0(\mathbb{R}^d)$. From Corollary 1.4.13 we conclude that $0 \leq \hat{f}(t) < 1$, $t \neq 0$. Consequently,

$$(\hat{f}(t))^n < \delta, \quad t \in \mathbb{R}^d \setminus B^o(\epsilon)$$

for some $n \in \mathbb{N}$. The function $p := f * \cdots * f$ (n fold convolution) has the desired properties. □

Definition 3.2.6. A function $f \in C^\infty(\mathbb{R}^d)$ is called *rapidly decreasing* if

$$\|f\|_{\alpha,\beta} := \sup_{x\in\mathbb{R}^d} |x^\alpha D^\beta f(x)| < \infty$$

for all $\alpha, \beta \in \mathbb{N}_0$.

We denote by $\mathcal{S}(\mathbb{R}^d)$ the linear space of all rapidly decreasing functions[4] on $\mathbb{R}^d$. If g is a complex-valued function on $\mathbb{R}^d$, then we write[5]

$$M_\alpha g(x) := x^\alpha g(x), \quad x \in \mathbb{R}^d$$

so that

$$\|f\|_{\alpha,\beta} = \|M_\alpha D^\beta f\|_\infty .$$

It is clear that $C_{00}^\infty(\mathbb{R}^d) \subset \mathcal{S}(\mathbb{R}^d)$. Another example of a rapidly decreasing function is given by

$$f(x) = e^{-\|x\|^2}, \quad x \in \mathbb{R}^d.$$

The next lemma is simple, we omit the proof.

Lemma 3.2.7. *If $f \in \mathcal{S}(\mathbb{R}^d)$, then also $M_\alpha D^\beta f \in \mathcal{S}(\mathbb{R}^d)$ for all $\alpha, \beta \in \mathbb{N}_0^d$.*

Theorem 3.2.8. *The Fourier transformation maps $\mathcal{S}(\mathbb{R}^d)$ onto $\mathcal{S}(\mathbb{R}^d)$ and*

$$M_\beta D^\alpha \hat{f} = (-1)^{|\alpha|}(D^\beta M_\alpha f)\hat{}, \quad f \in \mathcal{S}(\mathbb{R}^d). \tag{1}$$

Moreover, if $f_n, f \in \mathcal{S}(\mathbb{R}^d)$ and

$$\lim_{n\to\infty} \|f_n - f\|_{\alpha,\beta} \to 0$$

for all $\alpha, \beta \in \mathbb{N}_0^d$, then

$$\lim_{n\to\infty} \|\hat{f}_n - \hat{f}\|_{\alpha,\beta} \to 0$$

for all $\alpha, \beta \in \mathbb{N}_0^d$.

Proof. Equation (1) is a consequence of Theorem 1.2.1 and Theorem 1.2.9.
Let Q be a positive polynomial such that[6]

$$\int_{\mathbb{R}^d} \frac{1}{Q}\, d\lambda = 1.$$

---

[4] The function space $\mathcal{S}(\mathbb{R}^d)$ is usually called *Schwartz space*.

[5] Note that the symbol M_α is also used to denote moments of measures.

[6] We can take for example $Q(x) = c(1 + \|x\|^2)^d$ with a suitable constant $c > 0$.

If $g \in \mathcal{S}(\mathbb{R}^d)$ we have

$$(2\pi)^{d/2}|\hat{g}| \le \int_{\mathbb{R}^d} |g| \, d\lambda = \int_{\mathbb{R}^d} \frac{|gQ|}{Q} \, d\lambda \le \|gQ\|_\infty < \infty.$$

Replacing here g by $M_\beta D^\alpha \hat{f}$ and using equation (1) we see that $\hat{f} \in \mathcal{S}(\mathbb{R}^d)$ and that the second statement of the theorem is true.

In the same way, we see that $\check{f} \in \mathcal{S}(\mathbb{R}^d)$ whenever $f \in \mathcal{S}(\mathbb{R}^d)$. Using this, the fact that the Fourier transformation maps $\mathcal{S}(\mathbb{R}^d)$ onto $\mathcal{S}(\mathbb{R}^d)$ follows from the relation

$$f = (\check{f})\hat{}, \quad f \in \mathcal{S}(\mathbb{R}^d).$$

The proof is complete. □

3.3 Analytic characteristic functions of one variable

Definition 3.3.1. A complex-valued function f on $\mathbb{R}$ is said to be *analytic* if it is infinitely often differentiable and its Taylor series at 0 has positive convergence radius. If the convergence radius is equal to infinity, then f is said to be an *entire function.*

We note that f is analytic if and only if there exists a complex-valued function θ of one complex variable, which is holomorphic in an open disk around 0 and $f(t) = \theta(t)$ for all real t in this disk. If θ is holomorphic in the whole complex plane and $f(t) = \theta(t)$ for all $t \in \mathbb{R}$, then f is entire.

Characteristic functions that are not differentiable at 0, like $e^{-|t|}$ or $\max(1 - |t|, 0)$, are not analytic. The characteristic functions e^{-t^2} and $\cos t$ are entire, while $\frac{1}{1-it}$ and $\frac{1}{1+t^2}$ are analytic but not entire.

3.3.2. We have seen in Section 1.2 that the existence of derivatives of a characteristic function f is closely related to the existence of moments of the corresponding distribution μ. We will prove that analyticity is closely related to finiteness of integrals of the form $\int e^{-rs} \, d\mu(s)$ for certain $r \in \mathbb{R}$. We introduce the numbers

$$s_\mu^+ = s_f^+ = \sup \left\{ r \in \mathbb{R} : \int_{-\infty}^{\infty} e^{-rx} \, d\mu(x) < \infty \right\}$$

$$s_\mu^- = s_f^- = \inf \left\{ r \in \mathbb{R} : \int_{-\infty}^{\infty} e^{-rx} \, d\mu(x) < \infty \right\}$$

and the set

$$S_\mu = S_f = \{ z \in \mathbb{C} : s_\mu^- < \operatorname{Im} z < s_\mu^+ \}.$$

Note that $s_\mu^- \le 0 \le s_\mu^+$ and hence S_μ is empty if and only if $s_\mu^- = s_\mu^+ = 0$.

If the support of μ is bounded, then $S_\mu = \mathbb{R}$. If μ is the exponential distribution with parameter α, then $f(t) = \frac{\alpha}{\alpha - \mathrm{i}t}$ and $S_\mu = \{z \in \mathbb{C} : -\alpha < \operatorname{Im} z < \infty\}$.

Theorem 3.3.3. *Let μ be a probability distribution with characteristic function f. If S_μ is not empty, then the integral*

$$f(z) := \int_{-\infty}^{\infty} \mathrm{e}^{\mathrm{i}zx}\, \mathrm{d}\mu(x)$$

exists for all z in the strip S_μ and it represents there a holomorphic function. Moreover,

$$f^{(n)}(z) = \int_{-\infty}^{\infty} (\mathrm{i}x)^n \cdot \mathrm{e}^{\mathrm{i}zx}\, \mathrm{d}\mu(x), \quad n \in \mathbb{N} \tag{1}$$

for all $z \in S_\mu$.[7]

Proof. The fact that $f(z)$ exists follows from $|\mathrm{e}^{\mathrm{i}zx}| = \mathrm{e}^{-\operatorname{Im} z \cdot x}$. We now proceed in a similar way as in the proof of (1.2.1.i). Let $z \in S_\mu$ be fixed and choose a positive δ such that $z + h \in S_\mu$ for all $h \in \mathbb{C}$ with $|h| < \delta$. If, in addition, $h \neq 0$ then we have

$$\frac{f(z+h) - f(z)}{h} = \int_{-\infty}^{\infty} x \cdot \frac{\mathrm{e}^{\mathrm{i}hx} - 1}{hx} \cdot \mathrm{e}^{\mathrm{i}zx}\, \mathrm{d}\mu(x). \tag{2}$$

Theorem B.1.5 with $n = 0$ shows that the integrand times $\frac{\delta}{2}$ is dominated by

$$\frac{\delta |x|}{2} \cdot \mathrm{e}^{|hx|} \cdot \mathrm{e}^{-\operatorname{Im} z \cdot x}.$$

Using that $r < \mathrm{e}^r$, $r \in \mathbb{R}$, we see that this expression is less than

$$\mathrm{e}^{\delta |x|} \cdot \mathrm{e}^{-\operatorname{Im} z \cdot x}$$

whenever $|h| < \frac{\delta}{2}$. This function is μ-integrable with respect to x by our choice of δ. Taking the limit $h \to 0$ in (2) and applying Lebesgue's theorem on dominated convergence we obtain (1) for $n = 1$. The general case is obtained by repeating the arguments above. □

Corollary 3.3.4. *If the support of a distribution is bounded, then its characteristic function is entire.*

Corollary 3.3.5. *The characteristic function of μ is analytic if and only if*

$$s_\mu^- < 0 < s_\mu^+ .$$

In the sequel, we always assume that an analytic characteristic function f is defined in the strip S_μ.

---

[7] See also Lemma 3.4.1.

Theorem 3.3.6. *Let f be an analytic characteristic function.*

(i) *If g is an analytic characteristic function and f and g coincide on an interval $(-a,a) \subset \mathbb{R}$, then $f = g$ on $\mathbb{R}$.*

(ii) *If g is an entire function and f and g coincide on an interval $(-a,a) \subset \mathbb{R}$, then $f = g$ on $\mathbb{R}$.*[8]

(iii) *If two probability distributions have analytic characteristic functions and the same moments, then they are equal.*

Proof. (i) It follows from Theorem 3.3.3 that f and g have holomorphic extensions to the strip $S := S_f \cap S_g$. Since S is connected and $f = g$ on $(-a,a)$, the extensions are equal on S (cf. Theorem C.1.8). This implies that the corresponding distributions are equal, hence $S_f = S_g$.

(ii) We extend f to the strip S_f and consider g on the whole complex plane. Since S_f is simply connected and $f = g$ on $(-a,a)$, the extensions are equal on S_f.
(iii) By Theorem 1.2.1, the corresponding characteristic functions have the same Taylor expansion at zero. Hence (iii) follows from (i). □

Remark 3.3.7. A distribution is in general not uniquely determined by its moments. To see an example we consider the *log-normal distribution* which has the density p given by

$$p(x) = \frac{1}{x\sqrt{2\pi}} \cdot \mathrm{e}^{-(\log x)^2/2}, \quad x > 0$$

(see Figure 3.3). The moments M_k of this distribution are given by

$$M_k = \frac{1}{\sqrt{2\pi}} \int_0^\infty x^{k-1} \mathrm{e}^{-(\log x)^2/2}\,\mathrm{d}x, \quad k \in \mathbb{N}.$$

Substituting $t = \log x$, we get

$$\begin{aligned} M_k &= \frac{1}{\sqrt{2\pi}} \int_{-\infty}^{\infty} \mathrm{e}^{-t^2/2+kt}\,\mathrm{d}t \\ &= \mathrm{e}^{k^2/2} \frac{1}{\sqrt{2\pi}} \int_{-\infty}^{\infty} \mathrm{e}^{-(t-k)^2/2}\,\mathrm{d}t \\ &= \mathrm{e}^{k^2/2}. \end{aligned}$$

[8] See also Theorem 4.2.3.

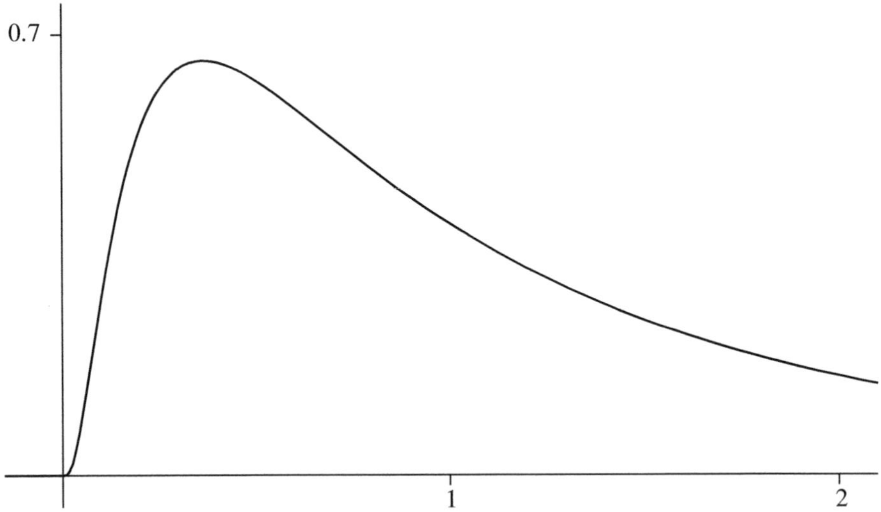

Figure 3.3. The function p from Remark 3.3.7.

The same argument as above shows that

$$\begin{aligned}\int_0^\infty x^k \sin(2\pi \log x) \cdot p(x)\,\mathrm{d}x &= \frac{1}{\sqrt{2\pi}} \int_{-\infty}^{\infty} \mathrm{e}^{-t^2/2+kt} \sin(2\pi t)\,\mathrm{d}t \\ &= \mathrm{e}^{k^2/2} \frac{1}{\sqrt{2\pi}} \int_{-\infty}^{\infty} \mathrm{e}^{-y^2/2} \sin(2\pi y)\,\mathrm{d}y \\ &= 0\end{aligned}$$

for all $k \in \mathbb{N}_0$. Consequently, all the functions p_a, $a \in [-1, 1]$, defined by

$$p_a(x) = p(x)\big(1 + a \sin(2\pi \log x)\big), \quad x > 0$$

are densities and they have the same moment sequence.[9]

Next we present an application of analytic characteristic functions.

Theorem 3.3.8. *If the random variables X and Y have analytic characteristic functions and the equation*

$$\mathbb{E}(X^k \cdot Y^l) = \mathbb{E}(X^k) \cdot \mathbb{E}(Y^l)$$

holds for all k and l in $\mathbb{N}$, then X and Y are independent.[10]

---

9 See, e.g., [1, 35, 55] or Chapter 2 in [6] for more information on this topic.

10 We refer to [7] for some related results.

Proof. Let f_X, f_Y, and g be the characteristic functions of X, Y, and the random vector (X, Y), respectively. If $s, t \in \mathbb{R}$ and $|s|$ and $|t|$ are small enough, then we have

$$\begin{aligned} f_X(s) \cdot f_Y(t) &= \sum_{k=0}^{\infty} \frac{\mathbb{E}(X^k) \cdot (\mathrm{i}s)^k}{k!} \cdot \sum_{l=0}^{\infty} \frac{\mathbb{E}(Y^l) \cdot (\mathrm{i}t)^l}{l!} \\ &= \sum_{k,l=0}^{\infty} \frac{\mathbb{E}(X^k) \cdot (\mathrm{i}s)^k \cdot \mathbb{E}(Y^l) \cdot (\mathrm{i}t)^l}{k! \cdot l!} \\ &= \sum_{k,l=0}^{\infty} \frac{\mathbb{E}(X^k Y^l) \cdot (\mathrm{i}s)^k \cdot (\mathrm{i}t)^l}{k! \cdot l!}. \end{aligned}$$

Using Theorem 1.2.1 we see that the last sum is the Taylor series of the function g at zero. Now fix s and t. The display above shows that $r \mapsto g(rs, rt)$, $r \in \mathbb{R}$, is an analytic characteristic function which is equal to the analytic characteristic function $r \mapsto f_X(rs) \cdot f_Y(rt)$ in a neighborhood of zero. By (3.3.6.ii), the equality holds on $\mathbb{R}$. In particular,

$$g(s, t) = f_X(s) \cdot f_Y(t).$$

Thus, X and Y are independent (cf. Theorem 1.3.10). □

Theorem 3.3.9. *Let f_1 and f_2 be characteristic functions. If the characteristic function $f = f_1 f_2$ is analytic, then so are f_1 and f_2. Moreover, the relations $S_{f_j} \supset S_f$ $(j = 1, 2)$ hold.*[11]

Proof. Denote by μ_1, μ_2 and μ the corresponding distributions. For all $A, B > 0$ and $s_\mu^- < r < s_\mu^+$ we have

$$\begin{aligned} \int_{-A}^{A} \mathrm{e}^{-rx}\, \mathrm{d}\mu_1(x) \int_{-B}^{B} \mathrm{e}^{-ry}\, \mathrm{d}\mu_2(y) &= \int_{-A}^{A} \int_{-B}^{B} \mathrm{e}^{-r(x+y)}\, \mathrm{d}\mu_1(x)\, \mathrm{d}\mu_2(y) \\ &\le \int_{-\infty}^{\infty} \int_{-\infty}^{\infty} \mathrm{e}^{-r(x+y)}\, \mathrm{d}\mu_1(x)\, \mathrm{d}\mu_2(y) \\ &= \int_{-\infty}^{\infty} \mathrm{e}^{-rs}\, \mathrm{d}(\mu_1 * \mu_2)(s) \\ &= \int_{-\infty}^{\infty} \mathrm{e}^{-rs}\, \mathrm{d}\mu(s) < \infty. \end{aligned}$$

This being true for all $A, B > 0$, we conclude that

$$\int_{-\infty}^{\infty} \mathrm{e}^{-rx}\, \mathrm{d}\mu_1(x) < \infty \quad \text{and} \quad \int_{-\infty}^{\infty} \mathrm{e}^{-rx}\, \mathrm{d}\mu_2(x) < \infty$$

whenever $s_\mu^- < r < s_\mu^+$. □

[11] See also Theorem 3.3.14.

Corollary 3.3.10. *Let f_1 and f_2 be characteristic functions. If $f_1 \cdot f_2$ is an entire characteristic function, then f_1 and f_2 are entire characteristic functions as well.*

We have seen in Theorem 1.1.2 that characteristic functions are positive definite. We now show that analytic characteristic functions are positive definite in the corresponding strip in the following sense.

Lemma 3.3.11. *Every analytic characteristic function f satisfies the inequality*

$$\sum_{j,k=1}^{n} f(z_j - \bar{z}_k) c_j \overline{c_k} \geq 0$$

for an arbitrary choice of $c_1, \ldots, c_n \in \mathbb{C}$ and $z_1, \ldots, z_n \in S_f$ such that $z_j - \bar{z}_k \in S_f$.

Proof. Denoting by μ the corresponding distribution, we have

$$f(z) = \int_{-\infty}^{\infty} \mathrm{e}^{\mathrm{i}zx} \, \mathrm{d}\mu(x), \quad z \in S_f$$

and hence

$$\begin{aligned} \sum_{j,k=1}^{n} f(z_j - \bar{z}_k) c_j \bar{c}_k &= \int_{-\infty}^{\infty} \sum_{j,k=1}^{n} \mathrm{e}^{\mathrm{i}(z_j - \bar{z}_k)x} \, c_j \overline{c_k} \, \mathrm{d}\mu(x) \\ &= \int_{-\infty}^{\infty} \sum_{j,k=1}^{n} \mathrm{e}^{\mathrm{i}z_j x} \, \overline{\mathrm{e}^{\mathrm{i}z_k x}} \, c_j \overline{c_k} \, \mathrm{d}\mu(x) \\ &= \int_{-\infty}^{\infty} \left| \sum_{j=1}^{n} \mathrm{e}^{\mathrm{i}z_j x} c_j \right|^2 \mathrm{d}\mu(x) \geq 0. \end{aligned}$$

□

Next we prove some simple but useful properties of analytic characteristic functions.

Theorem 3.3.12. *Let f be an analytic characteristic function.*

(i) *The function $x \mapsto f(x + \mathrm{i}y)$ is positive definite on $\mathbb{R}$ for all $y \in (s_f^-, s_f^+)$.*
(ii) $f(-\bar{z}) = \overline{f(z)}, \quad z \in S_f$.
(iii) $f(\mathrm{i}y) > 0, \quad \mathrm{i}y \in S_f$.
(iv) $|f(x + \mathrm{i}y)| \leq f(\mathrm{i}y), \quad x \in \mathbb{R}, \ \mathrm{i}y \in S_f$.
(v) *The function $y \mapsto \log f(\mathrm{i}y)$ is convex on the interval (s_f^-, s_f^+).*

Proof. The first statement is obtained from Lemma 3.3.11 with $z_j = x_j + \mathrm{i}y/2$ while (ii) and (iv) follow from (i) (see Lemma 1.4.8 and Theorem 1.4.12).[12] Let μ be the distribution corresponding to f. Then

$$f(\mathrm{i}y) = \int_{-\infty}^{\infty} \mathrm{e}^{-yx} \, \mathrm{d}\mu(x) > 0.$$

[12] Actually, these properties also follow easily from the integral representation of f.

By Lemma 3.3.11, the matrix $A = \big(f(z_j - \bar{z}_k)\big)_{j,k=1}^{n}$ is positive semidefinite and hence $\det A \geq 0$. In the special case where $n = 2$, $z_j = \mathrm{i}y_j$, this inequality gives

$$f(\mathrm{i}[y_1 + y_2])^2 \leq f(\mathrm{i}2y_1)f(\mathrm{i}2y_2), \quad s_f^- < y_1, y_2, y_1 + y_2 < s_f^+. \tag{1}$$

From (iii) we see that f is strictly positive on the imaginary axis. Inequality (1) and the continuity of f imply that the function $y \mapsto \log f(\mathrm{i}y)$ is convex. □

Lemma 3.3.13. *Let f be an analytic characteristic function of the form $f = \mathrm{e}^g$, where g is holomorphic in S_f and $g(0) = 0$. Then*

(i) *for $x, y \in \mathbb{R}$ such that $|y| \leq r < \min(-s_f^-, s_f^+)$ we have*

$$\operatorname{Re} g(x + \mathrm{i}y) \leq g(\mathrm{i}y) \leq \max\big(g(-\mathrm{i}r), g(\mathrm{i}r)\big);$$

(ii) $g(\mathrm{i}y) \geq \left.\frac{\mathrm{d}g(\mathrm{i}y)}{\mathrm{d}y}\right|_{y=0} \cdot y.$

Proof. By (3.3.12.v), the function $y \mapsto \log f(\mathrm{i}y)$ is convex and hence the inequality

$$g(\mathrm{i}y) \leq \max\big(g(-\mathrm{i}r), g(\mathrm{i}r)\big) \tag{1}$$

holds. Since $|f| = \mathrm{e}^{\operatorname{Re} g}$, we obtain (i) from (3.3.12.iv) and (1).

In view of $g(\mathrm{i}y) = \log f(\mathrm{i}y)$, the function g is convex on the imaginary axis. Convexity and the fact that $g(0) = 0$ imply (ii). □

The next theorem, which is essentially due to Yu. V. Linnik,[13] is a generalization of Theorem 3.3.9. If all the numbers α_k below are integers or rational numbers, then Theorem 3.3.14 follows easily from Theorem 3.3.9.

Theorem 3.3.14 (Linnik). *Let f be a holomorphic function on the open disk $D(0,\delta)$, $\delta > 0$, such that $f(-z) = \overline{f(z)}$ for all $z \in D(0,\delta)$. Further, let $f_1, \dots, f_r$ be characteristic functions and $\alpha_1, \dots, \alpha_r$ be positive real numbers such that*

$$\prod_{k=1}^{r} |f_k(t)|^{\alpha_k} = |f(t)|, \quad -\delta < t < \delta.$$

Then f_k is analytic and $S_{f_k} \supset D(0,\delta)$, $k = 1, \dots, r$.

Proof. As in the proof of Theorem 1.2.13 we write

$$g_k(t) = f_k(t)f_k(-t) = |f_k(t)|^2, \quad t \in \mathbb{R}$$

[13] Linnik's original formulation contains the conditions $f(z) \neq 0$, $z \in D(0,\delta)$, and

$$\prod_{k=1}^{r} f_k^{\alpha_k}(t) = f(t), \quad -\delta < t < \delta.$$

Our conditions are somewhat weaker and allow us to avoid the use of powers of complex numbers.

and $g(z) = f(z)f(-z) = |f(z)|^2$, $z \in D(0,\delta)$. Then g is an even holomorphic function on $D(0,\delta)$, the g_k's are characteristic functions and g, g_k are nonnegative on the interval $(-\delta,\delta)$. Moreover,

$$\prod_{k=1}^{r} g_k^{\alpha_k}(t) = g(t), \quad -\delta < t < \delta. \tag{1}$$

We will prove the statement of the theorem for the functions g_k. The statement for f_k follows then from Theorem 3.3.9. We choose $d \in (0,\delta]$ such that $g(z) \neq 0$ if $|z| < d$. From (1) we obtain

$$\sum_{k=1}^{r} \alpha_k \log \frac{1}{g_k(t)} = -h(t), \quad -d < t < d \tag{2}$$

where $h(t) = \log g(t)$.

By Theorem 1.2.13, the functions g_k are infinitely often differentiable. We prove first that they are analytic and $S_{g_k} \supset D(0,R)$ for some $R > 0$. We may suppose that $\alpha_k \geq 1$ $(k = 1,\dots,r)$. Indeed, if $N_0 \in \mathbb{N}$ is sufficiently large, then $N_0\alpha_k \geq 1$ for all k. Raising both sides of (1) to the power N_0 we obtain an equation of the same type containing factors of the form $g_k^{N_0\alpha_k}$.

We introduce the notation $\gamma_q = g^{2q}$ and $\gamma_{q,k}(t) = g_k^{2q\alpha_k}(t)$, $t \in (-\delta,\delta)$. Raising both sides of (1) to the power $2q$, differentiating the new equation $2q$-times using the Leibniz rule (cf. (B.1.13.1)), and setting $t = 0$ we obtain

$$\gamma_q^{(2q)}(0) = \sum_{l_1+\dots+l_r=2q} \frac{(2q)!}{l_1!\cdot\ldots\cdot l_r!}\gamma_{q,1}^{(l_1)}(0)\cdot\ldots\cdot\gamma_{q,r}^{(l_r)}(0). \tag{3}$$

To compute the derivatives on the right-hand side we apply Faà di Bruno's formula (B.1.14) which yields

$$\gamma_{q,k}^{(m)}(0) = \sum c(b,m)\cdot g_k'(0)^{b_1}\cdot g_k''(0)^{b_2}\cdot\ldots\cdot g_k^{(m)}(0)^{b_m} \tag{4}$$

where the sum is over all $b \in \mathbb{N}_0^m$ satisfying $b_1 + 2b_2 + \dots + mb_m = m$ and

$$c(b,m) = \frac{m!2q\alpha_k(2q\alpha_k - 1)\cdot\ldots\cdot(2q\alpha_k - |b| + 1)}{b_1!b_2!\cdot\ldots\cdot b_m!\,(1!)^{b_1}(2!)^{b_2}\cdot\ldots\cdot(m!)^{b_m}}.$$

Since $\alpha_k \geq 1$ and $|b| \leq m$, the number $c(b,m)$ is positive if $m \leq 2q$. Noting that $g_k^{(2l-1)}(0) = 0$ and the sign of $[g_k^{(2l)}(0)]^{b_{2l}}$ is $(-1)^{lb_{2l}}$, we see that the sign of the nonvanishing terms[14] on the right of (4) is equal to $(-1)^{m/2}$. From this we conclude that the sign of the nonvanishing terms on the right-hand side of (3) is $(-1)^q$. Since

[14] Note that m is even if there are nonvanishing terms.

the right-hand side of (3) contains the terms $2q\alpha_k g_k^{(2q)}$ we obtain

$$\left|\sum_{k=1}^{r} 2q\alpha_k g_k^{(2q)}(0)\right| \le |\gamma_q^{(2q)}(0)|$$

and hence

$$\left|g_k^{(2q)}(0)\right| \le \frac{1}{2q\alpha_k}|\gamma_q^{(2q)}(0)|, \quad k = 1, \dots, r. \tag{5}$$

Applying the inequality (C.1.13) to the Taylor expansion of $\gamma_q = g^{2q}$ with r replaced by $\delta_0 = \frac{\delta}{2}$ we get

$$\frac{|\gamma_q^{(2q)}(0)|}{(2q)!} \le \frac{2}{\delta_0^{2q}} \cdot \sup_{|z|=\delta_0} \left|\mathrm{Re}\big(g^{2q}(z)\big)\right| \le 2M^{2q} \tag{6}$$

where $M = \frac{1}{\delta_0} \cdot \sup_{|z|=\delta_0} |g(z)|$. The inequalities (5) and (6) show that

$$\limsup_{q\to\infty} \left(\frac{g_k^{(2q)}(0)}{(2q)!}\right)^{\frac{1}{2q}} \le M$$

and hence g_k is holomorphic on the disk $D(0, 1/M)$.

Denote by δ_k the radius of convergence of the Taylor expansion of g_k. We already know that $\delta_k > 0$. Without loss of generality we may assume that $\delta_1 \le \dots \le \delta_r$. To finish the proof we have to show that $\delta_1 = \delta$. Assume, on the contrary, that $\delta_1 < \delta$. The equations

$$G_k(t) = \sum_{j=0}^{\infty} \frac{g_k^{(2j)}(0)}{(2j)!}(-t)^j, \quad t \in (-\delta_k^2, \delta_k^2)$$

and

$$G(t) = \sum_{j=0}^{\infty} \frac{g^{(2j)}(0)}{(2j)!}(-t)^j, \quad t \in (-\delta^2, \delta^2)$$

define functions whose Taylor expansions have δ_k^2 and δ^2 as radii of convergence, respectively. We have $G_k(t^2) = g_k(it)$, $G(t^2) = g(it)$ and therefore

$$\prod_{k=1}^{r} G_k^{\alpha_k}(t) = G(t), \quad t \in (-\delta_1^2, \delta_1^2). \tag{7}$$

Relation (1.2.1.ii) shows that

$$G_k^{(l)}(0) \ge 0, \quad l \in \mathbb{N}_0. \tag{8}$$

Let $\delta_0 \in (0, \delta_1^2)$ be such that $G(\delta_0) \ne 0$ and define the functions H_k by

$$H_k(t) = G_k(\delta_0 + t),\ |\delta_0 + t| < \delta_1^2,\ k = 1, \dots, r.$$

By (7) we have

$$\prod_{k=1}^{r} H_k^{\alpha_k}(t) = G(\delta_0 + t), \quad |\delta_0 + t| < \delta_1^2,$$

hence

$$\prod_{k=1}^{r} \left(\frac{H_k(t)}{H_k(0)}\right)^{\alpha_k} = \frac{G(\delta_0 + t)}{G(\delta_0)}.$$

We raise this equation to the power n, differentiate the resulting equation n times and set $t = 0$. The left-hand side contains the term $n\alpha_1 H_1^{(n)}(0)/H_1(0)$. Since $H_k^{(l)}(0) \geq 0$ for all l (cf. (8)) we conclude that

$$n\alpha_1 H_1^{(n)}(0) \leq H_1(0) \cdot \frac{\mathrm{d}^n}{\mathrm{d}t^n} \frac{G^n(\delta_0 + t)}{G^n(\delta_0)}\bigg|_{t=0}.$$

We now choose arbitrary real numbers r_1 and r_2 such that $\delta_1^2 < r_1 < r_2 < \delta^2$. Applying the inequality (C.1.13) to the Taylor expansion of $G^n(\delta_0+t)$ with r replaced by $r_1 - \delta_0$ we get

$$\frac{1}{n!} \cdot \frac{\mathrm{d}^n}{\mathrm{d}t^n} G^n(\delta_0 + t)\bigg|_{t=0} \leq \frac{2}{(r_1 - \delta_0)^n} \cdot \sup_{|z|=r_1-\delta_0} \left|\mathrm{Re}\big(G^n(\delta_0 + z)\big)\right| \leq 2M^n,$$

where $M = \frac{1}{r_1-\delta_1} \cdot \sup_{|z|\leq r_2} |G(z)|$. Consequently,

$$\limsup_{n\to\infty} \left(\frac{H_1^{(n)}(0)}{n!}\right)^{\frac{1}{n}} \leq M$$

and therefore H_1 is holomorphic on the disk $D(\delta_0, 1/M)$. Since M does not depend on δ_0, we can choose δ_0 such that this disk contains δ_1^2. This shows that G_1 is bounded on $(-\delta_1^2 - \epsilon, \delta_1^2 + \epsilon)$ for some positive ϵ. On the other hand, the Taylor series of G_1 at 0 has nonnegative coefficients (cf. inequality (8)) and hence this series cannot converge for values larger than δ_1^2 (see Lemma B.1.15). This contradiction shows that $\delta_1 = \delta$. □

3.4 Holomorphic L^2 Fourier transforms

In this section we consider the Fourier transforms of square integrable functions as well as their extensions to certain regions of $\mathbb{C}$.[15] We will use the notation

$$\mathbb{C}_{\mathrm{im}}^+ := \{z \in \mathbb{C} : \operatorname{Im} z > 0\}, \quad \mathbb{C}_{\mathrm{im}}^- := \{z \in \mathbb{C} : \operatorname{Im} z < 0\}$$

and

$$\mathbb{C}_{\mathrm{re}}^+ := \{z \in \mathbb{C} : \operatorname{Re} z > 0\}, \quad \mathbb{C}_{\mathrm{re}}^- := \{z \in \mathbb{C} : \operatorname{Re} z < 0\}.$$

[15] For further reading on this topic we refer to Rudin's book [46].

Lemma 3.4.1. *If $F \in L^2(0,\infty)$, then the function*

$$f(z) = \int_0^\infty F(t)\mathrm{e}^{\mathrm{i}tz}\,\mathrm{d}t, \quad z \in \mathbb{C}^+_{\mathrm{im}}$$

is holomorphic. Moreover,

$$\int_{-\infty}^\infty |f(x+\mathrm{i}y)|^2\,\mathrm{d}x \le 2\pi \cdot \int_0^\infty |F(t)|^2\,\mathrm{d}t, \quad y > 0.$$

Proof. If $z = x + \mathrm{i}y$ where $y > 0$, then $|\mathrm{e}^{\mathrm{i}tz}| = \mathrm{e}^{-ty}$ showing that the integral above exists.

Assume that $y > \delta > 0$, $\operatorname{Im} z_n > \delta$, and $z_n \to z$. Using the Cauchy–Schwarz inequality we obtain

$$|f(z) - f(z_n)|^2 \le \int_0^\infty |F(t)|^2\,\mathrm{d}t \cdot \int_0^\infty \left|\mathrm{e}^{\mathrm{i}tz} - \mathrm{e}^{\mathrm{i}tz_n}\right|^2\,\mathrm{d}t.$$

The second integrand on the right is bounded by the integrable function $t \mapsto 4\mathrm{e}^{-2\delta t}$ and tends to zero as $n \to \infty$. The dominated convergence theorem shows that f is continuous.

Let γ be a closed path in $\mathbb{C}^+_{\mathrm{im}}$. By Fubini's theorem we have

$$\int_\gamma f(z)\,\mathrm{d}z = \int_0^\infty F(t) \cdot \int_\gamma \mathrm{e}^{\mathrm{i}tz}\,\mathrm{d}z\,\mathrm{d}t.$$

Cauchy's Theorem C.1.5 shows that

$$\int_\gamma \mathrm{e}^{\mathrm{i}tz}\,\mathrm{d}z = 0.$$

Thus, the integral of f over γ is zero and hence, by Morera's Theorem C.1.6, f is holomorphic on $\mathbb{C}^+_{\mathrm{im}}$.

To prove the last statement let $y > 0$ be fixed. By the definition of f, we have

$$\int_{-\infty}^\infty f(x+\mathrm{i}y)\,\mathrm{d}x = \int_{-\infty}^\infty \int_0^\infty \left[F(t)\mathrm{e}^{-ty}\right]\mathrm{e}^{\mathrm{i}tx}\,\mathrm{d}t\,\mathrm{d}x, \quad y > 0.$$

Applying Plancherel's Theorem 1.8.9 we obtain

$$\frac{1}{2\pi} \cdot \int_{-\infty}^\infty |f(x+\mathrm{i}y)|^2\,\mathrm{d}x = \int_0^\infty |F(t)|^2\mathrm{e}^{-2ty}\,\mathrm{d}t \le \int_0^\infty |F(t)|^2\,\mathrm{d}t. \qquad \square$$

Lemma 3.4.2. *Let $0 < a < \infty$ and $F \in L^2(-a,a)$. The function*

$$f(z) = \int_{-a}^a F(t)\mathrm{e}^{\mathrm{i}tz}\,\mathrm{d}t, \quad z \in \mathbb{C}$$

is entire and

$$|f(z)| \leq C e^{a|z|}, \quad z \in \mathbb{C}$$

with some constant $C \geq 0$.

Proof. The fact that f is entire can be proved by the same arguments as in the proof of Lemma 3.4.1 while the inequality follows from

$$|f(z)| \leq \int_{-a}^{a} |F(t)| e^{-ty} \, dt \leq e^{a|y|} \int_{-a}^{a} |F(t)| \, dt$$

since $|y| \leq |z|$. □

Definition 3.4.3. For $a > 0$ we denote by $E(a)$ the set of all entire functions of *exponential type* a, i.e., $F \in E(a)$ provided that F is entire and

$$|F(w)| \leq C e^{a|w|}, \quad w \in \mathbb{C}$$

for some $C < \infty$.

Theorem 3.4.4 (Paley–Wiener). *Let* a *and* C *be positive constants and* $f \in E(a)$ *be an entire function such that*

$$\int_{-\infty}^{\infty} |f(x)|^2 \, dx < \infty. \tag{1}$$

Then the L^2 *Fourier transform* F *of* $f|_{\mathbb{R}}$ *vanishes outside* $[-a, a]$ *and*

$$f(z) = \int_{-a}^{a} F(t) e^{itz} \, dt \tag{2}$$

for all $z \in \mathbb{C}$.

Proof. For $\epsilon > 0$ define the function f_ϵ by

$$f_\epsilon(x) := f(x) e^{-\epsilon|x|}, \quad x \in \mathbb{R}.$$

We will prove that

$$\lim_{\epsilon \to 0} \int_{-\infty}^{\infty} f_\epsilon(x) e^{-itx} \, dx = 0, \quad t \in \mathbb{R} \setminus [-a, a]. \tag{3}$$

First we show that this relation implies the theorem.

Considering the restriction of f to the real axis, we have

$$\lim_{\epsilon \to 0} \|f - f_\epsilon\|_2 = 0.$$

Plancherel's theorem (cf. Theorem 1.8.9) implies that the Fourier transforms of the f_ϵ's converge in $L^2(\mathbb{R})$ to the L^2 Fourier transform F of f. By equation (3) the function F vanishes outside $[-a, a]$. Since f is equal to the inverse Fourier transform of F we see that equation (2) holds for almost every $z \in \mathbb{R}$. As each side of (2) is an entire function (see Corollary 3.3.4), we conclude that (2) holds for all $z \in \mathbb{C}$.

To prove (3) let Γ_α be the path defined by

$$\Gamma_\alpha(s) = s\mathrm{e}^{\mathrm{i}\alpha}, \quad \alpha \in \mathbb{R},\ s \in [0, \infty). \tag{4}$$

Write

$$\Pi_\alpha = \{w \in \mathbb{C} : \operatorname{Re}(w\mathrm{e}^{\mathrm{i}\alpha}) > a\},$$

and for $w \in \Pi_\alpha$, define the function Ψ_α by[16]

$$\Psi_\alpha(w) = \int_{\Gamma_\alpha} f(z)\mathrm{e}^{-wz}\,\mathrm{d}z = \mathrm{e}^{\mathrm{i}\alpha}\int_0^\infty f(s\mathrm{e}^{\mathrm{i}\alpha})\mathrm{e}^{-ws\mathrm{e}^{\mathrm{i}\alpha}}\,\mathrm{d}s.$$

Using the fact that $f \in E(a)$ we see that the modulus of the integrand is at most

$$C\mathrm{e}^{-(\operatorname{Re}(w\mathrm{e}^{\mathrm{i}\alpha})-a)s}.$$

The same arguments as in the proof of Lemma 3.4.1 show that Ψ_α is holomorphic in the half plane Π_α. In the cases $\alpha = 0$ and $\alpha = \pi$, we extend the definition of Ψ_α to larger half planes by

$$\Psi_0(w) = \int_0^\infty f(x)\mathrm{e}^{-wx}\,\mathrm{d}x, \quad w \in \mathbb{C}^+_{\mathrm{re}}$$

and

$$\Psi_\pi(w) = -\int_{-\infty}^0 f(x)\mathrm{e}^{-wx}\,\mathrm{d}x, \quad w \in \mathbb{C}^-_{\mathrm{re}}.$$

Inequality (1) and Lemma 3.4.1 show that Ψ_0 and Ψ_π are holomorphic in the indicated half planes. Moreover, it is easy to check that

$$\int_{-\infty}^\infty f_\epsilon(x)\mathrm{e}^{-\mathrm{i}tx}\,\mathrm{d}x = \Psi_0(\epsilon + \mathrm{i}t) - \Psi_\pi(-\epsilon + \mathrm{i}t), \quad t \in \mathbb{R}.$$

Hence, it remains to prove that the right-hand side of the equation above tends to zero as $\epsilon \to 0$ if $|t| > a$.

Next we show that any two of the functions Ψ_α are equal in the intersection of their domains of definition.

Let $0 < \beta - \alpha < \pi$ and write

$$\gamma = \frac{\alpha + \beta}{2}, \quad \eta = \cos\frac{\beta - \alpha}{2}.$$

Note that η is positive. If $w = r\mathrm{e}^{-\mathrm{i}\gamma}$, where $r \in (a/\eta, \infty)$, then $w \in \Pi_\alpha \cap \Pi_\beta$. This follows from

$$\operatorname{Re}\left(w\mathrm{e}^{\mathrm{i}\alpha}\right) = \eta r = \operatorname{Re}\left(w\mathrm{e}^{\mathrm{i}\beta}\right).$$

[16] See Section C.1 for the definition of line integrals.

Let Γ be the curve given by $\Gamma(t) = R\mathrm{e}^{\mathrm{i}t}$, $R > 0$, $\alpha \le t \le \beta$. If $z = \Gamma(t)$ is a point of this curve and w is as above, then

$$\operatorname{Re}(-wz) = -rR\cos(t-\gamma) \le -rR\eta$$

and hence the modulus of the integrand of

$$I_R := \int_\Gamma f(z)\mathrm{e}^{-wz}\,\mathrm{d}z$$

is at most

$$C\mathrm{e}^{(a-r\eta)R}.$$

Since $r \in (a/\eta, \infty)$, we conclude that the integral I_R tends to zero as $R \to \infty$.

By Cauchy's Theorem C.1.5

$$\int_{[0,R\mathrm{e}^{\mathrm{i}\beta}]} f(z)\mathrm{e}^{-wz}\,\mathrm{d}z = I_R + \int_{[0,R\mathrm{e}^{\mathrm{i}\alpha}]} f(z)\mathrm{e}^{-wz}\,\mathrm{d}z.$$

Taking the limit $R \to \infty$ we see that

$$\Psi_\alpha(w) = \Psi_\beta(w), \quad w = r\mathrm{e}^{-\mathrm{i}\gamma},\ r > \frac{a}{\gamma}.$$

Theorem C.1.8 shows that Ψ_α and Ψ_β coincide in the intersection of their domains of definition. Thus, the difference on the right of equation (3) does not change if we replace Ψ_0 and Ψ_π by $\Psi_{\pi/2}$ if $t < -a$, and by $\Psi_{-\pi/2}$ if $t > a$. From this we infer that the difference tends to 0 as $\epsilon \to 0$. □

Theorem 3.4.5 (Rudin). *Suppose $F \in E(2a)$, F is even, nonnegative on the real axis, and*

$$\int_{-\infty}^{\infty} F^2(u)\,\mathrm{d}u < \infty.$$

Then there are even functions $G_j, H_j \in E(a)$ such that

$$F(u) = \sum_{j=1}^{\infty} |G_j(u)|^2 + u^2 \sum_{j=1}^{\infty} |H_j(u)|^2, \quad u \in \mathbb{R}. \tag{1}$$

Proof. We may assume that $a = 1$. If $F(0) = 0$, then $F(w) = w^{2k}F_0(w)$ with some positive integer k and an entire function F_0 having no zero at 0. It is easy to check that F_0 satisfies the conditions of the theorem. If we have shown that F_0 admits a decomposition analogous to (1), then we obtain the desired decomposition for F by multiplying G_j and H_j by w^k and changing the roles of G_j and H_j if k is odd. Therefore we may assume, without loss of generality, that $F(0) = 1$.

Let $\pm\lambda_1, \pm\lambda_2, \ldots$ be the zeros of F, counted with their multiplicities. Since $F \in E(2)$, Corollary C.1.16 shows that

$$n \leq \frac{\log C + 2|\lambda_n|}{\log 2}$$

with some constant $C > 0$ and hence $|\lambda_n| > cn$, for some $c > 0$. Thus,

$$\sum_{n=1}^{\infty} \frac{1}{|\lambda_n|^2} < \infty.$$

Using this and the facts that F is even and $F \in E(2)$, Hadamard's factorization Theorem C.1.19 shows that

$$F(z) = \prod_{n=1}^{\infty} \left(1 - \frac{z^2}{\lambda_n^2}\right), \quad z \in \mathbb{C}. \tag{2}$$

Let $t_1 \leq t_2 \leq \cdots$ be those positive numbers for which it_n is a zero of F and write

$$\begin{aligned} A_m(z) &= \prod_{n=1}^{m} \left(1 + \frac{z^2}{t_n^2}\right) \\ Q_m &= F/A_m \\ Q &= \lim_{m\to\infty} Q_m. \end{aligned}$$

We show that $Q \in E(2)$. It is clear that $Q_m \in E(2)$. The Paley–Wiener Theorem 3.4.4 shows that the Fourier transform of the restriction of Q_m to $\mathbb{R}$ vanishes outside $[-2, 2]$. By definition, $Q_m \to Q$ pointwise. Since $0 \leq Q_m \leq F$ on the real axis, Lebesgue's dominated convergence theorem shows that $Q_m \to Q$ in $L^2(\mathbb{R})$. Consequently, $\hat{Q}_m \to \hat{Q}$ in $L^2(\mathbb{R})$ in view of Theorem 1.8.9 and hence $\hat{Q}$ vanishes outside of $[-2, 2]$. Thus, $\hat{Q}$ is Lebesgue integrable. Taking the inverse Fourier transform of $\hat{Q}$ and using Lemma 3.4.2 we see that $Q \in E(2)$.

The function Q is nonnegative on the real axis, it is even and has no pure imaginary zeros. Therefore, the real zeros of Q have even multiplicity and the non-real zeros occur in conjugate pairs. Hence, we can write Q as $Q = F_1 F_2$ where F_1 and F_2 are subproducts of the product in (2), F_1 and F_2 are even, entire and $F_2(\overline{w}) = \overline{F_1(w)}$.

Next we show that $F_1 \in E(1)$ and $F_2 \in E(1)$. Since $|\lambda_n| > cn$, the subproducts of (2) are dominated by

$$\prod_{n=1}^{\infty} \left(1 + \left|\frac{z}{cn}\right|^2\right).$$

Using this and the product representation of the sine function from Remark C.1.20, we see that F_1 and F_2 are of exponential type. The equation $|F_1|^2 = Q$ holds on both the real and imaginary axes. On the real axis, the function Q is the inverse Fourier transform of the integrable function $\hat{Q}$, hence it is bounded. The function R defined by $R(w) := e^{iw} F_1(w)$ is obviously bounded on the real axis. Using that $Q \in E(2)$ we see that

$$|R(it)| = e^{-t}|F_1(it)|^2 = e^{-t} Q(it) \le e^{-t} C e^{2t}, \quad t > 0,$$

with some $C > 0$, i.e., R is bounded on the upper half of the imaginary axis. Since F_1 is of exponential type, R is of exponential type as well. The Phragmen–Lindelöf Theorem C.1.21 shows that R is bounded in the first and second quadrant. Applying the same argument to the lower half plane we conclude that $F_1 \in E(1)$ and $F_2 \in E(1)$.

By the definition of A_m and Q we have

$$\begin{aligned} F(z) &= Q(z) \prod_{n=1}^{\infty} \left(1 + \frac{z^2}{t_n^2}\right) \\ &= F_1(z) F_2(z) \sum_{k=0}^{\infty} c_k^2 z^{2k} \end{aligned}$$

with some nonnegative numbers c_k. If u is real, then $F_1(u)F_2(u) = |F_1(u)|^2$ and therefore

$$\begin{aligned} F(u) &= \sum_{k=0}^{\infty} |c_k u^k F_1(u)|^2 \\ &= \sum_{j=0}^{\infty} |c_{2j} u^{2j} F_1(u)|^2 + u^2 \sum_{j=0}^{\infty} |c_{2j+1} u^{2j} F_1(u)|^2, \quad u \in \mathbb{R}. \end{aligned}$$

Thus, the functions $G_j(z) := c_{2j} z^{2j} F_1(z)$ and $H_j(z) := c_{2j+1} z^{2j} F_1(z)$ have the desired properties. □

3.5 Further properties of Gaussian distributions

We now have all the necessary tools to prove further important properties of Gaussian distributions.

Theorem 3.5.1. *Let P be a polynomial on $\mathbb{R}^d$ with complex coefficients such that $P(0) = 0$. The function $f = e^P$ is the characteristic function of a d-dimensional random vector X if and only if*

$$P(t) = i(m,t) - \frac{1}{2}(Ct,t), \quad t \in \mathbb{R}^d \tag{1}$$

with some $m \in \mathbb{R}^d$ and a $d \times d$ positive semidefinite real matrix $C = (c_{ij})$.

Proof. First we show that if $f = e^P$ is a characteristic function, then the degree of P is at most two. Suppose first that $d = 1$. Since $P(0) = 0$ the constant term of P is 0. The case $P \equiv 0$ being trivial, assume that

$$P(t) = \sum_{k=1}^{n} c_k t^k, \quad c_k \in \mathbb{C},\ t \in \mathbb{R}$$

where $n \geq 1$ and $c_n \neq 0$. Since $|f(z)| = e^{\operatorname{Re} P(z)}$, $z \in \mathbb{C}$, the inequality (3.3.12.iv) implies that

$$\operatorname{Re}\left[\sum_{k=1}^{n} c_k z^k\right] \leq \sum_{k=1}^{n} c_k (\mathrm{i} \operatorname{Im} z)^k, \quad z \in \mathbb{C}.$$

Replacing here z by $s \cdot z$, $s \in \mathbb{R}$, dividing both sides by s^n and letting $s \to \infty$ we see that

$$\operatorname{Re}(c_n z^n) \leq c_n (\mathrm{i} \operatorname{Im} z)^n.$$

Putting all n-th roots $z_1, \dots z_n$ of $|c_n|/c_n$ into this inequality, we obtain

$$|c_n| \leq c_n (\mathrm{i} \operatorname{Im} z_j)^n, \quad j = 1, \dots, n.$$

Taking modulus and dividing by $|c_n|$ shows that $|\operatorname{Im} z_j| \geq 1$. Since $|z_j| = 1$, we conclude that $|\operatorname{Im} z_j| = 1$ for all $j = 1, \dots, n$. This is only possible if $n = 1$ or $n = 2$. Thus, the degree of P is at most 2. The case where d is arbitrary follows now from the fact that $s \mapsto f(s \cdot t)$ is a characteristic function of one variable for all $t \in \mathbb{R}^d$ (see, e.g., Remark 1.1.10) by using Lemma B.6.7.

Next we show that P has the form (1). We already know that

$$P(t) = \sum_{j=1}^{d} a_j t_j + \sum_{j,k=1}^{d} b_{jk} t_j t_k$$

with some a_j, $b_{jk} \in \mathbb{C}$. We may obviously assume that $b_{jk} = b_{kj}$. Theorem 1.2.9 shows that all moments of X exist. Hence, by Theorem 1.2.1,

$$\frac{\partial}{\partial t_j} f(0) = a_j = \mathrm{i}\mathbb{E}(X_j)$$

$$\frac{\partial^2}{\partial t_j \partial t_k} f(0) = a_j a_k + b_{jk} + b_{kj} = a_j a_k + 2b_{jk} = -\mathbb{E}(X_j X_k).$$

Thus, $m := -\mathrm{i}(a_1, \dots, a_d)$ is the expectation and $C := -2(b_{jk})_{j,k=1}^{d}$ is the covariance matrix of X. From this it follows that $m \in \mathbb{R}^d$ and that C is real and positive semidefinite (see page 349).

Finally, suppose that C is a positive semidefinite real matrix and $m \in \mathbb{R}^d$. Then f is the characteristic function of the random vector $Y = BX + m$ where X is standard Gaussian and B is a symmetric real matrix such that $B^2 = C$ (see 1.10.1). □

Theorem 3.5.2. *Let f be a characteristic function of a d-dimensional random vector X and P be a polynomial on $\mathbb{R}^d$. If $f = \mathrm{e}^P$ in a neighborhood of zero, then X is Gaussian.*

Proof. For fixed $t \in \mathbb{R}$ consider the characteristic function $x \mapsto f(xt)$, $x \in \mathbb{R}$. By (3.3.6.ii), $f(xt) = \mathrm{e}^{P(xt)}$ for all $x \in \mathbb{R}$ and hence $f = \mathrm{e}^P$. We can now apply Theorem 3.5.1. □

Remark 3.5.3. The function

$$f(t) = \mathrm{e}^{-\alpha|t|-|t|^\alpha}, \quad t \in \mathbb{R}$$

is a characteristic function for all $\alpha \in [2, \infty)$ (see Figure 3.4). To see this we show that f is convex on the interval $[0, \infty)$. The fact that f is a characteristic function follows then from Pólya's Theorem 3.9.11. The second derivative of f at $t > 0$ is

$$[\alpha^2(1 + t^{\alpha-1})^2 - \alpha(\alpha - 1)t^{\alpha-2}] \cdot f(t).$$

It suffices to prove that the expression in the brackets is positive which is equivalent to

$$t + t^\alpha > \sqrt{1 - 1/\alpha} \cdot t^{\frac{\alpha}{2}}.$$

This inequality holds because $t \geq t^{\frac{\alpha}{2}}$ if $0 \leq t \leq 1$ and $t^\alpha \geq t^{\frac{\alpha}{2}}$ if $t \geq 1$.

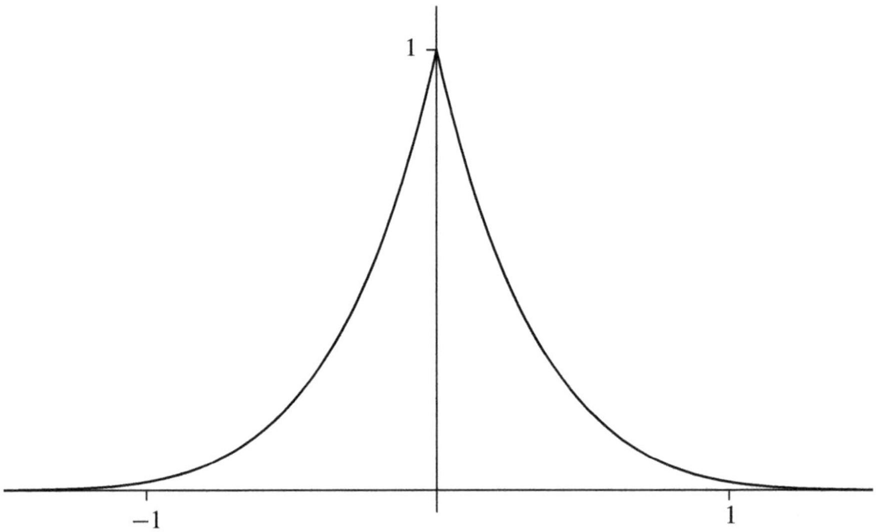

Figure 3.4. The function f from Remark 3.5.3 with $\alpha = 3$.

Theorem 3.5.4 (Lévy–Cramér). *If f_1 and f_2 are characteristic functions on $\mathbb{R}$ such that for some $a \in \mathbb{R}$ and $b > 0$*

$$f_1(t) f_2(t) = \mathrm{e}^{\mathrm{i}at - bt^2}, \quad t \in \mathbb{R}$$

then

$$f_j(t) = \mathrm{e}^{\mathrm{i}a_j t - b_j t^2}, \quad t \in \mathbb{R}$$

for some $a_j \in \mathbb{R}$ and $b_j \geq 0$, $j = 1, 2$.

Proof. Without loss of generality we may assume that $a = 0$ and $b = 1$. From Corollary 3.3.10 we infer that f_1 and f_2 are entire characteristic functions. The equation $\mathrm{e}^{-z^2} = f_1(z) f_2(z)$ holds clearly for every $z \in \mathbb{C}$. It follows that f_1 has no zeros and therefore, by Lemma C.1.11, $f_1 = \mathrm{e}^g$ for some entire function g of the form

$$g(z) = \sum_{k=1}^{\infty} c_k z^k, \quad c_k, z \in \mathbb{C}.$$

We may assume that $c_1 = g'(0) = 0$, for otherwise we replace $f_1(z)$ by $\mathrm{e}^{-c_1 z} f_1(z)$ and $f_2(z)$ by $\mathrm{e}^{c_1 z} f_2(z)$.

Applying inequality (3.3.13.ii) to the functions $f_1(z) = \mathrm{e}^{g(z)}$ and $f_2(z) = \mathrm{e}^{-g(z) - z^2}$ we obtain

$$g(\mathrm{i}y) \geq 0 \quad \text{and} \quad -g(\mathrm{i}y) + y^2 \geq 0, \quad y \in \mathbb{R}.$$

Applying now the inequality on the left of (3.3.13.i) to f_1 and f_2, we see that

$$\operatorname{Re} g(x + \mathrm{i}y) \leq g(\mathrm{i}y) \quad \text{and} \quad \operatorname{Re}\left(-g(x + \mathrm{i}y) - (x + \mathrm{i}y)^2\right) \leq -g(\mathrm{i}y) + y^2$$

for all $x, y \in \mathbb{R}$. These four inequalities imply that $|\operatorname{Re} g(z)| \leq |z|^2$, $z \in \mathbb{C}$. Using Corollary C.1.14 we obtain that $c_n = 0$ for each $n > 2$. Since f_1 is bounded on $\mathbb{R}$ and $f_1(-x) = \overline{f_1(x)}$, $x \in \mathbb{R}$, we must have $c_2 \leq 0$, completing the proof. □

Corollary 3.5.5. *If X and Y are independent d-dimensional random vectors and $X + Y$ is Gaussian, then X and Y are Gaussian as well.*

Proof. Let $a \in \mathbb{R}^d$ be arbitrary. By Lemma 1.10.3 the random variable $(a, X + Y) = (a, X) + (a, Y)$ is Gaussian. The previous theorem shows that (a, X) and (a, Y) are Gaussian. The statement follows now from Theorem 1.10.4. □

Theorem 3.5.6. *Let $f_1, \dots, f_r$ be characteristic functions on $\mathbb{R}^d$ and let δ, $\alpha_1, \dots, \alpha_r$ be positive real numbers such that*

$$\prod_{k=1}^{r} |f_k(t)|^{\alpha_k} = \mathrm{e}^{P(t)}, \quad t \in \mathbb{R}^d, \ \|t\| < \delta$$

holds for some real polynomial P. Then the distributions corresponding to the f_k's are Gaussian.

Proof. Suppose first that $d = 1$. By Theorem 3.3.14, the functions f_k are entire. As in the proofs of Theorems 3.3.14 and 1.2.13 we write

$$g_k(t) = f_k(t) f_k(-t) = |f_k(t)|^2, \quad t \in \mathbb{R}$$

and $Q(t) = P(t) + P(-t)$. Then g_k is an entire, nonnegative characteristic function. Moreover,

$$\prod_{k=1}^{r} g_k^{\alpha_k}(t) = \mathrm{e}^{Q(t)}, \quad t \in \mathbb{R}. \tag{1}$$

We will prove the statement of the theorem for the functions g_k. The statement for f_k follows then from Theorem 3.5.4.

By Lemma C.1.11 there exist entire functions h_k such that $g_k = \mathrm{e}^{h_k}$ and $h_k(0) = 0$. These functions satisfy the equation

$$\sum_{k=1}^{r} \alpha_k h_k(z) = Q(z), \quad z \in \mathbb{C}.$$

As h_k is even, its first derivative at 0 is equal to 0. Inequality (3.3.13.ii) shows that $h_k(\mathrm{i}t) \geq 0$, $t \in \mathbb{R}$. Consequently, $Q(\mathrm{i}t) \geq 0$ and $0 \leq h_k(\mathrm{i}t) \leq Q(\mathrm{i}t)$. From (3.3.13.i) and Corollary C.1.14 we conclude that h_k is a polynomial. That the distribution corresponding to g_k is Gaussian follows from Theorem 3.5.1.

Now let d be arbitrary. For each $a \in \mathbb{R}^d \setminus \{0\}$ the univariate characteristic functions $t \mapsto f_k(ta)$, $t \in \mathbb{R}$, satisfy the equation

$$\prod_{k=1}^{r} |f_k(ta)|^{\alpha_k} = \mathrm{e}^{P(ta)}, \quad t \in \mathbb{R}, \ |t| < \frac{\delta}{\|a\|}$$

hence they correspond to Gaussian distributions. The statement of the theorem follows now from Theorem 1.10.4. □

Theorem 3.5.7 (Darmois–Skitovitch). *Let $X = (X_1, \dots, X_d)$ be a random vector where $X_1, \dots, X_d$ are independent and let $a, b \in \mathbb{R}^d$. If the random variables*

$$Y_a = (a, X) = \sum_{k=1}^{d} a_k X_k, \quad Y_b = (b, X) = \sum_{k=1}^{d} b_k X_k$$

are independent, then X_k is Gaussian for all k with $a_k b_k \neq 0$.[17]

Proof. Since Y_a and Y_b are independent, Theorem 1.1.8 shows that

$$f_{(Y_a, Y_b)}(t, s) = f_{Y_a}(t) \cdot f_{Y_b}(s), \quad t, s \in \mathbb{R}.$$

[17] See also Theorem 1.10.10.

Using this and the independence of the X_k's we obtain

$$\prod_{k=1}^{d} f_k(a_k t + b_k s) = \prod_{k=1}^{d} f_k(a_k t) \prod_{k=1}^{d} f_k(b_k s), \quad t, s \in \mathbb{R}. \tag{1}$$

where f_k denotes the characteristic function of X_k. We may assume that a_k and b_k are not both zero. Changing the enumeration if necessary we also assume that $a_k b_k \neq 0$ if $1 \leq k \leq m$ and $a_k b_k = 0$ if $k > m$. If $k > m$ and $a_k \neq 0$, then the factor $f_k(a_k t)$ appears on both sides of (1). The same holds for $f_k(b_k s)$ if $k > m$ and $b_k \neq 0$. We choose $\delta > 0$ such that none of the factors in (1) is equal to zero if $t, s \in (-\delta, \delta)$. Then we have

$$\prod_{k=1}^{m} f_k(a_k t + b_k s) = \prod_{k=1}^{m} f_k(a_k t) \prod_{k=1}^{m} f_k(b_k s), \quad t, s \in (-\delta, \delta). \tag{2}$$

We show that this equation implies that X_k is Gaussian. By the same argument as in the proof of Theorem 3.5.6, we may assume that all the factors in (2) are positive. Setting $g_k = \log f_k$ we write (2) in the form

$$\sum_{k=1}^{m} g_k(a_k t + b_k s) = \sum_{k=1}^{m} g_k(a_k t) + \sum_{k=1}^{m} g_k(b_k s), \quad t, s \in (-\delta, \delta).$$

If the numbers b_k/a_k are all different, then Theorem C.9.5 shows that g_k is a polynomial in a neighborhood of 0 and hence, by Theorem 3.5.2, X_k is Gaussian. Assume that $b_1/a_1 = \cdots = b_{n_1}/a_{n_1}$ for some n_1 and $b_1/a_1 \neq b_k/a_k$ if $k > n_1$. We then write Y_a and Y_b in the form

$$Y_a = a_1 \cdot \left[X_1 + \frac{a_2}{a_1} X_2 + \cdots + \frac{a_{n_1}}{a_1} X_{n_1} \right] + \cdots$$

and

$$Y_b = b_1 \cdot \left[X_1 + \frac{b_2}{b_1} X_2 + \cdots + \frac{b_{n_1}}{b_1} X_{n_1} \right] + \cdots$$

The expressions in the brackets are equal. Applying the same arguments to the remaining X_j's, we obtain that the sums in the brackets are Gaussian. By Theorem 3.5.4 of Lévy and Cramér the summands are Gaussian as well. □

The next theorem is a multivariate analogue of the previous one.[18]

[18] We refer to [33] for further reading on results of this type and their applications.

Theorem 3.5.8. *Let $X_1, \dots, X_r$ be independent random vectors in $\mathbb{R}^d$ and let $A_k, B_k \in \mathbb{R}^{d\times d}$, $k = 1, \dots, r$, be nonsingular matrices. If the random vectors*

$$Y_A = \sum_{k=1}^{r} A_k X_k \quad \textit{and} \quad Y_B = \sum_{k=1}^{r} B_k X_k$$

are independent, then X_k is Gaussian for all k.

Proof. Proceeding in the same way as in the proof of Theorem 3.5.7, we obtain the functional equation

$$\sum_{k=1}^{r} g_k(A_k t + B_k s) = \sum_{k=1}^{r} g_k(A_k t) + \sum_{k=1}^{r} g_k(B_k s), \qquad t, s \in B^o(\delta)$$

with some $\delta > 0$, where $g_k = \log f_k$ and f_k is the characteristic function of X_k. Theorem C.9.5 shows that both sums on the right-hand side are polynomials of t and s, respectively. In view of Theorem 3.5.2, the random vectors Y_A and Y_B are Gaussian. Hence, by Theorem 3.5.4 of Lévy and Cramér, $A_k X_k$ is Gaussian as well. Since A_k is nonsingular, we infer that X_k is Gaussian. □

3.6 Fourier transformation of radial measures and functions

Radial measures and functions have the advantage of a simple structure. In this section we start exploiting radiality in more detail.

Definition 3.6.1. A function $g : \mathbb{R}^d \to \mathbb{C}$ is called *radial* if

$$g(t) = g(Ot), \quad t \in \mathbb{R}^d$$

for every orthogonal matrix $O \in \mathcal{O}(d)$. A measure $\mu \in \mathcal{M}(\mathbb{R}^d)$ is called *radial* if

$$\mu(B) = \mu(OB), \quad B \in \mathcal{B}(\mathbb{R}^d)$$

for all $O \in \mathcal{O}(d)$.

Setting $O = -I$ where I is the identity matrix, we see that radial positive definite functions are real-valued and even.

Lemma 3.6.2. *A complex-valued function g on $\mathbb{R}^d$ is radial if and only if there exists a complex-valued function h on $[0, \infty)$ such that*

$$g(t) = h(\|t\|), \quad t \in \mathbb{R}^d.$$

Proof. Assume that g is radial and define the function h by

$$h(r) := g(re_1), \quad r \in [0, \infty)$$

where $e_1 = (1, 0, \ldots, 0)$ is the first vector of the standard basis of $\mathbb{R}^d$. Let $t \in \mathbb{R}^d$ be arbitrary and choose an orthogonal matrix $O_t \in \mathcal{O}(d)$ such that $O_t t = \|t\| \cdot e_1$. We then have

$$g(t) = g(O_t t) = g(\|t\| \cdot e_1) = h(\|t\|).$$

To prove the converse statement assume that $g(t) = h(\|t\|)$ with some function h. Then $f(Ot) = h(\|Ot\|) = h(\|t\|) = f(t)$, $t \in \mathbb{R}^d$, i.e., f is radial. □

Lemma 3.6.3. *A measure $\mu \in \mathcal{M}^b(\mathbb{R}^d)$ is radial if and only if the function $\hat{\mu}$ is radial.*

Proof. If μ is radial, then

$$\int g(t)\,\mathrm{d}\mu(t) = \int g(Ot)\,\mathrm{d}\mu(t) \tag{1}$$

for all functions g of the form

$$g = \sum_{j=1}^{n} c_j \mathbf{1}_{B_j}, \quad n \in \mathbb{N},\ c_j \in \mathbb{C},\ B_j \in \mathcal{B}(\mathbb{R}^d)$$

and for all $O \in \mathcal{O}(d)$. Since these functions span a dense linear subspace of $L^1(\mu)$, we conclude that equation (1) holds for all continuous bounded functions g. In particular,

$$\hat{\mu}(Ot) = \int \mathrm{e}^{-\mathrm{i}(Ot,s)}\,\mathrm{d}\mu(s) = \int \mathrm{e}^{-\mathrm{i}(t,O^{-1}s)}\,\mathrm{d}\mu(s) = \hat{\mu}(t).$$

The converse statement can be proved in the same way by using the fact that functions of the form $t \mapsto \mathrm{e}^{\mathrm{i}(t,s)}$, $s \in \mathbb{R}^d$, span a dense linear subspace of $L^1(\mu)$. □

Next we consider radiality in terms of random vectors.

Theorem 3.6.4. *Let X be a d-dimensional random vector. The following statements are equivalent:*

(i) *X and OX have the same distribution for every orthogonal matrix $O \in \mathcal{O}(d)$;*

(ii) *the distribution of X is radial;*

(iii) *the characteristic function of X is radial;*

(iv) *for any $a \in \mathbb{R}^d$ the random variables (a, X) and $\|a\| X_1$ have the same distribution.*

Proof. The equivalence of (i) and (ii) is trivial while the equivalence of (ii) and (iii) follows from Lemma 3.6.3.

Assume that (i) holds and let $O_a \in \mathcal{O}(d)$ be such that $O_a a = \|a\| e_1$ where $e_1 = (1, 0, \ldots, 0) \in \mathbb{R}^d$. Then

$$(a, O_a^{-1} X) = (O_a a, X) = \|a\| X_1$$

so that (a, X) and $\|a\| X_1$ have the same distribution.

Finally, assume that (a, X) and $\|a\| X_1$ have the same distribution. We then have

$$f_X(a) = \mathbb{E}\left[\mathrm{e}^{\mathrm{i}(a,X)}\right] = \mathbb{E}\left[\mathrm{e}^{\mathrm{i}\|a\| X_1}\right] = f_{X_1}(\|a\|)$$

i.e., f_X is radial. □

Lemma 3.6.5. *Let $g \in \mathcal{L}^2(\mathbb{R}^d)$ be a radial function. Then the function*

$$f(t) := g * \tilde{g}(t) = \int_{\mathbb{R}^d} g(t + y) \overline{g(y)} \, \mathrm{d}y, \quad t \in \mathbb{R}^d$$

is radial, $f \in P^c(\mathbb{R}^d)$, and f vanishes at infinity.

Proof. In view of Lemma 1.5.4 it remains to prove that f is radial. Let $O \in \mathcal{O}(d)$ be an orthogonal matrix. Using the fact that the Lebesgue measure is invariant under orthogonal transformations we obtain

$$\begin{aligned} f(Ot) &= \int_{\mathbb{R}^d} g(Ot + y) \overline{g(y)} \, \mathrm{d}y \\ &= \int_{\mathbb{R}^d} g(Ot + Oy) \overline{g(Oy)} \, \mathrm{d}y \\ &= \int_{\mathbb{R}^d} g(t + y) \overline{g(y)} \, \mathrm{d}y = f(t), \quad t \in \mathbb{R}^d \end{aligned}$$

i.e., f is radial. □

Using Bessel functions, the Fourier transform of a radial function can be expressed as a univariate integral.

Theorem 3.6.6. *The Fourier transform of a radial function $f \in \mathcal{L}^1(\mathbb{R}^d)$ is given by*

$$\hat{f}(t) = \frac{(2\pi)^{d/2}}{\|t\|^{d/2-1}} \int_0^\infty r^{d/2} J_{d/2-1}(r\|t\|) f(r \cdot e) \, \mathrm{d}r, \quad t \in \mathbb{R}^d \setminus \{0\},$$

where $e \in \mathbb{R}^d$ is an arbitrary element of unit length.

Proof. Since $J_{-1/2}(x) = \sqrt{\frac{2}{\pi x}} \cos x$ the equation above is trivial if $d = 1$. Suppose that $d > 1$. By Lemma 3.6.3 the function $\hat{f}$ is radial. We may therefore assume that

$e = (1, 0, \dots, 0)$. Writing $h(r) := f(r \cdot e)$, $r \geq 0$, we have

$$\hat{f}(t) = \hat{f}(\|t\| \cdot e) = \int_{\mathbb{R}^d} \mathrm{e}^{\mathrm{i}\|t\| \cdot x_1} h(\|x\|) \, \mathrm{d}x$$
$$= \int_{\mathbb{R}} \mathrm{e}^{\mathrm{i}\|t\| \cdot x_1} \int_{\mathbb{R}^{d-1}} h\left(\sqrt{x_1^2 + \|y\|^2}\right) \mathrm{d}y \, \mathrm{d}x_1.$$

Using Corollary B.7.7 we see that the integral over $\mathbb{R}^{d-1}$ is equal to

$$\sigma(S^{d-2}) \int_0^\infty h\left(\sqrt{x_1^2 + u^2}\right) u^{d-2} \, \mathrm{d}u$$

where, in view of Proposition B.7.8,

$$\sigma(S^{d-2}) = \frac{2\pi^{(d-1)/2}}{\Gamma((d-1)/2)}.$$

Therefore, we have

$$\hat{f}(t) = \sigma(S^{d-2}) \int_{-\infty}^\infty \int_0^\infty \mathrm{e}^{\mathrm{i}\|t\| \cdot x_1} h\Big(\sqrt{x_1^2 + u^2}\Big) u^{d-2} \, \mathrm{d}u \, \mathrm{d}x_1.$$

Introducing polar coordinates (r, φ) by $x_1 = r \cos \varphi$ and $u = r \sin \varphi$, it follows

$$\hat{f}(t) = \sigma(S^{d-2}) \int_0^\infty h(r) r^{d-1} \left[\int_0^\pi \mathrm{e}^{\mathrm{i}\|t\| r \cos \varphi} \sin^{d-2} \varphi \, \mathrm{d}\varphi \right] \mathrm{d}r.$$

From Lemma C.5.3 we know that

$$J_\nu(x) = \frac{1}{\sqrt{\pi}\, \Gamma\left(\nu + \frac{1}{2}\right)} \left(\frac{x}{2}\right)^\nu \int_0^\pi \mathrm{e}^{\mathrm{i}x \cos \varphi} \sin^{2\nu} \varphi \, \mathrm{d}\varphi, \quad x \geq 0, \ \nu \geq 0$$

from which the assertion follows. □

Corollary 3.6.7. *If* $f \in \mathcal{L}^1(\mathbb{R}^3)$, *then*

$$\hat{f}(t) = \frac{4\pi}{\|t\|} \int_0^\infty r \sin(r\|t\|) f(r \cdot e) \, \mathrm{d}r, \quad t \in \mathbb{R}^3 \setminus \{0\}.$$

If f is a complex-valued function on $[0, \infty)$ such that

$$F_d(t) := f(\|t\|), \ t \in \mathbb{R}^d$$

is Lebesgue integrable, then the Fourier transform of F_d is radial. We will use the notation

$$\hat{F}_d(t) = \mathcal{F}_d(f)(\|t\|), \quad t \in \mathbb{R}^d.$$

Next we prove a recursion formula for the Fourier transformation for radial functions.[19]

[19] See [54] and [26] for more information on recursion formulae.

Theorem 3.6.8. *Let $f : [0,\infty) \to \mathbb{C}$ be such that the functions $t \mapsto f(\|t\|)$, $t \in \mathbb{R}^{d+2}$, and $t \mapsto f(\|t\|)$, $t \in \mathbb{R}^d$, are in $L^1(\mathbb{R}^{d+2})$ and $L^1(\mathbb{R}^d)$, respectively. Then we have*

$$\mathcal{F}_{d+2}(f)(r) = -\frac{1}{2\pi}\frac{1}{r}\frac{\mathrm{d}}{\mathrm{d}r}\mathcal{F}_d(f)(r), \quad r > 0. \tag{1}$$

Proof. In view of Theorem 3.6.6

$$\mathcal{F}_d(f)(s) = (2\pi)^{d/2}\int_0^\infty \tilde{J}_{d/2-1}(rs) f(r) r^{d-1}\,\mathrm{d}r, \quad s > 0 \tag{2}$$

where $\tilde{J}_\nu(x) := x^{-\nu} J_\nu(x)$. By the second equation in Proposition C.5.5 we have[20]

$$\frac{\mathrm{d}}{\mathrm{d}x}\tilde{J}_\nu(x) = -x\tilde{J}_{\nu+1}(x). \tag{3}$$

while Lemma C.5.2 shows that

$$|\tilde{J}_\nu(x)| \le C_\nu, \quad \nu \ge 0$$

with some positive constant C_ν. Using equation (3) and differentiating both sides of equation (2) with respect to s we obtain (1). The fact that interchanging differentiation and integration is permissible follows from

$$\begin{aligned}\int_0^\infty &\left|\frac{\mathrm{d}}{\mathrm{d}s}\tilde{J}_{d/2-1}(rs)\right| |f(r)| r^{d-1}\,\mathrm{d}r \\ &= s\int_0^\infty \left|\tilde{J}_{d/2}(rs)\right| |f(r)| r^{d+1}\,\mathrm{d}r \\ &\le s C_{d/2}\int_0^\infty |f(r)| r^{d+1}\,\mathrm{d}r < \infty.\end{aligned}$$

That the last integral is finite is a consequence of Corollary B.7.7 since, by assumption, the function $t \mapsto f(\|t\|)$, $t \in \mathbb{R}^{d+2}$, is in $L^1(\mathbb{R}^{d+2})$. □

Using induction on d. we immediately obtain the following corollary:

Corollary 3.6.9. *Let $f : [0,\infty) \to \mathbb{C}$ be such that the functions $t \mapsto f(\|t\|)$, $t \in \mathbb{R}^n$ are in $L^1(\mathbb{R}^n)$ for all $1 \le n \le 2d+2$. Then we have*

$$\mathcal{F}_{2d+1}(f)(r) = \frac{1}{(2\pi)^d}\sum_{j=1}^{d}\frac{(-1)^j(2d-j-1)!}{2^{d-j}(d-j)!(j-1)!}\frac{1}{r^{2d-j}}\left(\frac{\mathrm{d}}{\mathrm{d}r}\right)^j \mathcal{F}_1(f)(r)$$

and

$$\mathcal{F}_{2d+2}(f)(r) = \frac{1}{(2\pi)^d}\sum_{j=1}^{d}\frac{(-1)^j(2d-j-1)!}{2^{d-j}(d-j)!(j-1)!}\frac{1}{r^{2d-j}}\left(\frac{\mathrm{d}}{\mathrm{d}r}\right)^j \mathcal{F}_2(f)(r).$$

[20] See also Remark C.5.6.

3.7 Radial characteristic functions

Radial positive definite functions have many applications in probability and statistics. They occur for example as characteristic functions of radial distributions and as covariance functions of isotropic random fields. We have already shown (cf. Lemma 1.3.4) that the characteristic function of the radial density

$$\varphi(x) = \frac{1}{(2\pi)^{d/2}} \cdot e^{-\frac{1}{2}\|x\|^2}, \quad x \in \mathbb{R}^d$$

is given by

$$g(t) = e^{-\frac{1}{2}\|t\|^2}, \quad t \in \mathbb{R}^d.$$

We continue with further examples of radial characteristic functions.

Lemma 3.7.1. *Let U be a random vector that is uniformly distributed on the closed ball $B^c(R) \subset \mathbb{R}^d$. Then its characteristic function f_U (see Figure 3.5) is given by*

$$f_U(t) = \Gamma(d/2+1)\left(\frac{2}{R\|t\|}\right)^{d/2} J_{d/2}(R\|t\|), \quad t \in \mathbb{R}^d \setminus \{0\}.$$

Proof. The case $d = 1$ being simple, assume that $d > 1$. Theorem 3.6.6 and Corollary C.5.5 show that

$$\begin{aligned} f_U(t) &= \frac{(2\pi)^{d/2}}{\lambda_d(B^c(R))}\|t\|^{1-d/2} \int_0^R r^{d/2} J_{d/2-1}(r\|t\|)\,\mathrm{d}r \\ &= \frac{(2\pi)^{d/2}}{\lambda_d(B^c(R))}\|t\|^{-d} \int_0^{R\|t\|} x^{d/2} J_{d/2-1}(x)\,\mathrm{d}x \\ &= \frac{(2\pi)^{d/2}}{\lambda_d(B^c(R))}\|t\|^{-d} \left[x^{d/2} J_{d/2}(x)\right]_{x=0}^{R\|t\|} \end{aligned}$$

from which the lemma follows. □

Recall that a d-dimensional random vector is said to be *uniformly distributed on the d-dimensional sphere of radius R* if the distribution of X/R is equal to the measure σ_d introduced in Theorem B.7.5.

Lemma 3.7.2. *Let U be a random vector that is uniformly distributed on the d-dimensional sphere of radius R. Then its characteristic function f_U is given by*

$$f_U(t) = \Gamma(d/2)\left(\frac{2}{R\|t\|}\right)^{d/2-1} J_{d/2-1}(R\|t\|), \quad t \in \mathbb{R}^d \setminus \{0\}.$$

Proof. To simplify the notation we may assume that $R = 1$. By Lemma 3.6.3 the function f_U is radial and hence $f_U(t) = f_U(\|t\|e_1)$ where $e_1 = (1, 0, \ldots, 0) \in \mathbb{R}^d$.

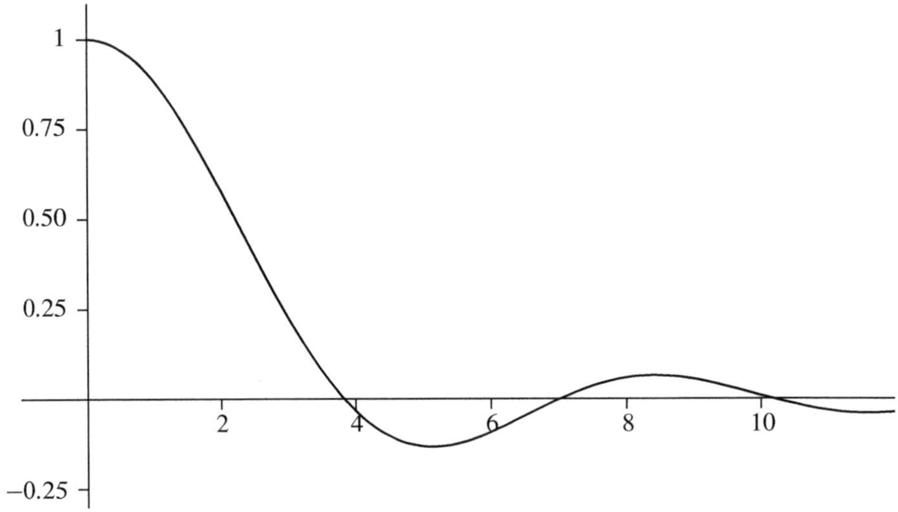

Figure 3.5. The characteristic function f_U from Lemma 3.7.1 with $d = 2$ and $R = 1$ (shown as a function of the norm).

Using this and equation (B.7.11.1) we obtain

$$\begin{aligned} f_U(t) &= \frac{1}{\sigma_{d-1}(S^{d-1})} \int_{S^{d-1}} e^{itu} \, d\sigma_{d-1}(u) \\ &= \frac{1}{\sigma_{d-1}(S^{d-1})} \int_{S^{d-1}} e^{i\|t\| u_1} \, d\sigma_{d-1}(u) \\ &= \frac{1}{\sigma_{d-1}(S^{d-1})} \int_{S^{d-2}} \int_{-1}^{1} (1-s^2)^{(d-3)/2} e^{i\|t\| s} \, ds \, d\sigma_{d-2}(v). \end{aligned}$$

The definition of $J_{d/2-1}$ (cf. Definition C.5.1 and Remark C.5.6) and Proposition B.7.8 lead immediately to the stated formula. □

Lemma 3.7.3. *The function f defined by*

$$f(t) = e^{-\|t\|}, \; t \in \mathbb{R}^d, \; d \in \mathbb{N}$$

is a characteristic function and

$$\hat{f}(t) = 2^d \pi^{(d-1)/2} \Gamma((d+1)/2) \cdot (1 + \|t\|^2)^{-(d+1)/2}, \quad t \in \mathbb{R}^d. \tag{1}$$

Proof. Theorem 3.6.6 shows that

$$\hat{f}(t) = (2\pi)^{d/2} \|t\|^{1-d/2} \int_0^\infty r^{d/2} J_{d/2-1}(r\|t\|) e^{-r} \, dr, \quad t \in \mathbb{R}^d \setminus \{0\}.$$

Using the formula

$$J_{d/2-1}(x) = \left(\frac{x}{2}\right)^{d/2-1} \sum_{k=0}^{\infty} \frac{(-1)^k}{k!\,\Gamma(k+d/2)} \cdot \left(\frac{x}{2}\right)^{2k}$$

(cf. Theorem C.5.4) we obtain

$$\hat{f}(t) = 2\pi^{d/2} \sum_{k=0}^{\infty} \frac{(-1)^k \|t\|^{2k}}{2^{2k} k!\Gamma(k+d/2)} \int_0^{\infty} r^{2k+d-1} \mathrm{e}^{-r}\, \mathrm{d}r$$

for all t where the power series above is absolutely convergent. The definition of the Gamma function and Legendre's duplication formula (cf. Proposition C.4.5) show that the integral above is equal to

$$\frac{1}{\sqrt{\pi}} 2^{2k+d-1} \Gamma(k+d/2)\Gamma(k+(d+1)/2)$$

and hence

$$\hat{f}(t) = 2^d \pi^{(d-1)/2} \sum_{k=0}^{\infty} \frac{(-1)^k \Gamma(k+(d+1)/2)}{k!} \cdot \|t\|^{2k}, \quad \|t\| < 1.$$

Using this and Proposition B.1.3 we see that equation (1) holds whenever $\|t\| < 1$ while Theorem 3.3.6 implies that it holds for all $t \in \mathbb{R}^d$.

The fact that f is a characteristic function follows from $\hat{f} \geq 0$. □

Corollary 3.7.4. *Let P be a polynomial with real coefficients such that $P(0) = 1$ and $P(z) \neq 0$ if $\operatorname{Re} z \geq 0$. Then*

$$t \mapsto 1/P(\|t\|)^b, \quad t \in \mathbb{R}^d$$

is a characteristic function for all $b > 0$ and all $d \in \mathbb{N}$.

Proof. Since P has no zero with nonnegative real part, it is the product of polynomials of the form $Q(t) = 1 + pt + qt^2$ where $p, q \geq 0$. Thus, it suffices to show that $t \mapsto 1/Q(\|t\|)^b$ is positive definite. We have

$$\Gamma(b) = \int_0^{\infty} x^{b-1} \mathrm{e}^{-x}\, \mathrm{d}x = s^b \int_0^{\infty} u^{b-1} \mathrm{e}^{-su}\, \mathrm{d}u, \quad b, s > 0.$$

Setting $s = Q(\|t\|)$ we obtain

$$\frac{1}{Q(\|t\|)^b} = \frac{1}{\Gamma(b)} \int_0^{\infty} u^{b-1} \mathrm{e}^{-Q(\|t\|)u}\, \mathrm{d}u, \quad b > 0,\ t \in \mathbb{R}^d. \tag{1}$$

Using Lemma 3.7.3 and Lemma 1.3.4 we see that

$$t \mapsto \mathrm{e}^{-Q(\|t\|)u} = \mathrm{e}^{-u} \mathrm{e}^{-up\|t\|} \mathrm{e}^{-uq\|t\|^2}, \quad t \in \mathbb{R}^d$$

is positive definite if $u \geq 0$. Therefore, the Corollary follows from Lemma 1.5.7. □

Taking the inverse Fourier transform in equation (3.7.3.1), we see that the function $t \mapsto (1 + \|t\|^2)^{-(d+1)/2}$ is the characteristic function of an absolutely continuous distribution. The next theorem states more.

Theorem 3.7.5. *The function*

$$f(t) = \frac{1}{(1 + \|t\|^2)^b}, \quad t \in \mathbb{R}^d$$

is a characteristic function for all $b > 0$. If $b > \frac{d}{2}$, then[21]

$$\hat{f}(x) = \frac{2^{1-b}}{\Gamma(b)} \|x\|^{b-d/2} K_{b-d/2}(\|x\|), \quad x \in \mathbb{R}^d \setminus \{0\}.$$

Proof. The first statement follows from Corollary 3.7.4.

If $b > d/2$, then f is Lebesgue integrable (cf. Corollary B.7.7). By equation (3.7.4.1) we have

$$f(t) = \frac{1}{\Gamma(b)} \int_0^\infty u^{b-1} e^{-(1+\|t\|^2)u}\, du.$$

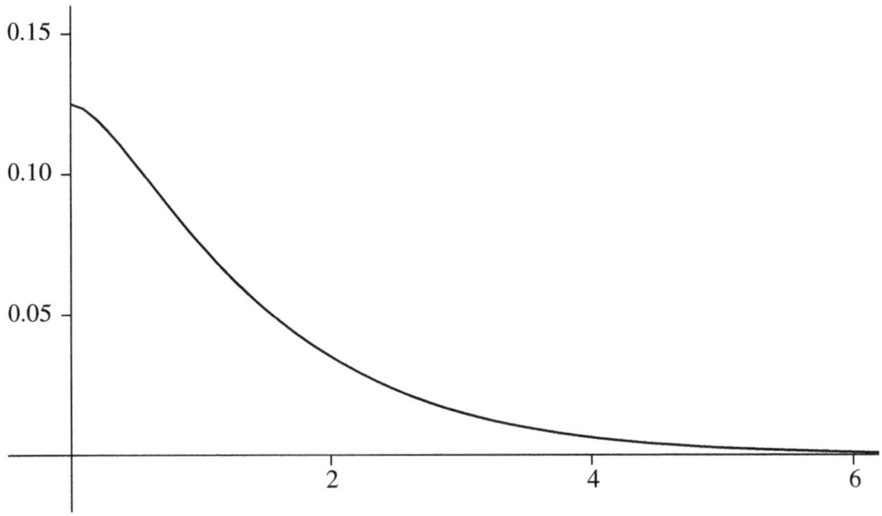

Figure 3.6. The function $\hat{f}$ from Theorem 3.7.5 with $b = 3$ and $d = 4$ (shown as a function of the norm).

[21] Recall that K_ν denotes the modified Bessel function (cf. Definition C.5.7). The function $\hat{f}$ is shown in Figure 3.6.

Consequently,

$$\begin{aligned}\hat{f}(x) &= (2\pi)^{-d/2} \int_{\mathbb{R}^d} f(t) \mathrm{e}^{-\mathrm{i}(t,x)} \, \mathrm{d}t \\ &= (2\pi)^{-d/2} \frac{1}{\Gamma(b)} \int_{\mathbb{R}^d} \int_0^\infty u^{b-1} \mathrm{e}^{-(1+\|t\|^2)u} \mathrm{e}^{-\mathrm{i}(t,x)} \, \mathrm{d}u \, \mathrm{d}t \\ &= (2\pi)^{-d/2} \frac{1}{\Gamma(b)} \int_0^\infty u^{b-1} \mathrm{e}^{-u} \int_{\mathbb{R}^d} \mathrm{e}^{-\|t\|^2 u} \mathrm{e}^{-\mathrm{i}(t,x)} \, \mathrm{d}t \, \mathrm{d}u.\end{aligned}$$

Using Lemma 1.3.4 we obtain

$$\hat{f}(x) = \frac{1}{2^{d/2}\Gamma(b)} \int_0^\infty u^{b-d/2-1} \mathrm{e}^{-u} \mathrm{e}^{-\|x\|^2/(4u)} \, \mathrm{d}u.$$

From Lemma C.5.8 we know that

$$K_\nu(r) = \frac{1}{2a^\nu} \int_0^\infty \mathrm{e}^{-\frac{r}{2}\cdot(\frac{u}{a}+\frac{a}{u})} u^{\nu-1} \, \mathrm{d}u, \quad a, r > 0, \ \nu \in \mathbb{R}.$$

Setting $r = \|x\|$, $a = \|x\|/2$, and $\nu = b - d/2$ leads to

$$\begin{aligned}K_{b-d/2}(\|x\|) &= \frac{1}{2} \left(\frac{\|x\|}{2}\right)^{d/2-b} \int_0^\infty \mathrm{e}^{-u} \mathrm{e}^{-\|x\|^2/(4u)} u^{b-d/2-1} \, \mathrm{d}u \\ &= 2^{b-1}\Gamma(b) \|x\|^{d/2-b} \hat{f}(x), \quad x \neq 0,\end{aligned}$$

from which the second statement follows. □

Next we compute the self-convolution of the indicator function of a ball.

Lemma 3.7.6 (Euclid's hat). *The function*[22]

$$g_d := \frac{1}{\lambda_d(B^c(1/2))} \cdot \mathbf{1}_{B^c(1/2)} * \mathbf{1}_{B^c(1/2)}$$

is a radial characteristic function on $\mathbb{R}^d$ *and we have*

$$g_d(t) = \begin{cases} \dfrac{d\Gamma(d/2)}{\sqrt{\pi}\,\Gamma((d+1)/2)} \displaystyle\int_{\|t\|}^1 (1-x^2)^{(d-1)/2} \, \mathrm{d}x, & t \in B^c(1) \\ 0, & t \in \mathbb{R}^d \setminus B^c(1). \end{cases}$$

The distribution corresponding to g_d *is absolutely continuous with density (see Figure 3.7)*

$$p_d(x) = \frac{\Gamma(d/2+1)}{\pi^{d/2}} \frac{J_{d/2}^2(\|x\|/2)}{\|x\|^d}, \quad x \in \mathbb{R}^d \setminus \{0\}.$$

---

[22] Scale mixtures of Euclid's hat have been investigated in [20].

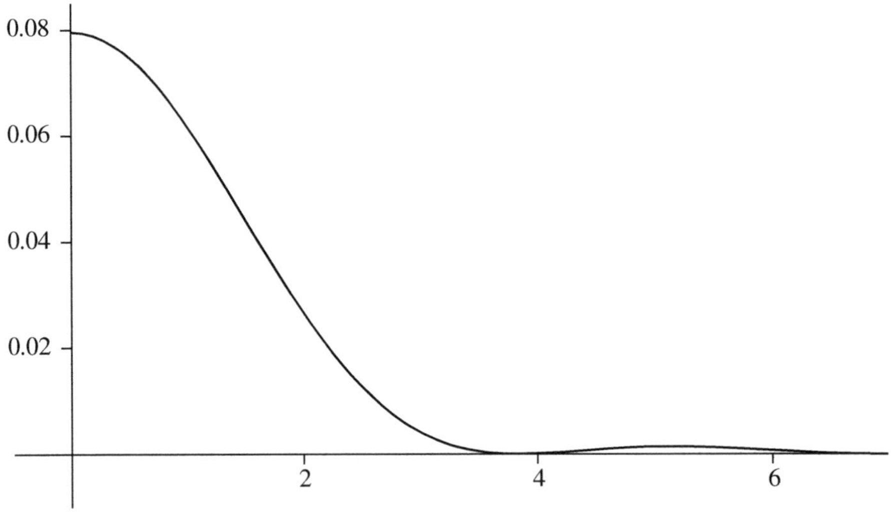

Figure 3.7. The density p_d from Lemma 3.7.6 with $d = 2$ (shown as a function of the norm).

Proof. It follows from Lemma 3.6.5 that g_d is a radial characteristic function. To prove the remaining statements we may assume that $d \geq 2$.[23] We have

$$q(t) := \mathbf{1}_{B^c(1/2)} * \mathbf{1}_{B^c(1/2)}(t) = \int_{\mathbb{R}^d} \mathbf{1}_{B^c(1/2)}(t+y)\mathbf{1}_{B^c(1/2)}(y)\,\mathrm{d}y$$
$$= \lambda_d(B^c(1/2) \cap (B^c(1/2) - t)).$$

If $\|t\| > 1$, then the measure above is zero. Assume that $\|t\| \leq 1$ and let $e := (1, 0, \ldots, 0)$. Since q is radial, we have $q(t) = q(-\|t\| \cdot e)$. Using the fact that the set $B^c(1/2) \cap (B^c(1/2) + \|t\| \cdot e)$ is symmetric with respect to the hyperplane $\{y \in \mathbb{R}^d : y_1 = \|t\|/2\}$ we obtain

$$q(t) = 2\lambda_d\left(\left\{y \in \mathbb{R}^d : \|t\|/2 \leq y_1 \leq 1/2,\ \|y\| \leq 1/2\right\}\right)$$
$$= 2\int_{\|t\|/2}^{1/2} \lambda_{d-1}\left(B\left(\sqrt{\frac{1}{4} - s^2}\right)\right) \mathrm{d}s.$$

By Proposition B.7.9

$$q(t) = \frac{\pi^{(d-1)/2}}{2^{d-2}\Gamma((d+1)/2)} \int_{\|t\|/2}^{1/2} (1-4s^2)^{\frac{d-1}{2}}\,\mathrm{d}s$$
$$= \frac{\pi^{(d-1)/2}}{2^{d-1}\Gamma((d+1)/2)} \int_{\|t\|}^{1} (1-x^2)^{\frac{d-1}{2}}\,\mathrm{d}x$$

from which the representation of g_d follows.

[23] We already considered the case $d = 1$ in Remark 1.5.5.

From Lemma 3.7.1 we conclude that the Fourier transform of $\mathbf{1}_{B^c(1/2)}$ is given by

$$\hat{\mathbf{1}}_{B^c(1/2)}(x) = \frac{J_{d/2}(\|x\|/2)}{(2\|x\|)^{d/2}}.$$

Thus, the last statement is an immediate consequence of Theorem 1.3.6 and equation (1.8.2.iii). □

Remark 3.7.7. Write the function g_d in the form $g_d(t) = h_d(\|t\|)$. For $0 \le s \le 1$ we then have $h_1(s) = 1 - s$ and

$$h_2(s) = \frac{2}{\pi}(\arccos s - s(1-s^2)^{1/2})$$

(see Figure 3.8).

Partial integration of the integral in Lemma 3.7.6 gives the recursion formula[24]

$$h_d(s) = h_{d-2}(s) - \frac{\Gamma(d/2)}{\sqrt{\pi}\,\Gamma((d+1)/2)} s(1-s^2)_+^{(d-1)/2}, \quad s \in \mathbb{R}, \quad d \ge 3.$$

The derivative

$$h_n'(s) = -\frac{d\,\Gamma(d/2)}{\sqrt{\pi}\,\Gamma((d+1)/2)}(1-s^2)_+^{(d-1)/2}$$

is nonpositive and nondecreasing, so that h_d is convex and nonincreasing for all $d \in \mathbb{N}$.[25]

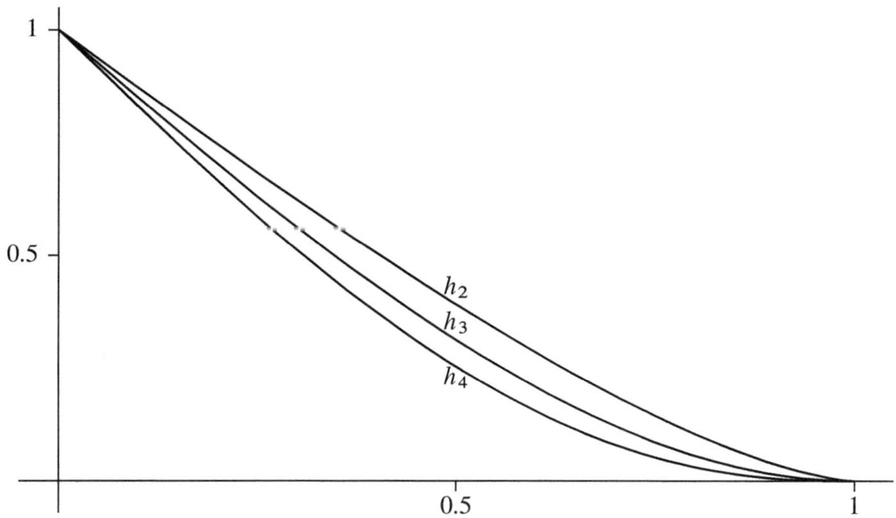

Figure 3.8. The functions h_2, h_3 and h_4 from Remark 3.7.7.

[24] See also the proof of Proposition C.1.10.

[25] More information on positive definite functions related to Euclid's hat can be found in [20].

Corollary 3.7.8. *The function φ_d defined by*

$$\varphi_d(t) = (1 - \|t\|)_+^{\lfloor \frac{d+2}{2} \rfloor}, \quad t \in \mathbb{R}^d$$

is a characteristic function.

Proof. If d is even, then $\varphi_d(t) = \varphi_{d+1}(t, 0)$, $t \in \mathbb{R}^d$. Therefore, it suffices to consider the case where $d = 2n + 1$, $n \in \mathbb{N}$. Lemma 3.7.6 shows that the function

$$g_d(t) = c\mathbf{1}_{B^c(1)}(t) P_n(\|t\|), \quad t \in \mathbb{R}^d$$

is positive definite, where c is some positive constant and

$$P_n(s) = \int_s^1 (1 - x^2)^n \, dx, \quad s \in \mathbb{R}.$$

By Proposition C.1.10

$$P_n(s) = (1 - s)^{n+1} Q_n(s),$$

where Q_n is a polynomial of degree n, having only zeros with negative real part. Using the definition of P_n we see that

$$(1 - \|t\|)_+^{n+1} = \frac{1}{c Q_n(\|t\|)} \cdot g_d(t), \quad t \in \mathbb{R}^d.$$

Thus, the statement follows from Corollary 3.7.4. □

3.8 Schoenberg's theorems on radial characteristic functions

In this section we first characterize radial characteristic functions on $\mathbb{R}^d$ (Theorem 3.8.2). Then we investigate what happens if the dimension d tends to infinity (Theorem 3.8.5).

3.8.1. We will use the notation

$$\Omega_d(r) = \Gamma(d/2) \left(\frac{2}{r}\right)^{d/2-1} J_{d/2-1}(r), \quad r > 0 \tag{1}$$

and $\Omega_d(0) = 1$ (see Figure 3.9). By Lemma 3.7.2, $t \mapsto \Omega_d(\|t\|)$ is the characteristic function of a random vector that is uniformly distributed on S^{d-1}.

Theorem 3.8.2 (Schoenberg). *A real-valued function f on $\mathbb{R}^d$ is a radial characteristic function if and only if $f(t) = \psi(\|t\|)$ where*

$$\psi(r) = \int_0^\infty \Omega_d(rs) \, d\mu(s) \tag{1}$$

for some probability measure μ on $[0, \infty)$.

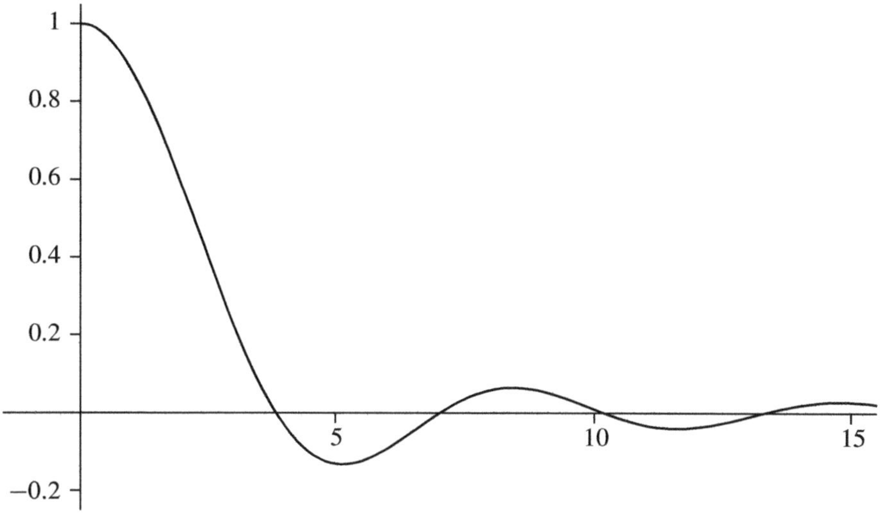

Figure 3.9. The function Ω_4 from equation (3.8.1.1).

Proof. Since $t \mapsto \Omega_d(\|t\|s)$ is a positive definite function on $\mathbb{R}^d$ for all s, the function $f(t) := \psi(\|t\|)$, where ψ is defined by (1), is positive definite as well. Thus, f is a characteristic function by Bochner's Theorem 1.7.3. The other direction follows from the fact that f is the characteristic function of the random vector RX where R is a random variable with distribution μ and X is a random vector uniformly distributed on S^{d-1} and independent of R (cf. Theorem 1.1.6). □

As the proof of the preceding theorem shows, we have the following characterization of random vectors with a radial distribution.

Corollary 3.8.3. *A d-dimensional random vector Y has a radial distribution if and only if there exist a nonnegative random variable R and a random vector X which is uniformly distributed on S^{d-1} such that R and X are independent and RX has the same distribution as Y.*

The set of all characteristic functions ϕ on $\mathbb{R}$ such that $t \mapsto \phi(\|t\|)$ is a characteristic function on $\mathbb{R}^d$ is denoted by Φ_d. If $\phi \in \Phi_d$ for all d, then we write $\phi \in \Phi_\infty$. Notice that

$$\Phi_\infty \subset \cdots \subset \Phi_2 \subset \Phi_1 .$$

Lemma 3.8.4. *The relation*

$$\lim_{d\to\infty} \Omega_d(r\sqrt{d}\,) = \mathrm{e}^{-r^2/2}$$

holds uniformly for $r \in \mathbb{R}$.

Proof. Using the definition of the Bessel function $J_{d/2-1}$ (see Definition C.5.1) it is easy to see that $r \mapsto \Omega_d(r\sqrt{d}\,)$, $r \in \mathbb{R}$, is the characteristic function corresponding to the density

$$p_d(x) = \frac{\Gamma(d/2)}{\sqrt{\pi d}\,\Gamma((d-1)/2)} \left(1 - \frac{x^2}{d}\right)^{(d-3)/2} \mathbf{1}_{(-\sqrt{d},\sqrt{d})}, \quad x \in \mathbb{R}.$$

In view of Stirling's formula (cf. Corollary C.4.10)

$$\lim_{d\to\infty} \frac{\Gamma(d/2)}{\sqrt{\pi d}\,\Gamma((d-1)/2)} = 1.$$

Using this and the relation

$$\lim_{d\to\infty} \left(1 - \frac{x^2}{d}\right)^{(d-3)/2} = \mathrm{e}^{-x^2/2}$$

we see that

$$\lim_{d\to\infty} p_d(x) = p_\infty(x) := \frac{1}{\sqrt{2\pi}} \mathrm{e}^{-x^2/2}, \quad x \in \mathbb{R}.$$

The characteristic function corresponding to the density p_∞ is $r \mapsto \mathrm{e}^{-r^2/2}$. Therefore, the lemma follows from Proposition 1.6.6. □

Theorem 3.8.5 (Schoenberg). *A characteristic function ϕ on $\mathbb{R}$ belongs to Φ_∞ if and only if*

$$\phi(r) = \int_0^\infty \mathrm{e}^{-r^2 s^2/2}\, \mathrm{d}\mu(s), \quad r \in \mathbb{R}$$

for some probability measure μ on $[0, \infty)$.

Proof. Since $t \mapsto \mathrm{e}^{-\|t\|^2 s^2/2}$ is a characteristic function on $\mathbb{R}^d$ for all s and for all d (cf. Lemma 1.3.4), the same argument as in the proof of Theorem 3.8.2 shows that $\phi \in \Phi_\infty$.

Next assume that $\phi \in \Phi_d$ for all d. Then for each d there exists a probability measure μ_d on $[0, \infty)$ such that

$$\phi(r) = \int_0^\infty \Omega_d(rs\sqrt{d}\,)\, \mathrm{d}\mu_d(s), \quad r \in \mathbb{R}. \tag{1}$$

In view of Theorem E.1.13, the sequence $\{\mu_d\}_{d=1}^\infty$ contains a subsequence $\{\mu_{d_k}\}_{k=1}^\infty$ converging vaguely to some finite nonnegative measure μ satisfying $\mu(\mathbb{R}) \le 1$.

Let $\epsilon > 0$ and $t \in \mathbb{R} \setminus \{0\}$ be arbitrary. Equation (1) and Lemma 3.8.4 imply

$$\phi(r) = \int_0^\infty \mathrm{e}^{-r^2 s^2/2}\, \mathrm{d}\mu_d(s) + \sigma_d(r)$$

where $|\sigma_d| < \epsilon$ for sufficiently large values of d. This implies that

$$\phi(r) = \int_0^a e^{-r^2 s^2/2} \, d\mu_d(s) + \rho_{d,a}(r)$$

where $|\rho_{d,a}| < 2\epsilon$ whenever both a and d are sufficiently large. If a is a continuity point of μ, letting $d \to \infty$ we get

$$\phi(r) = \int_0^a e^{-r^2 s^2/2} \, d\mu(s) + \delta_a(r)$$

where $|\delta_a| < 2\epsilon$. Finally, letting $a \to \infty$ and then $\epsilon \to 0$ we obtain the desired representation of ϕ.[26] □

3.9 Convex and completely monotone functions

As we will see in this section, sufficient conditions for positive definiteness can be formulated in terms of convexity. Convexity conditions can be checked efficiently by taking derivatives and checking for nonnegativity. Using numerical algorithms this leads to simple and powerful tests for positive definiteness. One of the main results of the present section is Theorem 3.9.8 which describes the connection between radial characteristic functions and completely monotone functions.

Definition 3.9.1. For $a \in \mathbb{R}$ and $n \in \mathbb{N}_0$ we denote by C_n^a the set of all functions $g : [a, \infty) \to \mathbb{R}$ having the following properties:

(i) g is bounded;

(ii) g is n-times differentiable;[27]

(iii) $g^{(n)}$ is monotone.

Note that C_0^a is the set of functions which are bounded and monotone.

Lemma 3.9.2. *For all $n \in \mathbb{N}$ we have*

$$C_n^a \subset C_{n-1}^a \subset \cdots \subset C_0^a.$$

Moreover, the derivatives $g^{(k)}$, $1 \le k \le n$, of a function $g \in C_n^a$ are alternately nonnegative and decreasing or nonpositive and increasing.

Proof. Assume that $g \in C_n^a$ for some $n \ge 1$. Without loss of generality we may further assume that $g^{(n)}$ is decreasing. We prove that $g^{(n)}$ is nonnegative and $g^{(n-1)}$ is nonpositive. The lemma follows then by induction on n.

[26] See also the proof of Bernstein's Theorem 3.9.6 for a similar argument.

[27] We consider differentiability from the right at a.

Since $g^{(n)}$ is monotone the limit

$$\delta_n := \lim_{x\to\infty} g^{(n)}(x)$$

exists. We show that $\delta_n = 0$. If $\delta_n > 0$, then there exists $A_n > a$ such that $g^{(n)}(x) > \delta_n/2$ whenever $x \geq A_n$. Thus, the function $g^{(n-1)}$ is increasing on the interval $[A_n, \infty)$. Let $x \geq A_n$. By the mean value theorem there exists $\xi \in [x, x+1]$ such that

$$g^{(n-1)}(x+1) = g^{(n-1)}(x) + g^{(n)}(\xi) > g^{(n-1)}(x) + \delta_n/2.$$

This relation shows that $\lim_{x\to\infty} g^{(n-1)}(x) = \infty$. Repeating this argument we obtain that $\lim_{x\to\infty} g(x) = \infty$, contradicting our assumption. In the same way we see that $\delta_n < 0$ is not possible and hence $\delta_n = 0$. Using this and the fact that $g^{(n)}$ is decreasing, we conclude that $g^{(n)}$ must be nonnegative on $[a, \infty)$. This implies that $g^{(n-1)}$ is increasing. The same argument as above shows that $\lim_{x\to\infty} g^{(n-1)}(x) = 0$ and hence $g^{(n-1)}$ is nonpositive. □

Lemma 3.9.3. *If $n \in \mathbb{N}$ and $g \in C_n^a$, then $\lim_{x\to\infty} x^k g^{(k)}(x) = 0$ for all $k = 1, \ldots, n$.*

Proof. Lemma 3.9.2 shows that g is monotone and hence the limit $\delta_0 := \lim_{x\to\infty} g(x)$ exists. Replacing g by $g - \delta_0$ we may suppose that $\delta_0 = 0$. Without loss of generality we may further assume that g is decreasing. Then $g \geq 0,\ g' \leq 0$ and g' is increasing. We consider first the case $n = 1$. For $0 < c < 1$ we have

$$g(cx) = -\int_{cx}^{\infty} g'(u)\,\mathrm{d}u \geq -\int_{cx}^{x} g'(u)\,\mathrm{d}u \geq -(1-c)xg'(x) \geq 0 \qquad (1)$$

whenever $cx \geq a$, showing that $\lim_{x\to\infty} xg'(x) = 0$.

Assume that the lemma is true for some n and let $g \in C_{n+1}^a$. Since $C_{n+1}^a \subset C_n^a$ we have $g^{(n)} \in C_1^a$. Applying the relation (1) for $g^{(n)}$ instead of g we see that $\lim_{x\to\infty} x^{n+1} g^{(n+1)}(x) = 0$. □

Theorem 3.9.4. *Let $n \in \mathbb{N}_0$ and $g : [0, \infty) \to \mathbb{R}$ be an n-times differentiable bounded function such that $g^{(n)}$ is convex or concave. Then*[28]

$$\lim_{x\to\infty} x^k g^{(k)}(x) = 0, \quad k = 1, \ldots, n+1 \qquad (1)$$

[28] This result is essentially due to P. Lévy [40] though his formulation and proof is slightly different from ours.

where $g^{(n+1)}$ denotes the right-hand derivative.[29] *Moreover, the function*

$$F(u) := \frac{(-1)^{n+1}}{(n+1)!} \int_u^\infty v^{n+1} \, \mathrm{d}g^{(n+1)}(v), \quad u \in [0,\infty)$$

is monotone, bounded and

$$g(x) = \int_0^\infty \left(1 - \frac{x}{u}\right)_+^{n+1} \mathrm{d}F(u) + \lim_{u\to\infty} g(u), \quad x \in [0,\infty).$$

Proof. It suffices to consider the case where $g^{(n)}$ is convex. The same argument as in the proof of Lemma 3.9.3 shows that we may suppose $\lim_{u\to\infty} g(u) = 0$. By Theorem B.4.4 there exists $a \geq 0$ such that $g^{(n)}$ is monotone on $[a,\infty)$. Consequently, by Lemma 3.9.3, the relation (1) holds if $k \leq n$.

Since $g^{(n)}$ is convex it is absolutely continuous and $g^{(n+1)}$ is increasing (cf. Theorem B.4.7 and Theorem B.4.3). Consequently, we may replace g by $g^{(n)}$ in relation (3.9.3.1). This leads to the inequality

$$g^{(n)}(cx) \geq -(1-c)xg^{(n+1)}(x) \geq 0, \quad cx \geq a$$

from which (1) with $k = n+1$ follows.

Using the relation (1) repeated integration by parts (cf. relation (B.5.6.1)) gives

$$g(0) = -\int_0^\infty g'(v)\,\mathrm{d}v = \cdots = \frac{(-1)^n}{(n+1)!} \int_0^\infty v^{n+1} \, \mathrm{d}g^{(n+1)}(v).$$

This shows that the integral on the right-hand side converges. Thus, the integral in the definition of F converges as well. It is clear that F is monotone and bounded. Integrating by parts again and using Theorem B.5.4 we obtain

$$\begin{aligned} g(x) = -\int_x^\infty g'(u)\,\mathrm{d}u &= \frac{(-1)^n}{(n+1)!} \int_x^\infty (u-x)^{n+1} \, \mathrm{d}g^{(n+1)}(u) \\ &= \int_x^\infty (u-x)^{n+1} \frac{1}{u^{n+1}} \, \mathrm{d}F(u) \\ &= \int_x^\infty \left(1 - \frac{x}{u}\right)^{n+1} \mathrm{d}F(u) \\ &= \int_0^\infty \left(1 - \frac{x}{u}\right)_+^{n+1} \mathrm{d}F(u). \end{aligned}$$

□

Definition 3.9.5. An infinitely differentiable function $g : (0,\infty) \to \mathbb{R}$ is called *completely monotone* if

$$(-1)^n g^{(n)}(x) \geq 0$$

for all $n \in \mathbb{N}_0$ and $x > 0$. A continuous function $g : [0,\infty) \to \mathbb{R}$ is called *completely monotone* if $g|_{(0,\infty)}$ is completely monotone.

[29] That the right-hand derivative exists follows from Theorem B.4.3.

The function $x \mapsto e^{-ux}$ is completely monotone for all $u \in [0, \infty)$. The next theorem shows that this function is the prototype of completely monotone functions.

Theorem 3.9.6 (Hausdorff–Bernstein–Widder). *An infinitely differentiable function $g : [0, \infty) \to \mathbb{R}$ is completely monotone if and only if it admits the representation*

$$g(x) = \int_0^\infty e^{-ux} \, d\mu(u), \quad x \in [0, \infty)$$

where μ is a finite nonnegative measure on $[0, \infty)$.

Proof. It is clear that g is completely monotone if it admits the representation above. It follows from Theorem 3.9.4 that for all $n \in \mathbb{N}$ there exists a finite nonnegative measure on $[0, \infty)$ such that

$$g(x) = \int_0^\infty \left(1 - \frac{xv}{n}\right)_+^n \, d\mu_n(v), \quad x \geq 0. \tag{1}$$

Since $g(0) = \mu_n([0, \infty))$, some subsequence $\{\mu_{n_k}\}$ converges weakly to a finite nonnegative measure μ (cf. Theorem E.1.13 and Theorem E.1.12). Moreover,

$$\lim_{n \to \infty} \left(1 - \frac{xv}{n}\right)_+^n = e^{-xv}$$

where for fixed x the convergence is uniform in v. Using this and taking the limit along the subsequence $\{\mu_{n_k}\}$ in equation (1) we obtain the desired representation of g.[30] □

Corollary 3.9.7. *An infinitely differentiable function $g : (0, \infty) \to \mathbb{R}$ is completely monotone if and only if it admits the representation*

$$g(x) = \int_0^\infty e^{-ux} \, d\mu(u), \quad x \in [0, \infty)$$

where μ is a nonnegative measure on $[0, \infty)$.

Proof. It is easy to show that g is completely monotone if it admits the representation above (cf. Lemma C.7.4).

To prove the converse, assume that g is completely monotone on $(0, \infty)$. Then for each $a > 0$ the function $x \mapsto g(x + a)$ is completely monotone on $[0, \infty)$. Hence, by Theorem 3.9.6, there is a finite nonnegative measure μ_a on $[0, \infty)$ such that

$$g(x + a) = \int_0^\infty e^{-ux} \, d\mu_a(u), \quad x \in [0, \infty).$$

[30] We used the same idea as in the proof of Schoenberg's Theorem 3.8.5. Here the situation is simpler since we have uniform convergence of the integrands.

We define the measure μ by $\mathrm{d}\mu(x) := \mathrm{e}^{ax}\,\mathrm{d}\mu_a(x)$. We then have

$$g(x) = \int_0^\infty \mathrm{e}^{-ux}\,\mathrm{d}\mu(u) \tag{1}$$

for all $x > a$. Using this and the uniqueness of the Laplace transform (cf. Theorem C.7.5) we conclude that $\mathrm{e}^{ax}\,\mathrm{d}\mu_a(x) = \mathrm{e}^{bx}\,\mathrm{d}\mu_b(x)$ for all $b > 0$. Thus, μ does not depend on a and therefore equation (1) remains valid for all $x > 0$. □

Combining the Hausdorff–Bernstein–Widder Theorem 3.9.6 and Schoenberg's Theorem 3.8.5 we obtain the following characterization.

Theorem 3.9.8. *For a continuous function $g : [0,\infty) \to \mathbb{R}$ the following properties are equivalent:*

(i) $g(\|\cdot\|)$ *is positive definite on every* $\mathbb{R}^d$*;*

(ii) $g(\sqrt{\cdot})$ *is completely monotone on* $[0,\infty)$*;*

(iii) g *admits the representation*

$$g(r) = \int_0^\infty \mathrm{e}^{-r^2 s}\,\mathrm{d}\mu(s), \quad r \in [0,\infty),$$

where μ is finite nonnegative measure on $[0,\infty)$.

Theorem 3.9.9 (Askey). *Let $d \in \mathbb{N}$, $k := \lfloor d/2 \rfloor$ and let $g : [0,\infty) \to \mathbb{R}$ be a continuous function such that*

(i) $g(0) = 1$*;*

(ii) $\lim_{r\to\infty} g(r) = 0$*;*

(iii) $(-1)^k g^{(k)}$ *is convex.*

Then the function f defined by

$$f(t) := g(\|t\|), \quad t \in \mathbb{R}^d$$

is a characteristic function in $\mathbb{R}^d$. Moreover, there exists a probability measure ν on $(0,\infty)$ such that

$$f(t) = \int_0^\infty \varphi_d(rt)\,\mathrm{d}\nu(r), \quad t \in \mathbb{R}^d$$

where φ_d is the characteristic function defined in Corollary 3.7.8.[31]

Proof. The function g is obviously bounded. Since $(-1)^k g^{(k)}$ is convex, the function $(-1)^k g^{(k+1)}$ is increasing where $g^{(k+1)}$ denotes the right-hand derivative. The statements of the theorem follow immediately from Theorem 3.9.4 by setting $\mathrm{d}\nu(r) := \mathrm{d}(-F(1/r))$ and noting that the function F is increasing. □

[31] Generalizations of Askey's theorem as well as further references can be found in [22].

Definition 3.9.10. An even continuous function $f : \mathbb{R} \to [0, \infty)$ with $f(0) = 1$ is said to be of *Pólya-type* if f is convex on $(0, \infty)$ and $\lim_{t\to\infty} f(t) = 0$.

The case $d = 1$ of the previous theorem has been considered by G. Pólya in [43].

Theorem 3.9.11 (Pólya). *Every function of the Pólya type is the characteristic function of an absolutely continuous distribution.*

3.9.12. Below we give a more detailed statement. For its proof we will need the following facts. Every function f of the Pólya type admits the representation

$$f(t) = \int_0^\infty (1 - |t/y|)_+ \, \mathrm{d}\nu(y), \quad t \in \mathbb{R}$$

where $\mathrm{d}\nu(y) = \mathrm{d}[1 - f(y) + yf'(y)]$ is a probability distribution on $(0, \infty)$ and f' denotes the right-hand derivative. This follows from Theorem 3.9.4 but it can also be proved directly using the relation

$$f(t) = -t \int_t^\infty \frac{\mathrm{d}}{\mathrm{d}y}\left[\frac{f(y)}{y}\right] \mathrm{d}y = \int_t^\infty (1 - t/y)\, \mathrm{d}[1 - f(y) + yf'(y)], \quad t > 0$$

where $\frac{\mathrm{d}}{\mathrm{d}y}$ denotes the right-hand derivative. The fact that the function

$$F(y) := 1 - f(y) + yf'(y) \tag{1}$$

is increasing is illustrated by Figure 3.10. For $y \in (0, \infty)$ and $x \in \mathbb{R}$ let

$$K(x, y) := \frac{2}{\pi} \frac{\sin^2 yx/2}{yx^2} = \frac{y}{2\pi}\left(\frac{\sin yx/2}{yx/2}\right)^2$$

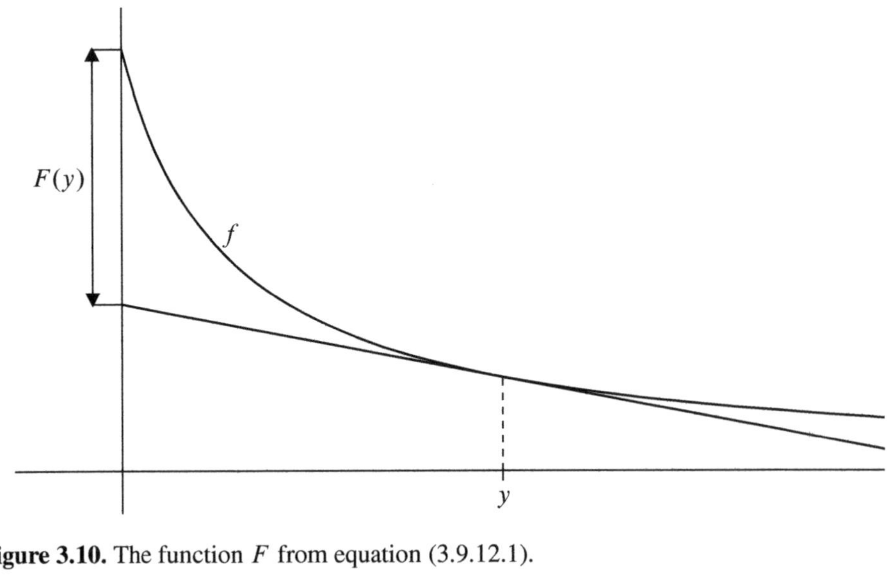

Figure 3.10. The function F from equation (3.9.12.1).

if $x \neq 0$ and $K(0, y) := y/2\pi$. By Remark 1.5.5 the function $x \mapsto K(x, y)$ is a density for any fixed y and the corresponding characteristic function is given by $t \mapsto (1 - |t/y|)_+$. That is,

$$(1 - |t/y|)_+ = \int_{\mathbb{R}} K(x, y)\, \mathrm{e}^{\mathrm{i}tx}\, \mathrm{d}x. \tag{2}$$

Theorem 3.9.13. *Let f be a Pólya-type function. Then we have*

(i) *f is the characteristic function of an absolutely continuous distribution with density*

$$p(x) = \frac{2}{\pi} \int_0^\infty \frac{\sin^2(yx/2)}{yx^2}\, \mathrm{d}\nu(y)$$

where $\mathrm{d}\nu(y) = \mathrm{d}[1 - f(y) + yf'(y)]$ is a probability distribution on $(0, \infty)$;

(ii) *p is finite and continuous on $\mathbb{R} \setminus \{0\}$;*

(iii) *p is bounded if and only if*

$$\int_0^\infty y\, \mathrm{d}\nu(y) < \infty.$$

Proof. (i) We have seen in Theorem 3.9.12 that ν is a probability distribution. By Fubini's theorem,

$$\begin{aligned} 1 = \int_0^\infty 1\, \mathrm{d}\nu(y) &= \int_0^\infty \int_{-\infty}^\infty K(x, y)\, \mathrm{d}x\, \mathrm{d}\nu(y) \\ &= \int_{-\infty}^\infty \int_0^\infty K(x, y)\, \mathrm{d}\nu(y)\, \mathrm{d}x \\ &= \int_{\mathbb{R}\times(0,\infty)} K(x, y)\, \mathrm{d}\lambda \times \nu(x, y) \end{aligned}$$

showing that $K \in L^1(\lambda \times \nu)$. Moreover, the function p defined by

$$p(x) := \int_0^\infty K(x, y)\, \mathrm{d}\nu(y), \quad x \in \mathbb{R}$$

is a probability density. Since $K \in L^1(\lambda \times \nu)$, applying Fubini's theorem and (3.9.12.2) we obtain

$$\begin{aligned} \int_{-\infty}^\infty p(x)\, \mathrm{e}^{\mathrm{i}tx}\, \mathrm{d}x &= \int_{-\infty}^\infty \int_0^\infty K(x, y)\, \mathrm{d}\nu(y)\, \mathrm{e}^{\mathrm{i}tx}\, \mathrm{d}x \\ &= \int_0^\infty \int_{-\infty}^\infty K(x, y)\, \mathrm{e}^{\mathrm{i}tx}\, \mathrm{d}x\, \mathrm{d}\nu(y) \\ &= \int_0^\infty (1 - |t/y|)_+\, \mathrm{d}\nu(y) = f(t). \end{aligned}$$

(ii) Using the inequalities

$$K(x,y) \le \frac{1}{2\pi}, \quad 0 < y \le 1 \quad \text{and} \quad K(x,y) \le \frac{2}{\pi x^2}, \quad y \ge 1$$

we see that $p(x) < \infty$, $x \neq 0$. Moreover, in view of these inequalities, Lebesgue's dominated convergence theorem shows that p is continuous on $\mathbb{R} \setminus \{0\}$.

(iii) If $\int y \, d\nu(y) < \infty$ then, using the inequality

$$K(x,y) \le y/2\pi = K(0,y)$$

we conclude that $p(x) \le p(0) < \infty$ for every x.

Assume that p is bounded. Then, by Fatou's lemma,

$$\begin{aligned} \infty > \liminf_{x\to 0} p(x) &= \liminf_{x\to 0} \int_0^\infty K(x,y)\, d\nu(y) \\ &\ge \int_0^\infty \liminf_{x\to 0} K(x,y)\, d\nu(y) \\ &= \frac{1}{2\pi}\int_0^\infty y\, d\nu(y). \end{aligned}$$

This completes the proof. □

Remark 3.9.14. The fact that a function f of Pólya-type is positive definite can also be proved in the following way. For $m \in \mathbb{N}$ let the function $g_m : \mathbb{R} \to [0,\infty)$ be defined by the relations (see Figure 3.11)

(i) $g_m(j/m) = f(j/m), \quad j = -m^2, \dots, m^2;$

(ii) g_m is linear on the intervals $[j/m, (j+1)/m], \quad j = -m^2, \dots, m^2 - 1;$

(iii) $g_m(x) = f(m)$ if $|x| > m$.

It is not hard to see that g_m can be written as

$$g_m = f(m) + \sum_{j=-m^2}^{m^2} p_j \psi_{j/m}$$

where $\psi_a(x) = (1 - |x/a|)_+$ and $p_j \ge 0$. This shows that g_m is positive definite. On the other hand, $g_m \to f$ uniformly on $\mathbb{R}$.

Example 3.9.15. We have shown in Remark 3.5.3 that the function

$$f_\alpha(t) = e^{-\alpha|t| - |t|^\alpha}, \quad t \in \mathbb{R}$$

is of Pólya-type for all $\alpha \in [2,\infty)$. It is easy to check that the functions

$$g_\alpha(t) = e^{-|t|^\alpha} \quad \text{and} \quad h_\alpha(t) = \frac{1}{1+|t|^\alpha}, \quad t \in \mathbb{R}$$

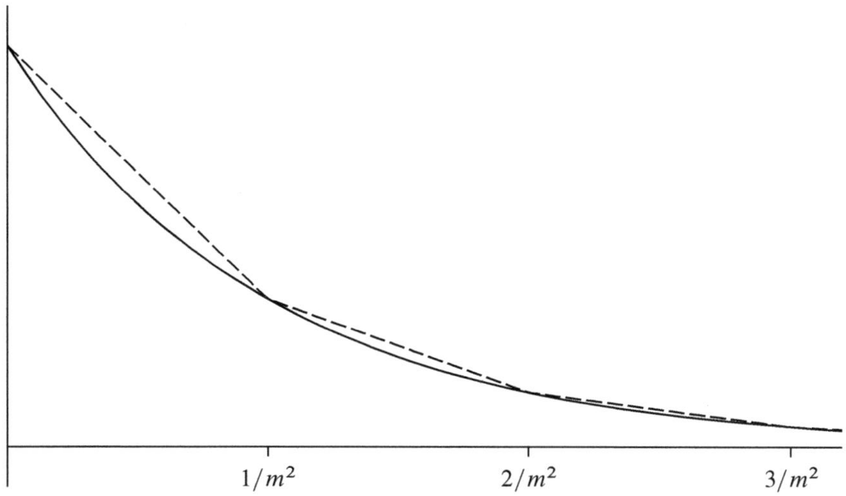

Figure 3.11. The functions f (continuous line) and g_m from Remark 3.9.14.

are of the Pólya type if $0 < \alpha \le 1$. Note that g_α and h_α are characteristic functions also for $1 < \alpha \le 2$. To prove this we show that these functions are positive definite. As to g_α, we may assume that $0 < \alpha < 2$ and write

$$D_\alpha = \int_0^\infty \frac{1 - \cos s}{s^{1+\alpha}}\, \mathrm{d}s.$$

The substitution $s = ty$ leads to the formula

$$-|t|^\alpha = C_\alpha \int_{-\infty}^\infty \frac{\cos ty - 1}{|y|^{1+\alpha}}\, \mathrm{d}y, \quad t \in \mathbb{R}$$

where $C_\alpha = 1/2D_\alpha > 0$. Replacing the integral above by $\int_{|t|\ge 1/m}$, $m \in \mathbb{N}$, we see that g_α is the pointwise limit of functions of the form $q_m = \mathrm{e}^{\varphi_m + c_m}$ where φ_m is positive definite and $c_m \in \mathbb{R}$. By Lemma 1.4.17 the function q_m is positive definite.

The positive definiteness of h_α follows from the equation

$$h_\alpha(t) = \int_0^\infty \mathrm{e}^{-y(1+|t|^\alpha)}\, \mathrm{d}y$$

and from the positive definiteness of g_α.

3.10 Convolution roots with compact support

If f is a characteristic function and $f = h * \tilde{h}$ with some square integrable function h, then h is called a *convolution root* of f. By Theorem 1.8.16, a characteristic function has a convolution root if and only if its spectral measure is absolutely continuous. In this section we assume that f has compact support and investigate the existence of compactly supported convolution roots. This problem is closely related to factorization of entire functions (cf. Corollary 3.10.3). Applications of results of this type range from probability theory, time series, spatial statistics, to optics, crystallography, and signal processing.[32]

We consider first the discrete one-dimensional case.

Theorem 3.10.1. *Let* $N \in \mathbb{N}$ *and* $f \in P(\mathbb{Z})$ *be such that* $\operatorname{supp} f \subset [-2N, 2N]$. *Then there exists a complex-valued function* h *on* $\mathbb{Z}$ *such that* $\operatorname{supp} h \subset [-N, N]$ *and*

$$f(n) = h * \tilde{h}(n) = \sum_{k=-N}^{N} h(n+k)\overline{h(k)}, \quad n \in \mathbb{Z}.$$

Proof. Define the polynomial p by

$$p(z) = \sum_{n=-2N}^{2N} f(n) z^n, \quad z \in \mathbb{C}.$$

It follows from Theorem 1.9.6 that p is nonnegative on $\mathbb{T}$. By Theorem B.1.4, p can be written in the form

$$p(z) = q(z)\overline{q(1/\overline{z})}, \quad z \in \mathbb{C} \setminus \{0\}$$

where

$$q(z) = \sum_{k=0}^{2N} b_k z^k, \quad z, b_k \in \mathbb{C}.$$

Setting $b_k := 0$ if $k \in \mathbb{Z} \setminus [0, 2N]$ and equating the coefficients we obtain

$$f(n) = \sum_{k=0}^{2N} b_{n+k}\overline{b_k}, \quad n \in \mathbb{Z}.$$

The function h defined by $h(k) := b_{k+N}$ has the desired properties. □

[32] See [15] for more details and references.

Theorem 3.10.2 (Boas–Kac, Kreĭn). *Let $f \in P^c(\mathbb{R})$ be such that the support of f is contained in $[-2r, 2r]$. Then there is a square integrable function h vanishing outside $[-r, r]$ such that*

$$f(x) = h * \tilde{h}(x) = \int_{-r}^{r} h(x+y)\overline{h(y)}\,\mathrm{d}y, \quad x \in \mathbb{R}.$$

Proof. We may suppose that $r = \frac{1}{2}$. For all $m \in \mathbb{N}$ the sequence $\{f(n/m)\}_{n\in\mathbb{Z}}$ is positive definite and vanishes outside $\{-m, -m+1, \dots, m\}$. Hence, there exists a sequence $b^{(m)} = \left\{b_n^{(m)}\right\}_{n=-\infty}^{\infty}$ vanishing outside $\{0, 1, \dots, m\}$ and satisfying

$$f(n/m) = \sum_{k=0}^{m} b_{n+k}^{(m)} \cdot \overline{b_k^{(m)}}, \quad n \in \mathbb{Z}. \tag{1}$$

(see the proof of Theorem 3.10.1). For each m let[33]

$$h_m := \sqrt{m} \cdot b^{(m)} * \mathbf{1}_{[-1/(2m),1/(2m)]} \quad \text{and} \quad f_m(x) := h_m * \tilde{h}_m.$$

It follows from (1) and from the definition of h_m that f_m takes the same values as f at the points n/m and is linear in between. Since f is uniformly continuous, we see that $\lim_{m\to\infty} f_m(t) = f(t)$ uniformly on $\mathbb{R}$. As $\|h_m\|_2^2 = f_m(0) = f(0)$, there exists a subsequence $\{h_{m_k}\}$ converging weakly in $L^2(\mathbb{R})$ to some square integrable function h (see Theorem D.5.6). If $g \in L^2(\mathbb{R})$ and $\operatorname{supp}(g) \subset \mathbb{R} \setminus [-1, 1]$ then

$$\int_{-\infty}^{\infty} g(x)h(x)\,\mathrm{d}x = \lim_{k\to\infty} \int_{-\infty}^{\infty} g(x)h_{m_k}(x)\,\mathrm{d}x = 0$$

and hence $\operatorname{supp}(h) \subset [-1, 1]$. Therefore, h is integrable. Using that $h_{m_k} \xrightarrow{w} h$ and the fact that the supports of h and h_{m_k} are contained in $[-2, 2]$ we see that

$$\lim_{k\to\infty} \hat{h}_{m_k}(t) = \hat{h}(t), \; t \in \mathbb{R}.$$

Consequently,

$$\hat{f}(t) = \lim_{k\to\infty} \hat{f}_{m_k}(t) = |h_{m_k}(t)|^2 = |h(t)|^2,$$

i.e., $f = h * \tilde{h}$. □

Corollary 3.10.3. *If F is an entire function of finite exponential type $2r$ which is nonnegative and integrable on $\mathbb{R}$, then there exists an entire function H of exponential type r such that*

$$F(z) = H(z)\overline{H(\overline{z})}, \quad z \in \mathbb{C}. \tag{1}$$

[33] The first convolution below is defined by considering $b^{(m)}$ as a complex measure in $\mathcal{M}^f(\mathbb{R})$.

Proof. Since F is integrable and nonnegative on $\mathbb{R}$, the Fourier transform f of $F|_{\mathbb{R}}$ is a continuous positive definite function. The Paley–Wiener Theorem 3.4.4 shows that f vanishes outside $[-2r, 2r]$. By Theorem 3.10.2 there is a square integrable function h vanishing outside $[-r, r]$ such that $f = h * \tilde{h}$. Setting

$$H(z) := \int_{-r}^{r} h(t) e^{itz} \, dt, \quad z \in \mathbb{C}$$

Lemma 3.4.2 shows that H is an entire function of exponential type r. Moreover, it is easy to check that equation (1) holds. □

Recall that $B^o(r) \subset \mathbb{R}^d$ denotes the open ball with center 0 and radius $r \geq 0$.

Theorem 3.10.4 (Rudin). *Let $f \in P^c(\mathbb{R}^d)$ be an infinitely differentiable radial function such that the support of f lies in $B^o(2r)$, $r > 0$. Then f is the sum of a uniformly convergent series*[34]

$$f = \sum_{k=1}^{\infty} f_k * \tilde{f_k}$$

where each f_k is infinitely differentiable and vanishes outside $B^o(r)$.

Proof. Define the function F by

$$F(w) = \int_{\mathbb{R}^d} e^{-iwt_1} f(t) \, dt, \quad w \in \mathbb{C}.$$

Since f is even, F is even as well. Note that F can be written as

$$F(w) = \int_{-2r}^{2r} e^{-iwt_1} g(t_1) \, dt_1 \tag{1}$$

where $g = f$ if $d = 1$ and

$$g(t_1) = \int_{\mathbb{R}^{d-1}} f(t_1, t_2, \dots, t_d) \, dt_2 \cdots dt_d, \quad t_1 \in \mathbb{R}$$

if $d > 1$. In both cases g is an infinitely differentiable positive definite function vanishing outside $[-2r, 2r]$. By Theorem 3.2.8 the function $F|_{\mathbb{R}}$ is rapidly decreasing. Using the representation (1), Lemma 3.4.2 and Theorem 1.5.9 show that F is an entire function of exponential type $2r$ which is nonnegative on the real axis. By Theorem 3.4.5 there are even functions $G_j, H_j \in E(r)$ such that

$$F(u) = \sum_{j=1}^{\infty} |G_j(u)|^2 + u^2 \sum_{j=1}^{\infty} |H_j(u)|^2, \quad u \in \mathbb{R}. \tag{2}$$

[34] The paper [14] contains an analogue of Rudin's result where f is not supposed to be infinitely differentiable and $d \geq 3$. See also [15] where this analogue is stated without proof.

Since $|G_j(u)|^2 \leq F$ and $|H_j(u)|^2 \leq F$, the functions $G_j|_{\mathbb{R}}$ and $H_j|_{\mathbb{R}}$ are rapidly decreasing. Thus, $s \mapsto G_j(\|s\|)$ and $s \mapsto H_j(\|s\|)$ are integrable on $\mathbb{R}^d$. We define the function g_j by

$$g_j(t) := \int_{\mathbb{R}^d} e^{i(t,s)} G_j(\|s\|)\, ds. \tag{3}$$

By Lemma 3.6.3 the function g is radial while Theorem 3.2.8 shows that g_j is rapidly decreasing. Since G_j is even and entire, for fixed $t_2, \ldots, t_d \in \mathbb{R}$ there is an entire function $Q_j = Q_j^{t_2,\ldots,t_d}$ such that

$$Q_j(t_1) = G_j(\|(t_1, t_2, \ldots, t_d)\|), \quad t_1 \in \mathbb{R}.$$

Using that $G_j \in E(r)$ it is easy to check that Q_j belongs to $E(r)$. By the Paley–Wiener Theorem 3.4.4 the Fourier transform of Q_j vanishes outside $[-r, r]$. Hence,

$$g_j(t_1, 0, \ldots, 0) = \int_{\mathbb{R}^{d-1}} \left[\int_{\mathbb{R}} e^{it_1 x} Q_j(x)\, dx \right] dt_2 \cdots dt_d = 0$$

whenever $|t_1| \geq r$ (if $d = 1$, then the relation above does not contain the outer integral). Since g_j is radial we conclude that g_j vanishes outside $B^o(r)$.

Next, we have

$$|u|^2 |H_j(|u|)|^2 = \sum_{k=1}^{d} |u_k H_j(|u|)|^2.$$

Associate the function h_j with H_j as the function g_j was associated with G_j by equation (3). Then h_j is rapidly decreasing. Moreover, if h_{kj} denotes the partial derivative of h_j with respect to t_k, then $(2\pi)^{d/2} t_k H_j(\|t\|)$ is the Fourier transform of h_{kj}.

From equation (2) we see that

$$\hat{f}(t) = \sum_{j=1}^{\infty} |\hat{g}_j(t)|^2 + \sum_{k=1}^{d} \sum_{j=1}^{\infty} |\hat{h}_{kj}(t)|^2$$

where the series converges in $L^1(\mathbb{R}^d)$. Taking inverse Fourier transforms we obtain the uniformly convergent representation

$$f = \sum_{k=1}^{\infty} g_k * \tilde{g}_k + \sum_{k=1}^{d} \sum_{j=1}^{\infty} h_{kj} * \tilde{h}_{kj}$$

from which the theorem follows. □

3.11 Infinitely divisible characteristic functions

In this section we study a special class of characteristic functions on $\mathbb{R}^d$ known as infinitely divisible characteristic functions. This class plays an important role in the study of decomposition of distributions, in the theory of processes with independent increments (cf. Theorem 2.1.3), as well as in the study of limit theorems.

Definition 3.11.1. A characteristic function f on $\mathbb{R}^d$ is called *infinitely divisible* if for every positive integer n there exists a characteristic function f_n such that

$$f = (f_n)^n.$$

Equivalently, a probability measure μ is said to be infinitely divisible if for every $n \in \mathbb{N}$ there exists a probability measure μ_n such that

$$\mu = \underbrace{\mu_n * \cdots * \mu_n}_{n \text{ times}}.$$

The characteristic functions

$$t \mapsto e^{i(a,t)-q\|t\|^2} \quad \text{and} \quad t \mapsto \frac{1}{(1+\|t\|^2)^q}, \qquad t \in \mathbb{R}^d$$

where $a \in \mathbb{R}^d$, $q \geq 0$, are easily seen to be infinitely divisible (cf. Theorem 3.7.5).

The next lemma follows immediately from Definition 3.11.1. We omit the proof.

Lemma 3.11.2. *If the characteristic functions f and g are infinitely divisible, then so are $\overline{f}$, $|f|^2$ and fg.*

The next theorem shows that characteristic functions having zeros cannot be infinitely divisible.

Theorem 3.11.3. *Let f be an infinitely divisible characteristic function on $\mathbb{R}^d$. Then f has no zeros.*

Proof. For $n \in \mathbb{N}$ let f_n be a characteristic function such that $f = (f_n)^n$. The function $|f|^{2/n} = |f_n|^2$ is a characteristic function for all n. Define g by

$$g(t) := \lim_{n\to\infty} |f_n(t)|^2 = \lim_{n\to\infty} |f(t)|^{2/n}, \qquad t \in \mathbb{R}^d.$$

If $f(t) \neq 0$, then $g(t) = 1$, otherwise $g(t) = 0$. From this we conclude that g is equal to 1 in a neighborhood of 0. Since g is positive definite, Corollary 1.5.2 and Lemma 1.5.1 show that $g(t) = 1$ and hence $f(t) \neq 0$ for all $t \in \mathbb{R}^d$. □

Remark 3.11.4. Let $g : \mathbb{R}^d \to \mathbb{C} \setminus \{0\}$ be a continuous function such that $g(0) > 0$. By Theorem C.8.7 there exists a unique continuous function $\theta : \mathbb{R}^d \to \mathbb{R}$ such that $\theta(0) = 0$ and

$$g(t) = |g(t)| \cdot e^{i\theta(t)}, \qquad t \in \mathbb{R}^d.$$

We define the functions $\ln g$ and g^p, $p \geq 0$, by the relations

$$\ln g(t) = \ln |g(t)| + \mathrm{i}\theta(t),$$
$$g^p(t) = |g(t)|^p \cdot \mathrm{e}^{\mathrm{i}p\theta(t)}, \quad t \in \mathbb{R}^d.$$

Theorem 3.11.5. *Let $\{f_k\}_{k=1}^{\infty}$ be a sequence of infinitely divisible characteristic functions converging pointwise on $\mathbb{R}^d$ to a continuous function f. Then f is an infinitely divisible characteristic function.*

Proof. Since f_k is infinitely divisible, $|f_k|^{2/n}$ is a characteristic function for all positive integers k and n. Lévy's continuity Theorem 1.6.3 and the relation

$$|f(t)|^{2/n} = \lim_{k\to\infty} |f_k(t)|^{2/n}, \quad t \in \mathbb{R}^d$$

show that f and $|f|^{2/n}$, $n \in \mathbb{N}$, are characteristic functions. Thus, $|f|^2$ is an infinitely divisible characteristic function and therefore it has no zeros. Using this we obtain

$$f^{1/n}(t) = \mathrm{e}^{\frac{1}{n}\ln f(t)} = \lim_{k\to\infty} \mathrm{e}^{\frac{1}{n}\ln f_k(t)} = \lim_{k\to\infty} f_k^{1/n}(t), \quad t \in \mathbb{R}^d.$$

This relation shows that f is infinitely divisible. □

Corollary 3.11.6. *If f is an infinitely divisible characteristic function, then so are f^p and $|f|^p$ for all $p \geq 0$.*

Proof. If p is rational, then the statement follows easily from Definition 3.11.1. Using this, the general case is a consequence of Theorem 3.11.5. □

In the next section we give an integral representation for the continuous logarithm of infinitely divisible characteristic functions.

3.12 Conditionally positive definite functions

In the present section we generalize the concept of positive definiteness and show that the new concept is closely related to infinite divisibility. We prove an integral representation and give a useful sufficient condition for a radial function to be conditionally positive definite.[35]

[35] Conditionally positive definite functions and their generalizations have many applications, see for example [9, 58] and also the notes in Remark 3.12.16.

Definition 3.12.1. A complex-valued Hermitian function f on $\mathbb{R}^d$ is called *conditionally positive definite* if the inequality

$$\sum_{i,j=1}^{n} f(t_i - t_j)c_i\overline{c_j} \geq 0$$

holds for all $n \in \mathbb{N}$, $t_1, \dots, t_n \in \mathbb{R}^d$ and for all complex numbers $c_1, \dots, c_n$ such that $c_1 + \dots + c_n = 0$.

Note that the inequality above is satisfied if f is constant. Therefore, we cannot conclude from this inequality alone that f is Hermitian.

Positive definite functions are obviously conditionally positive definite. Let $a \in \mathbb{R}^d$ be arbitrary and write

$$l(t) := \mathrm{i}(t, a), \quad t \in \mathbb{R}^d.$$

If $a \neq 0$, then l is not bounded and hence it is not positive definite. The function l is Hermitian and we have

$$\begin{aligned}\sum_{i,j=1}^{n} l(t_i - t_j)c_i\overline{c_j} &= \sum_{i,j=1}^{n} [l(t_i) - l(t_j)]c_i\overline{c_j} \\ &= \Big(\sum_{i=1}^{n} l(t_i)c_i\Big)\Big(\sum_{j=1}^{n} \overline{c_j}\Big) - \Big(\sum_{j=1}^{n} l(t_j)\overline{c_j}\Big)\Big(\sum_{i=1}^{n} c_i\Big) = 0\end{aligned}$$

whenever $\sum c_i = 0$. Thus, l is conditionally positive definite.

The next lemma states elementary properties of conditionally positive definite functions which follow immediately from Definition 3.12.1. We omit the proof.

Lemma 3.12.2. *Let f and g be conditionally positive definite functions on $\mathbb{R}^d$. Then so are the functions $\overline{f}$, $\mathrm{Re} f$ and $pf + qg$ for all $p, q \geq 0$. Moreover, the pointwise limit of conditionally positive definite functions is conditionally positive definite as well.*

Note that the product of conditionally positive definite functions need not be conditionally positive definite. A simple example is given by the product of the functions $t \mapsto -1$ and $t \mapsto e^{\mathrm{i}t}$, $t \in \mathbb{R}$.

Lemma 3.12.3. *If f is conditionally positive definite, then the Hermitian matrix*

$$A = \big(f(t_i - t_j)\big)_{i,j=1}^{n}$$

has at most one negative eigenvalue[36] *for all finite systems $t_1, \dots, t_n$ of elements of $\mathbb{R}^d$.*

---

[36] See also Remark 3.12.16.

Proof. The set

$$\{x \in \mathbb{C}^n : x_1 + \cdots + x_n = 0\}$$

is an $(n-1)$-dimensional nonnegative subspace of A (cf. Definition D.2.5). We conclude that A has no negative subspace of dimension greater than 1. Hence, the statement follows from Theorem D.2.6. □

Theorem 3.12.4. *Let $f : \mathbb{R}^d \to \mathbb{C}$ be a Hermitian function with $f(0) \le 0$. Then f is conditionally positive definite if and only if the matrix*

$$B = \Big(f(t_i - t_j) - f(t_i) - \overline{f(t_j)}\Big)_{i,j=1}^n$$

is positive semidefinite for all finite systems $t_1, \dots, t_n \in \mathbb{R}^d$.

Proof. To prove the "if part" let $t_1, \dots, t_n \in \mathbb{R}^d$ be arbitrary and let $c_1, \dots, c_n \in \mathbb{C}$ be such that $\sum_{i=1}^n c_i = 0$. Then we have

$$\sum_{i,j=1}^n f(t_i - t_j)c_i\overline{c_j} = \sum_{i,j=1}^n (f(t_i - t_j) - f(t_i) - \overline{f(t_j)})c_i\overline{c_j} \ge 0.$$

Thus, f is conditionally positive definite.

Conversely, suppose that f is conditionally positive definite. Let $t_1, \dots, t_n \in \mathbb{R}^d$ and $c_1, \dots, c_n \in \mathbb{C}$ be arbitrary and write $t_{n+1} := 0$, $c_{n+1} := -\sum_{i=1}^n c_i$. Then $\sum_{i=1}^{n+1} c_i = 0$ and we have

$$\begin{aligned}
\sum_{i,j=1}^n (f(t_i - t_j) - f(t_i) - \overline{f(t_j)})c_i\overline{c_j} &= \sum_{i,j=1}^n f(t_i - t_j)c_i\overline{c_j} \\
&\quad + \sum_{i=1}^n f(t_i)c_i\overline{c_{n+1}} + \sum_{j=1}^n f(-t_j)c_{n+1}\overline{c_j} \\
&= \sum_{i,j=1}^{n+1} f(t_i - t_j)c_i\overline{c_j} - f(0)|c_{n+1}|^2 \\
&\ge 0.
\end{aligned}$$

□

The next theorem establishes an important connection between infinitely divisible characteristic functions and conditionally positive definite functions.

Theorem 3.12.5. *A Hermitian function f on $\mathbb{R}^d$ is conditionally positive definite if and only if the function*

$$t \mapsto e^{pf(t)}, \quad t \in \mathbb{R}^d$$

is positive definite for all $p > 0$.

Proof. Suppose that $e^{pf(\cdot)}$ is positive definite for all $p > 0$. Then $(e^{pf(\cdot)} - 1)/p$ is conditionally positive definite for all $p > 0$. From the relation

$$f(t) = \lim_{p \to +0} \frac{e^{pf(t)} - 1}{p}, \quad t \in \mathbb{R}^d$$

we conclude that the function f is conditionally positive definite.

To prove the converse statement, suppose that f is conditionally positive definite. Then so is $f - f(0)$ and therefore we may suppose that $f(0) = 0$. Since pf is conditionally positive definite if $p > 0$, it suffices to consider the case $p = 1$. If (a_{ij}) is a positive semidefinite matrix, then so is $(\exp(a_{ij}))$ (cf. Corollary D.2.13). Using this, it follows from Theorem 3.12.4 that the matrix

$$\left(\exp(f(t_i - t_j) - f(t_i) - \overline{f(t_j)})\right)_{i,j=1}^n$$

is positive semidefinite. We have

$$\begin{aligned}
&\sum_{i,j=1}^n \exp(f(t_i - t_j))c_i\overline{c_j} \\
&\quad = \sum_{i,j=1}^n \exp(f(t_i - t_j) - f(t_i) - \overline{f(t_j)}) \exp(f(t_i)) \exp(\overline{f(t_j)})c_i\overline{c_j} \\
&\quad = \sum_{i,j=1}^n \exp(f(t_i - t_j) - f(t_i) - \overline{f(t_j)})d_i\overline{d_j} \geq 0
\end{aligned}$$

where $d_i = \exp(\overline{f(t_i)})c_i$. □

As a corollary we obtain that the function

$$t \mapsto a + \mathrm{i}(x, t) - (Ct, t), \quad t \in \mathbb{R}^d$$

is conditionally positive definite for all $a \in \mathbb{R}$, $x \in \mathbb{R}^d$ and all $d \times d$ positive semidefinite real matrices C.

Lemma 3.12.6. *A Hermitian function f on $\mathbb{R}^d$ is conditionally positive definite if and only if the inequality*

$$\nu * \tilde{\nu} * f(0) \geq 0$$

holds for all finitely supported complex measures ν on $\mathbb{R}^d$ such that $\hat{\nu}(0) = 0$.

Proof. If

$$\nu = \sum_{j=1}^n c_j \delta_{t_j}$$

then $\hat{\nu}(0) = c_1 + \cdots + c_n$. Using this the statement can be proved in the same way as Lemma 1.4.3. □

Lemma 3.12.7. *Let f be a conditionally positive definite function on $\mathbb{R}^d$ and let μ be a complex measure with finite support on $\mathbb{R}^d$. Then the function*

$$g = \mu * \tilde{\mu} * f$$

is conditionally positive definite. If $\hat{\mu}(0) = 0$, then g is positive definite.

In particular, the function

$$t \mapsto 2f(t) - f(t+y) - f(t-y), \quad t \in \mathbb{R}^d$$

is positive definite for all $y \in \mathbb{R}^d$.

Proof. Let ν be an arbitrary finitely supported complex measure on $\mathbb{R}^d$. The first two statements of the lemma follow from the equation

$$\nu * \tilde{\nu} * f(0) = (\nu * \mu) * (\nu * \mu)^{\tilde{}} * f(0)$$

using Lemma 3.12.6 and the fact that $(\nu * \mu)\hat{}(0) = \hat{\nu}(0)\hat{\mu}(0)$. Setting $\mu = \delta_0 - \delta_y$ we obtain the last statement. □

The previous simple lemma has several useful corollaries.

Corollary 3.12.8. *Let $t \mapsto f(t)$, $t \in \mathbb{R}^d$, be a conditionally positive definite function. If f is twice continuously differentiable, then the function*

$$-\frac{\partial^2}{\partial t_j^2} f$$

is positive definite for all $j = 1, \ldots, d$.

Proof. Let $e_1, \ldots, e_d$ be the standard orthonormal basis in $\mathbb{R}^d$. Then we have

$$-\frac{\partial^2}{\partial t_j^2} f(t) = \lim_{\epsilon \to 0} \frac{2f(t) - f(t + \epsilon e_j) - f(t - \epsilon e_j)}{\epsilon^2}, \quad t \in \mathbb{R}^d.$$

Hence, the statement follows from Lemma 3.12.7. □

Corollary 3.12.9. *Let $h : [0, \infty) \to \mathbb{R}$ be a continuous function which is twice continuously differentiable on $(0, \infty)$. If $h(\|\cdot\|^2)$ is conditionally positive definite on $\mathbb{R}^{d+1}$ for some $d \in \mathbb{N}$, then the function $-h'(\|\cdot\|^2)$ is positive definite on $\mathbb{R}^d$.*

Proof. For $\epsilon \in \mathbb{R}$ we consider the function

$$t \mapsto 2h(\|t\|^2) - h(\|t + \epsilon e_{d+1}\|^2) - h(\|t - \epsilon e_{d+1}\|^2), \quad t \in \mathbb{R}^{d+1}$$

where $e_1, \dots, e_{d+1}$ is the standard basis of $\mathbb{R}^{d+1}$. Lemma 3.12.7 shows that this function is positive definite. Its restriction to $\mathbb{R}^d$ is given by

$$s \mapsto 2\,[h(\|s\|^2) - h(\|s\|^2 + \epsilon^2)], \quad s \in \mathbb{R}^d.$$

Dividing by ϵ^2 and letting ϵ tend to zero we obtain the statement of the lemma. □

Corollary 3.12.10. *Let f be a conditionally positive definite function and let $\mu, \nu \in \mathcal{M}^f(\mathbb{R}^d)$ be such that $\hat{\mu}(0) = \hat{\nu}(0) = 0$. Then there exists a complex measure σ such that $\mu * \nu * f = \check{\sigma}$.*

Proof. The corollary follows from the identity

$$\mu * \nu = \frac{1}{4} \sum_{l=0}^{3} \mathrm{i}^l (\mu + \mathrm{i}^l \tilde{\nu}) * (\mu + \mathrm{i}^l \tilde{\nu})^{\tilde{}}$$

using Lemma 3.12.7 and the fact that $(\mu + \mathrm{i}^l \tilde{\nu})\hat{}(0) = 0$. □

Lemma 3.12.11. *For every continuous conditionally positive definite function f there exists a finite radial measure μ_0 such that*

(i) *the convolution $f * \mu_0$ exists and $f * \mu_0$ is a continuous positive definite function;*

(ii) $\hat{\mu}_0(t) > 0, \quad t \in \mathbb{R}^d \setminus \{0\}$.

Proof. In view of Lemma 3.2.5 for every $n \in \mathbb{N}$ there exists a radial probability measure ν_n with compact support satisfying

$$0 \le \hat{\nu}_n(t) \le 1/2, \quad \|t\| \ge 1/n.$$

Setting $\mu_n := \delta_0 - \nu_n$ we have $\hat{\mu}_n(0) = 0$ and

$$\hat{\mu}_n(t) \ge \frac{1}{2}, \quad \|t\| \ge 1/n. \tag{1}$$

By Lemma 3.12.7, the function $\mu_n * \tilde{\mu}_n * f$ is positive definite. Consequently, writing $r_n := \mu_n * \tilde{\mu}_n * f(0)$, we have

$$|\mu_n * \tilde{\mu}_n * f(t)| \le r_n, \quad t \in \mathbb{R}^d. \tag{2}$$

Since $\|\mu_n\| \le 2$, the equation

$$\mu_0 = \sum_{n=1}^{\infty} \frac{1}{2^n \max(1, r_n)} \cdot \mu_n * \tilde{\mu}_n \tag{3}$$

defines a finite radial measure. Using inequality (2) we see that the convolution $f * \mu_0$ exists and $f * \mu_0 \in P^c(\mathbb{R}^d)$.

Property (ii) follows from inequality (1). □

Theorem 3.12.12 (Lévy–Khinchin formula). *Let f be a continuous Hermitian function on $\mathbb{R}^d$ and let $h : \mathbb{R}^d \to \mathbb{R}^d$ be a bounded continuous function such that $h(t) = t$ in some neighborhood of zero and $h(-t) = -h(t)$, $t \in \mathbb{R}^d$. The function f is conditionally positive definite if and only if*

$$f(t) = a + \mathrm{i}(t, x) - (Ct, t) + \int_{\mathbb{R}^d} \left(\mathrm{e}^{\mathrm{i}(t,y)} - 1 - \mathrm{i}(t, h(y)) \right) \mathrm{d}\sigma(y), \quad t \in \mathbb{R}^d \quad (1)$$

where $a \in \mathbb{R}$, $x \in \mathbb{R}^d$, C is a $d \times d$ positive semidefinite real matrix and σ is a nonnegative measure on $\mathbb{R}^d$ such that $\sigma(\{0\}) = 0$, $\sigma(\mathbb{R}^d \setminus U) < \infty$ and

$$\int_U \|y\|^2 \,\mathrm{d}\sigma(y) < \infty \quad (2)$$

for every neighborhood U of zero. The representation (1) is unique.

Proof. Assume first that f admits the integral representation (1). That f is conditionally positive definite follows immediately from the fact that the functions

$$t \mapsto a + \mathrm{i}(t, x) - (Ct, t) \quad \text{and} \quad t \mapsto \mathrm{e}^{\mathrm{i}(t,y)} - 1 - \mathrm{i}(t, h(y))$$

are conditionally positive definite.

Next suppose that f is conditionally positive definite and let μ_0 be the measure from Lemma 3.12.11. Since $\mu_0 * f$ is positive definite it can be represented as

$$\mu_0 * f = \check{\sigma}_{\mu_0}$$

with some nonnegative measure σ_{μ_0}. We define the measure σ by $\sigma(\{0\}) := 0$ and

$$\mathrm{d}\sigma(y) := \frac{1}{\hat{\mu}_0(y)} \,\mathrm{d}\sigma_{\mu_0}(y), \quad y \neq 0.$$

Then σ is nonnegative and (3.12.11.ii) implies that

$$\sigma(\mathbb{R}^d \setminus U) < \infty$$

for every neighborhood U of zero. By Corollary 3.12.10, for $\mu, \nu \in \mathcal{M}^f(\mathbb{R}^d)$ with $\hat{\mu}(0) = \hat{\nu}(0) = 0$ there is a complex measure $\sigma_{\mu*\nu}$ such that

$$\mu * \nu * f = \check{\sigma}_{\mu*\nu}.$$

It follows from $\mu_0 * (\mu * \nu * f) = \mu * \nu * (\mu_0 * f)$ that

$$\hat{\mu}_0(y) \,\mathrm{d}\sigma_{\mu*\nu}(y) = (\mu * \nu)\hat{}(y) \,\mathrm{d}\sigma_{\mu_0}(y).$$

Using this relation and the definition of σ we obtain

$$\mathrm{d}\sigma_{\mu*\nu}(y) = (\mu * \nu)\hat{}(y) \,\mathrm{d}\sigma(y), \quad y \neq 0. \quad (3)$$

In particular, the right-hand side represents a finite measure. The special case $\mu_x := \delta_0 - \delta_x$ and $\nu := \tilde{\mu}_x$, gives

$$\int_{\mathbb{R}^d} |\hat{\mu}_x(y)|^2 \, d\sigma(y) = \int_{\mathbb{R}^d} |1 - e^{-i(x,y)}|^2 \, d\sigma(y)$$
$$= 2 \int_{\mathbb{R}^d} [1 - \cos((x,y))] \, d\sigma(y) < \infty, \quad x \in \mathbb{R}^d$$

from which the relation (2) readily follows.

Now we put

$$p(t) := f(t) - \int_{\mathbb{R}^d} \left(e^{i(t,y)} - 1 - i(t, h(y)) \right) d\sigma(y), \quad t \in \mathbb{R}^d.$$

That the integral above exists follows from (2) using the inequality in Lemma B.1.5 and the fact that $h(y) = y$ in some neighborhood of zero. Next we show that $\mu * \nu * p$ is a constant for all $\mu, \nu \in \mathcal{M}^f(\mathbb{R}^d)$ with $\hat{\mu}(0) = \hat{\nu}(0) = 0$. Writing $\chi := \mu * \nu$ for short, and applying Proposition C.9.6 we obtain

$$\chi * p(t) = \chi * f(t) - \int_{\mathbb{R}^d} \hat{\chi}(y) e^{i(t,y)} \, d\sigma(y)$$
$$= \check{\sigma}_\chi(t) - \int_{\mathbb{R}^d} \hat{\chi}(y) e^{i(t,y)} \, d\sigma(y), \quad t \in \mathbb{R}^d.$$

From equation (3) we see that the last integral is equal to

$$\int_{\mathbb{R}^d \setminus \{0\}} e^{i(t,y)} \, d\sigma_\chi(y) = \check{\sigma}_\chi(t) - \sigma_\chi(\{0\})$$

and hence

$$\chi * p(t) = \sigma_\chi(\{0\}), \quad t \in \mathbb{R}^d.$$

In particular, $(\delta_0 - \delta_x) * (\delta_0 - \delta_s) * p$ is constant for all $x, s \in \mathbb{R}^d$. Applying Lemma C.9.4 twice we see that p is a polynomial of degree at most 2. Setting $\nu := \tilde{\mu}$, we obtain that

$$\mu * \tilde{\mu} * p(0) = \sigma_{\mu * \tilde{\mu}}(\{0\}) \geq 0$$

and therefore, by Lemma 3.12.6, the function p is conditionally positive definite. Thus, e^p is positive definite in view of Theorem 3.12.5. The special form of p follows from (the simple part of) Theorem 3.5.1.

To prove uniqueness of the representation (1) assume that

$$p_1(t) + \int_{\mathbb{R}^d} \left(e^{i(t,y)} - 1 - i(t, h(y)) \right) d\sigma_1(y)$$
$$= p_2(t) + \int_{\mathbb{R}^d} \left(e^{i(t,y)} - 1 - i(t, h(y)) \right) d\sigma_2(y), \quad t \in \mathbb{R}^d$$

where p_j is a polynomial of degree at most 2 and σ_j is a measure having the properties of σ, $j = 1, 2$. Let $x \in \mathbb{R}^d$ be arbitrary. Convolving both sides of the equation above with $\chi := (\delta_0 - \delta_x) * (\delta_0 - \delta_x) * (\delta_0 - \delta_x)$ and applying Proposition C.9.6 we obtain

$$\int_{\mathbb{R}^d} \hat{\chi}(y) e^{i(t,y)} \, d\sigma_1(y) = \int_{\mathbb{R}^d} \hat{\chi}(y) e^{i(t,y)} \, d\sigma_2(y), \quad t \in \mathbb{R}^d.$$

Since $\hat{\chi}(y) \, d\sigma_j(y) = \left[1 - e^{-i(x,y)}\right]^3 d\sigma_j(y)$ is a finite measure, in view of property (2), we conclude that

$$\left[1 - e^{-i(x,y)}\right]^3 d\sigma_1(y) = \left[1 - e^{-i(x,y)}\right]^3 d\sigma_2(y), \quad x \in \mathbb{R}^d.$$

Taking into account that $\sigma_1(\{1\}) = \sigma_2(\{1\}) = 0$ we infer that $\sigma_1 = \sigma_2$. Hence we also have $p_1 = p_2$. □

Remark 3.12.13. If f is radial, then the measure σ in Theorem 3.12.12 is radial as well. This follows immediately from the construction of σ. In this case, $x = 0$ and C is a nonnegative multiple of the identity matrix.

Theorem 3.12.14 (Micchelli). *Let $h : [0, \infty) \to \mathbb{R}$ be a continuous function which is infinitely differentiable on $(0, \infty)$. Then the function $f := h(\|\cdot\|^2)$ is conditionally positive definite on $\mathbb{R}^d$ for all d if and only if $-h'$ is completely monotone on $(0, \infty)$.*[37]

Proof. If $-h'$ is completely monotone on $(0, \infty)$ then, by Corollary 3.9.7, it admits the representation

$$-h'(r) = \int_0^\infty e^{-rs} \, d\mu(s), \quad r \in [0, \infty)$$

where μ is a nonnegative measure on $[0, \infty)$. Let $\epsilon > 0$ and define the function h_ϵ by $h_\epsilon(r) := h(r + \epsilon)$, $r \in (-\epsilon, \infty)$. Then we have

$$\begin{aligned} h_\epsilon(r) &= h_\epsilon(0) + \int_0^r h'_\epsilon(t) \, dt \\ &= h_\epsilon(0) - \int_0^r \int_0^\infty e^{-(t+\epsilon)s} \, d\mu(s) \, dt \\ &= h_\epsilon(0) - \int_0^\infty e^{-\epsilon s} \int_0^r e^{-ts} \, dt \, d\mu(s) \\ &= h_\epsilon(0) + \int_0^\infty e^{-\epsilon s} \frac{e^{-rs} - 1}{s} \, d\mu(s) \end{aligned}$$

Since the function $e^{-\|\cdot\|^2 s} - 1$ is conditionally positive definite on $\mathbb{R}^d$ for all $s \in [0, \infty)$ and for all d, the function $h_\epsilon(\|\cdot\|^2)$ is conditionally positive definite on $\mathbb{R}^d$ for all d. Letting ϵ tend to zero, we conclude that the same is true for h.

[37] This result can easily be generalized for conditionally positive definite functions of order m (see [58]).

Next assume that $h(\|\cdot\|^2)$ is conditionally positive definite on every $\mathbb{R}^d$. Corollary 3.12.9 shows that the function $-h'(\|\cdot\|^2)$ is positive definite on every $\mathbb{R}^d$. From Schoenberg's Theorem 3.8.5 we conclude that $-h'$ is completely monotone. □

Example 3.12.15. Let $a > 0,\ 0 < b < 1$, and write $g(r) := -(a+r)^b,\ r \geq 0$. From the relation

$$g^{(k)}(r) = -b(b-1)\cdots(b-k+1)(a+r)^{b-k}, \quad k \in \mathbb{N}$$

we see that $-g'$ is completely monotone. Thus, $g(\|\cdot\|^2)$ is conditionally positive definite on every $\mathbb{R}^d$.

Remark 3.12.16. M. G. Kreĭn (see the historical remarks in [49]) generalized the concept of positive definiteness in the following way: A complex-valued Hermitian function f on $\mathbb{R}^d$ is said to have k *negative squares* if the Hermitian matrix

$$A = \big(f(t_i - t_j)\big)_{i,j=1}^{n}$$

has at most k negative eigenvalues (counted with their multiplicities) for any choice of n and $t_1,\ldots,t_n \in \mathbb{R}^d$, and for some choice of n and $t_1,\ldots,t_n$ the matrix A has exactly k negative eigenvalues. Denote by $P_k(\mathbb{R}^d)$ the set of all functions on $\mathbb{R}^d$ with k negative squares. As we have seen in Lemma 3.12.3, conditionally positive definite functions belong to $P_0(\mathbb{R}^d) \cup P_1(\mathbb{R}^d)$. Further examples are given by the functions

$$\begin{aligned}
&t \mapsto t^k e^t + (-t)^k e^{-t} \in P_{k+1}(\mathbb{R}),\\
&t \mapsto (-1)^k t^{2k} \in P_k(\mathbb{R}),\\
&t \mapsto (-1)^{k+1} t^{2k} \in P_{k+1}(\mathbb{R}),\\
&t \mapsto (-1)^k |t|^a \in P_k(\mathbb{R}), \quad a \in (2k-2, 2k],\ k > 0,
\end{aligned}$$

(see [49] for proofs).

In [36] M. G. Kreĭn proved that every continuous function $f \in P_k(\mathbb{R})$ is *definitizable* in the following sense: there exists a polynomial Q of degree k such that the inequality

$$\int_{-\infty}^{\infty}\int_{-\infty}^{\infty} f(x-y)Q\Big(-\mathrm{i}\frac{\mathrm{d}}{\mathrm{d}y}\Big)h(y)\,\overline{Q\Big(-\mathrm{i}\frac{\mathrm{d}}{\mathrm{d}x}\Big)h(x)}\,\mathrm{d}y\,\mathrm{d}x \geq 0$$

holds for every infinitely differentiable function h with compact support. He obtained the integral representation

$$f(x) = p(x) + \int_{-\infty}^{\infty} \frac{e^{\mathrm{i}tx} - S(x,t)}{|Q_0(t)|^2}\,\mathrm{d}\mu(t)$$

where p is a Hermitian solution of the differential equation

$$\overline{Q}\Big(-\mathrm{i}\frac{\mathrm{d}}{\mathrm{d}x}\Big)Q\Big(-\mathrm{i}\frac{\mathrm{d}}{\mathrm{d}x}\Big)p(x) = 0$$

where $\overline{Q}(t) := \overline{Q(\bar{t})}$, Q_0 is a polynomial that is obtained by deleting the non-real zeros of Q, S is a regularizing correction compensating for the real zeros of Q, and μ is a nonnegative measure satisfying

$$\int_{-\infty}^{\infty} \frac{1}{(1+t^2)^m}\, \mathrm{d}\mu(t) < \infty$$

where m denotes the degree of Q_0.

The concepts of negative squares and definitizability, as well as Kreĭn's integral representation have been generalized to commutative groups. Functions with k negative squares on a commutative group are definitizable in the sense that certain linear combinations of their translates are positive definite. We refer to [49] for more information on this topic. Definitizable functions are closely related to certain random fields which are generalizations of stationary processes.[38]

To see multivariate examples of functions with negative squares, denote by s_r^d, $1 \le r \le d$, the *elementary symmetric polynomial* of degree r in d real variables:

$$\begin{aligned}
s_1^d(t) &= t_1 + t_2 + \cdots + t_d \\
s_2^d(t) &= t_1 t_2 + t_1 t_3 + \cdots + t_{d-1} t_d \\
&\ \vdots \\
s_r^d(t) &= t_1 \cdots t_r + \cdots + t_{n-r+1} \cdots t_d \\
&\ \vdots \\
s_d^d(t) &= t_1 t_2 \cdots t_d, \quad t \in \mathbb{R}^d.
\end{aligned}$$

Then we have

$$\begin{aligned}
&\pm \mathrm{i} s_1^d \in P_1(\mathbb{R}^d), && s_2^d \in P_2(\mathbb{R}^d), && -s_2^d \in P_d(\mathbb{R}^d), \\
&\pm \mathrm{i} s_3^d \in P_{d+1}(\mathbb{R}^d), && \pm \mathrm{i}^d s_d^d \in P_{2^{d-1}}(\mathbb{R}^d)
\end{aligned}$$

(see [51] for proofs).

[38] See, among others, [4, 9, 42, 52, 53] and the references therein.

Chapter 4

The extension problem

In this chapter, we study the extendability of positive definite functions defined on a certain subset V of a commutative group. After proving some general results we consider the groups $\mathbb{R}^d$ and $\mathbb{Z}^d$.

4.1 General results

Throughout this section G is a discrete commutative group with character group $\hat{G}$ and V denotes a symmetric subset[1] of G containing 0.

Definition 4.1.1. A complex-valued function f on V is called *positive definite* if the inequality

$$\sum_{i,j=1}^{n} f(x_i - x_j) c_i \bar{c}_j \geq 0 \tag{1}$$

holds for all $c_1, \ldots, c_n \in \mathbb{C}$ and all $x_1, \ldots, x_n \in V$ such that $x_i - x_j \in V$, $i, j = 1, \ldots, n$.

We denote by $P(V)$ the set of all positive definite functions on V and write

$$P_0(V) := \{f : f \in P(V),\ f(0) \leq 1\}.$$

Remark 4.1.2. If $f \in P(V)$, then inequality (4.1.1.1) holds for all $x_1, \ldots, x_n \in G$ such that $x_i - x_j \in V$. Indeed, setting $y_i := x_i - x_1$ we have $y_i \in V$ and $y_i - y_j = x_i - x_j$.

If $V = T - T$ with some $T \subset G$, then every $f \in P(V)$ is T-positive definite (cf. Definition 2.9.1). Now let $G = \mathbb{R}$ and $V = (-2a, 2a)$ with some $a > 0$. Then $V = T - T$ where $T = (-a, a)$. In this case a complex-valued function f on V is T-positive definite if and only if f is positive definite on V. To see this, let $x_1, \ldots, x_n \in V$ be such that $x_i - x_j \in V$ for all i and j. Then the diameter of the set $\{x_1, \ldots, x_n\}$ is less than $2a$ and hence there exists $x_0 \in \mathbb{R}$ such that $x_i \in (x_0 - a, x_0 + a)$ for all i. Setting $y_i := x_i - x_0$ we have $y_i \in T$ and $y_i - y_j = x_i - x_j$. Consequently, f is positive definite on V.

Example 4.1.4 shows that there exist a finite set $T \subset \mathbb{R}^2$ and a function f on $T - T$ such that f is T-positive definite but not positive definite on $T - T$.

[1] A subset V of G is called *symmetric* if $V = -V$.

Example 4.1.3. For $a \in (0, \pi/2)$ let

$$V_a := \{e^{it} : -a < t < a\} \subset G := \mathbb{T}$$

and let $r \in \mathbb{R}$ be arbitrary.[2] Then the function

$$\gamma_r(e^{it}) := e^{irt}, \quad -a < t < a$$

is positive definite on V_a. Indeed, if $x, y, x - y \in V_a$ then $\gamma_r(x - y) = \gamma_r(x)\overline{\gamma_r(y)}$ and hence we can repeat the computations in Example 1.4.6:

$$\sum_{i,j=1}^{n} \gamma_r(x_i - x_j) c_i \overline{c_j} = \left| \sum_{i=1}^{n} c_i \gamma_r(x_i) \right|^2 \geq 0.$$

Example 4.1.4. Write[3]

$$T := \left\{ \left(\cos \frac{2\pi k}{6}, \sin \frac{2\pi k}{6} \right) : k = 0, 1, \ldots, 6 \right\}$$

and $V := T - T$. We show that there is a function $\psi : [0, 2] \to [-1, 1]$ such that the function $f(x) = \psi(\|x\|)$ is T-positive definite but $f \notin P(V)$.

Note that V consists of 19 points, of which 6 have norm 1, 6 have norm $\sqrt{3}$, and 6 have norm 2 (see Figure 4.1). Thus, apart from $\psi(0) = 1$, it suffices to specify three numbers:

$$\psi(1), \quad \psi(\sqrt{3}) \quad \text{and} \quad \psi(2).$$

Choose any

$$\gamma \in [-1, -1/2), \quad a \in [-1/2, 1], \quad \text{and} \quad b \in [-1, 1/2]$$

and put

$$\psi(1) := \alpha := \frac{1}{2} \cdot [a(1+\gamma) + b(1-\gamma)], \quad \psi(\sqrt{3}) := \beta := \frac{1}{2} \cdot [a(1+\gamma) - b(1-\gamma)]$$

and $\psi(2) := \gamma$. To see that f is T-positive definite, write

$$t_k = (\cos 2k\pi/6, \sin 2k\pi/6)$$

so we have to show that the matrix

$$A = (f(t_i - t_j))_{j,k=0}^{5}$$

is positive semidefinite. Writing down A, we see that it is the Toeplitz matrix corresponding to the function $(1, \alpha, \beta, \gamma, \beta, \alpha)$ on the group $\{0, 1, 2, 3, 4, 5\} = \mathbb{Z}/6\mathbb{Z}$ (factor

[2] Below we use additive notation for the group operation in $\mathbb{T}$ which is actually multiplication.

[3] This example is due to T. M. Bisgaard.

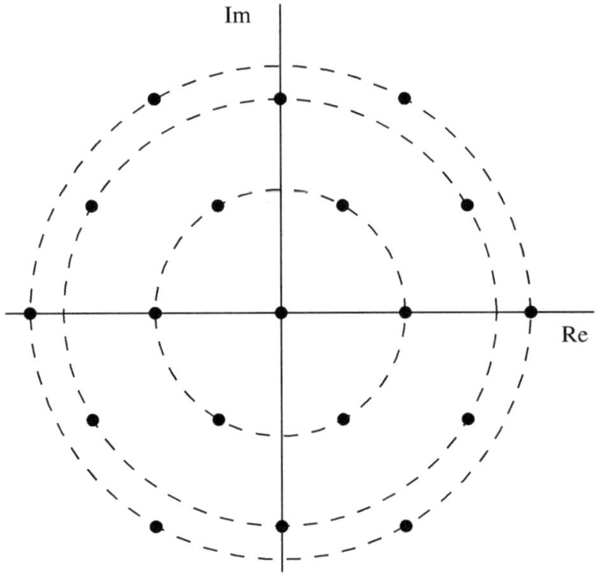

Figure 4.1. The set $V = T - T$ from Example 4.1.4.

group). By (1.9.6.iii) it suffices to show that the Fourier transform of this function is nonnegative. So one has to verify

$$1 + \alpha \cdot (z^k + \overline{z}^k) + \beta \cdot (z^{2k} + \overline{z}^{2k}) + \gamma \cdot z^{3k} \geq 0$$

for all $k = 0, \ldots, 5$, where $z = \mathrm{e}^{\mathrm{i}2\pi/6}$. This is readily done.[4]

To see that f is not positive definite on V, write $R = \{0, 2t_0, 2t_1\}$ and note that $R - R \subset V$. The corresponding 3×3 matrix is

$$B = \begin{bmatrix} 1 & \gamma & \gamma \\ \gamma & 1 & \gamma \\ \gamma & \gamma & 1 \end{bmatrix}.$$

Setting $v := [1, 1, 1]^{\mathrm{T}}$ we have $(Bv, v) = 3(1 + 2\gamma) < 0$. Thus, B is not positive semidefinite.

In the next theorem we list basic properties of positive definite functions on V which can be proved in the same way as in Section 1.4 and Section 1.5.

Theorem 4.1.5.

(i) *If $f \in P(V)$, then f is Hermitian.*

---

[4] By symmetry of the function it suffices to consider the values $k = 0, 1, 2, 3$.

(ii) *Let* $f_1, f_2 \in P(V)$. *Then the functions* $\overline{f_1}$, $\operatorname{Re} f_1$, $|f_1|^2$, *and* $f_1 \cdot f_2$ *are positive definite. Moreover,* $p_1 f_1 + p_2 f_2$ *is positive definite for all* $p_1, p_2 \geq 0$.

(iii) *The set* $P(V)$ *is a convex cone closed in the topology of pointwise convergence, while* $P_0(V)$ *is a compact convex set.*

(iv) *The inequalities of Theorem 1.4.12 hold for a function* $f \in P(V)$ *whenever the used arguments of* f *lie in* V.

(v) *Let* $G = \mathbb{R}^d$ *or* $G = \mathbb{T}^d$, *let* V *be open and* $f \in P(V)$. *If* $\operatorname{Re} f$ *is continuous at the neutral element, then* f *is uniformly continuous on* V.

(vi) *Let* $V \subset \mathbb{R}^d$ *and assume that* V *is an open ball with center* 0. *If* $f \in P^c(V)$ *is such that* $|f(t)| = 1$ *for all* t *from a neighborhood of* 0, *then*

$$f(t) = \mathrm{e}^{\mathrm{i}(x,t)}, \quad t \in V$$

for some $x \in \mathbb{R}^d$.

The next proposition can be proved in the same way as Lemma 1.5.8. We omit the proof.

Proposition 4.1.6. *Let* $r > 0$ *and* $f \in P^c(B^o(2r))$. *Then the inequality*

$$\int_{B^o(r)} \int_{B^o(r)} f(x-y)c(x)\overline{c(y)}\,\mathrm{d}x\,\mathrm{d}y = \int_{B^o(2r)} f(x)(c * \tilde{c})(x)\,\mathrm{d}x \geq 0$$

holds for every continuous function c *on* $\mathbb{R}^d$ *whose support lies in* $B^o(r)$.

Notation 4.1.7. Recall that the *support* $\operatorname{supp}(\mu)$ of a measure $\mu = \sum_j c_j \delta_{x_j} \in \mathcal{M}^f(G)$ is given by

$$\operatorname{supp}(\mu) = \{x_j : c_j \neq 0\}.$$

We write

$$\begin{aligned} T_V &:= \{\hat{\mu} : \mu \in \mathcal{M}^f(G), \operatorname{supp}(\mu) \subset V\} \\ T_V^r &:= \{p \in T_V : p \text{ is real-valued}\} \\ T_V^+ &:= \{p \in T_V : p \geq 0\} \end{aligned}$$

Further, let T_V^2 denote the set of all $q \in T_V^+$ of the form

$$q = \sum_{j=1}^n |q_j|^2 \quad \text{for some } n \in \mathbb{N}$$

where

$$q_j = \hat{\mu}_j \in T_V \quad \text{and} \quad \operatorname{supp}(\mu_j) - \operatorname{supp}(\mu_j) \subset V. \tag{1}$$

It is clear that

$$T_V^2 \subset T_V^+ \subset T_V^r \subset T_V$$

T_V is a complex linear space, T_V^r is a real linear space and the sets T_V^2 and T_V^+ are convex cones. Note that T_V^2 contains the constant character and hence $T_V^2 \neq \emptyset$. By (1.8.3.vi) a function $p = \hat{\mu} \in T_V$ belongs to T_V^r if and only if $\mu = \tilde{\mu}$. On T_V and T_V^r we will use the supremum norm

$$\|p\|_\infty = \sup\{|p(\gamma)| : \gamma \in \hat{G}\}.$$

Assume that V is finite. Then both of the linear spaces T_V and T_V^r are finite dimensional and we have $\dim(T_V) = \dim(T_V^r) = d$, where d is the number of points of V. For a complex-valued Hermitian function f on V we set

$$L_f(p) := \sum_{x \in V} f(x)\check{p}(x) = \int_V f \,\mathrm{d}\mu, \quad p = \hat{\mu} \in T_V^r.$$

Then L_f is a real linear functional on T_V^r. Conversely, if L is a real linear functional on T_V^r, then $L = L_f$ with some Hermitian function f on V.

Lemma 4.1.8. *If V is finite, then the cone T_V^2 is closed in T_V^r with respect to the norm $\|\cdot\|_\infty$.*

Proof. Let d denote the dimension of T_V^r. We prove that every $q \in T_V^2$ is a sum of d squares $|q_j|^2$ with some q_j satisfying (4.1.7.1). Suppose that

$$q = \sum_{j=1}^{m} |q_j|^2 \tag{1}$$

where $m > d$. Since $m > \dim(T_V^r)$, there exist real numbers r_j, not all zero, such that

$$\sum_{j=1}^{m} r_j |q_j|^2 = 0.$$

We may assume, without loss of generality, that $|r_1| \geq |r_2| \geq \cdots \geq |r_m|$ and $r_1 \neq 0$. Solving the above equation for $|q_1|^2$ and substituting the solution into (1) we obtain

$$q = \sum_{j=2}^{m} (1 - r_j/r_1)|q_j|^2$$

which is a sum of $m-1$ squares. Repeating these arguments we see that every $q \in T_V^2$ is a sum of d squares.

Suppose that a sequence $\{q_n\}_1^\infty \subset T_V^2$ converges to $q \in T_V^r$ with respect to the norm $\|\cdot\|_\infty$. By the first part of the proof there exist functions $q_{j,n}$ satisfying (4.1.7.1) such that

$$q_n = \sum_{j=1}^{d} |q_{j,n}|^2. \tag{2}$$

The sequence $\{q_n\}_1^\infty$ being convergent, there exists a constant K such that $\|q_n\|_\infty \le K$ for all n. It follows from equation (2) that $\|q_{j,n}\|_\infty^2 \le K$ for all j and n. The normed space T_V being finite dimensional, there is a sequence $n_i \to \infty$ such that q_{j,n_i} converges to some $q'_j \in T_V$, $j = 1, \dots, d$. Since the corresponding measures μ_{j,n_i} converge weakly, q'_j satisfies the condition on the support in (4.1.7.1). Obviously we have

$$q = \sum_{j=1}^{d} |q'_j|^2.$$

Thus, $q \in T_V^2$, showing that T_V^2 is closed. □

Lemma 4.1.9. *Let $f : V \to \mathbb{C}$ be a Hermitian function. Then f is positive definite on V if and only if*[5]

$$L_f(q) \ge 0 \quad for\ all\ q \in T_V^2. \tag{1}$$

Moreover, f can be extended to a positive definite function on G if and only if

$$L_f(p) \ge 0 \quad for\ all\ p \in T_V^+.$$

Proof. To prove the first statement assume that $f \in P(V)$ and let $\mu = \sum_{i=1}^n c_i \delta_{x_i}$ be such that $x_i \in V$ and $x_i - x_j \in V$ for all i, j. Then $\operatorname{supp}(\mu) - \operatorname{supp}(\mu) \subset V$. Since $(\mu * \tilde{\mu})\hat{} = |\hat{\mu}|^2$ we have

$$L_f(|\hat{\mu}|^2) = \int_V f \, \mathrm{d}(\mu * \tilde{\mu}) = \sum_{i,j=1}^{n} f(x_i - x_j) c_i \overline{c_j}.$$

This relation shows that f is positive definite if and only if (1) holds.

To prove the second statement. suppose that f can be extended to a positive definite function $\phi \in P(G)$. If $p \in T_V^+$, then $\check{p}$ is a positive definite function on G with $\operatorname{supp}(\check{p}) \subset V$. Using the fact that $\phi\check{p} \in P(G)$ and applying Theorem 1.9.6 we obtain

$$L_f(p) = \sum_{x \in V} f(x)\check{p}(x) = \sum_{x \in G} \phi(x)\check{p}(x) = (\phi\check{p})\hat{}(0) \ge 0.$$

To prove the converse statement, assume that $L_f(p) \ge 0$ for every $p \in T_V^+$. Without loss of generality we may suppose that $f(0) = 1$. If $p \le q$, $p, q \in T_V^r$, then $q - p \in T_V^+$ and hence $L_f(p) \le L_f(q)$. From this we conclude that

$$|L_f(p)| \le L_f(1) = 1$$

whenever $p \in T_V^r$ and $-1 \le p \le 1$. Thus, L_f is a linear functional of norm[6] 1 on T_V^r. By the Hahn–Banach theorem (cf. Theorem D.4.4), L_f can be extended to a

[5] See Notation 4.1.7 for the definition of L_f.

[6] Note that we consider the supremum norm on $C^r(\hat{G})$.

linear functional L of norm 1 on the linear space $\mathrm{C}^r(\hat{G})$. Since

$$1 = L(1) \le \|L\| = 1$$

L is nonnegative (cf. Lemma E.1.5). The Riesz representation theorem (see Theorem E.1.3) shows the existence of a (nonnegative) measure $\mu \in \mathcal{M}(\hat{G})$ such that

$$L(p) = L_f(p) = \int_{\hat{G}} p(\gamma)\, \mathrm{d}\mu(\gamma), \quad p \in T_V^r. \tag{2}$$

Let $x \in V$ be arbitrary. Since V is symmetric, the function p defined by $p(\gamma) := \gamma(x) + \gamma(-x)$ belongs to T_V^r. Substituting this p into (2) we obtain

$$f(x) + f(-x) = \check{\mu}(x) + \check{\mu}(-x).$$

If we define p by $p(\gamma) := \mathrm{i}(\gamma(x) - \gamma(-x))$, then (2) gives

$$f(x) - f(-x) = \check{\mu}(x) - \check{\mu}(-x).$$

Thus, $f(x) = \check{\mu}(x)$ for all $x \in V$, i.e., $\check{\mu}(x)$ is an extension of f. Since μ is nonnegative the function $\check{\mu}$ is positive definite. The proof is complete. □

Theorem 4.1.10. *Let V be finite. The following two conditions are equivalent:*

(i) *every function $f \in P(V)$ can be extended to a function in $P(G)$;*

(ii) $T_V^+ = T_V^2$.

Proof. Suppose that $T_V^+ = T_V^2$ and let $f \in P(V)$. By the first statement of Lemma 4.1.9 the inequality $L_f(p) \ge 0$ holds for all $p \in T_V^2 = T_V^+$. It follows from the second statement of Lemma 4.1.9 that f admits a positive definite extension.

By Lemma 4.1.8, the convex cone T_V^2 is closed. If $T_V^2 \ne T_V^+$ then there exists a real linear functional L on T_V^r such that L is nonnegative on T_V^2 and $L(p_0) < 0$ for some $p_0 \in T_V^+$ (cf. Corollary D.4.3 with $F = T_V^2$ and $C = \{p_0\}$). Let f be the Hermitian function on V for which $L = L_f$. Lemma 4.1.9 shows that f is positive definite on V but has no positive definite extension. □

Example 4.1.11. Let V_a and γ_r be as in Example 4.1.3 and assume that γ_r can be extended to a function $f_r \in P(\mathbb{T})$. By (4.1.5.v), the function f_r is continuous. Arguing as in the proof of Corollary 1.4.13, we see that f_r must be a character of $\mathbb{T}$. Thus, $f_r(t) = \mathrm{e}^{\mathrm{i}nt}$ with some $n \in \mathbb{Z}$. Consequently, γ_r can be extended to a positive definite function on $\mathbb{T}$ if and only if r is an integer.

In the rest of this section G_1 and G_2 are commutative groups, V_2 is a symmetric subset of G_2 containing 0 and G denotes the product group $G_1 \times G_2$. We identify G_1 and G_2 with the subgroups $G_1 \times \{0\}$ and $\{0\} \times G_2$, respectively. Doing so, every element $x \in G$ has a unique decomposition $x = x_1 + x_2$ with $x_i \in G_i$, $i = 1, 2$.

We will investigate positive definite functions on the set $V := G_1 \times V_2 \subset G$.

Lemma 4.1.12. *Let $f \in P(V)$, $y \in G_1$, and $c \in \mathbb{C}$. Then the function*

$$f_c^y(x) := (1 + |c|^2) f(x) + \overline{c} f(x + y) + c f(x - y), \quad x \in V$$

is positive definite on V.

Proof. It is easy to see that $x - y,\ x + y \in V$ if $x \in V$ and $y \in G_1$. Thus, the function f_c^y is well defined on V. Writing $h_1 := 0,\ h_2 := y,\ d_1 := 1$ and $d_2 := \overline{c}$, we have

$$f_c^y(x) = \sum_{k,l=1}^{2} f(x + h_l - h_k) d_l \overline{d_k}, \quad x \in V.$$

Now let $c_1, \ldots, c_n \in \mathbb{C}$ be arbitrary and $x_1, \ldots, x_n \in V$ be such that $x_i - x_j \in V$. Setting $x_{j,k} := x_j + h_k,\ c_{j,k} := c_j d_k$, using the equation above and the fact that $f \in P(V)$, we obtain

$$\sum_{i,j=1}^{n} f_c^y(x_i - x_j) c_i \overline{c_j} = \sum_{k,l=1}^{2} \sum_{i,j=1}^{n} f(x_{i,l} - x_{j,k}) c_{i,l} \overline{c_{j,k}} \geq 0$$

i.e., $f_c^y \in P(V)$. □

Lemma 4.1.13. *If f is an extremal point of the convex set $P_0(V)$, then*

$$f(x + y) = f(x) f(y)$$

holds for all $x \in G_1$ and $y \in V$.

Proof. This lemma can be proved by the same argument that was used in the first part of the proof of Theorem 1.4.20. We omit the details. □

Theorem 4.1.14. *If every positive definite function on V_2 can be extended to a positive definite function on G_2, then every positive definite function on V has a positive definite extension to G.*

Proof. First we prove that every extremal point f of $P_0(V)$ can be extended to a positive definite function on G. By the previous lemma we have

$$f(x + y) = f(x) f(y), \quad x \in G_1,\ y \in V.$$

The function f_2 defined by $f_2(y) := f(y),\ y \in V_2$, is positive definite on V_2. By assumption, it has an extension $\phi_2 \in P(G_2)$. Setting

$$\phi(x + y) := f(x) \phi_2(y), \quad x \in G_1,\ y \in G_2$$

the function ϕ is positive definite on G (cf. Lemma 1.4.16). Moreover, ϕ is an extension of f.

The set $P_0(V)$ is compact in the topology of pointwise convergence. Kreĭn–Milman's Theorem D.4.5 shows that every function $g \in P_0(V)$ is the pointwise limit of a net $\{f_\alpha\}$, where f_α is a finite convex combination of extremal points. By what we have proved, f_α admits an extension $\phi_\alpha \in P(G)$. Since $P_0(V)$ is compact, the net $\{\phi_\alpha\}$ has a subnet converging pointwise to some complex-valued function ϕ on G. It is clear that ϕ is a positive definite extension of g. □

4.2 The cases $\mathbb{R}^d$ and $\mathbb{Z}^d$

We start with the one-dimensional case.

Theorem 4.2.1. *Let N be a nonnegative integer. If $V = \{n \in \mathbb{Z} : |n| \leq N\}$, then every positive definite function on V can be extended to a positive definite function on $\mathbb{Z}$.*

Proof. By Theorem 4.1.10 it suffices to show that $T_V^+ = T_V^2$. The character group of $\mathbb{Z}$ is $\mathbb{T}$ and each function $p \in T_V$ can be written as

$$p(z) = \sum_{n=-N}^{N} c_n z^n, \quad z \in \mathbb{T}$$

with some complex numbers c_n. The equality $T_V^+ = T_V^2$ follows now at once from Theorem B.1.4. □

Theorem 4.2.2. *Let $a \in (0, \infty)$ and set $V := (-a, a)$.*

(i) *Every function $f \in P(V)$ can be extended to a function $\varphi \in P(\mathbb{R})$.*

(ii) *If f is measurable (continuous) on V, then every positive definite extension of f is measurable (continuous) on $\mathbb{R}$.*

Proof. (i) Let g be a positive definite function on $\mathbb{R}$ such that the support of g is a finite subset of V. We show that the function h, which is equal to fg on V and is equal to zero on $\mathbb{R} \setminus V$, is positive definite on $\mathbb{R}$. For this we prove that the matrix $\big(h(x_i - x_j)\big)_{i,j=1}^n$ is positive semidefinite for all $x_1, \dots, x_n \in \mathbb{R}$. Without loss of generality assume that $x_1 \leq x_2 \leq \cdots \leq x_n$. We set

$$S := \{(i,j) : x_i - x_j \in V\} \quad \text{and} \quad c_{ij} := f(x_i - x_j), \quad (i,j) \in S.$$

Then S and c_{ij} satisfy the conditions of Theorem D.2.19 (see also the beginning of Remark 4.1.2). Thus, there exists a positive semidefinite matrix $\big(a_{ij}\big)_{i,j=1}^n$ such that

$a_{ij} = c_{ij}$ for $(i, j) \in S$. Schur's Theorem D.2.12 shows that the matrix

$$\left(a_{ij}\, g(x_i - x_j)\right)_{i,j=1}^{n}$$

is positive semidefinite. Since the support of g is contained in V, we have $g(x_i - x_j) = 0$ for $(i, j) \notin S$ and hence

$$h(x_i - x_j) = a_{ij}\, g(x_i - x_j), \quad i, j = 1, \ldots, n.$$

Thus, h is positive definite and hence (cf. Theorem 1.9.6)

$$\hat{h}(\mathbf{1}) = \sum_{x \in V} f(x) g(x) \geq 0. \tag{1}$$

Setting[7]

$$L_f(q) := \sum_{x \in V} f(x) \check{q}(x), \quad q \in T_V^r$$

inequality (1) shows that L_f is nonnegative on T_V^+. Considering $\mathbb{R}$ as a discrete group and arguing as in the second part of the proof of (4.1.9.ii), we see that there exists a nonnegative finite measure on the character group of $\mathbb{R}$ such that $f(x) = \check{\mu}(x)$, $x \in V$. Consequently, $\check{\mu}$ is a positive definite extension of f.

(ii) The statements concerning measurability and continuity of φ follow from Theorem 2.11.8 and Corollary 1.5.2. □

As a corollary we obtain the following interesting result.

Theorem 4.2.3. *If an analytic function $g : \mathbb{R} \to \mathbb{C}$ is positive definite on an interval $(-a, a)$, then it is positive definite on $\mathbb{R}$.*

Proof. By the previous theorem, the restriction of g to $(-a, a)$ can be extended to a continuous positive definite function f on $\mathbb{R}$. It follows from (3.3.6.ii) that $f = g$. □

Remark 4.2.4. In a series of papers M. G. Kreĭn investigated the problem of describing all positive definite continuations to $\mathbb{R}$ of a given positive definite function on a finite interval.[8] To give details would go beyond the scope of this book. Below we present an example which was studied in [38].

We consider the function

$$f(t) = 1 - |t|, \quad t \in \mathbb{R}$$

and its restrictions $f_a := f|_{(-2a,2a)}$, $a > 0$. Using Pólya's Theorem 3.9.11 we see that f_a is positive definite if $a \leq 1$. If $a > 1$, then f_a is not positive definite since its

7 We use the Notation 4.1.7 with $G = \mathbb{R}$.

8 See, e.g., [38] or [25] for a list of references.

modulus is not majorized by $f(0)$. If $a \le 1$, then the periodic extension $\tilde{f}_a$ of f_a is positive definite. This can be seen from its Fourier series

$$\tilde{f}_a(t) = 1 - a + \sum_{k=-\infty}^{\infty} \frac{a}{(k+\frac{1}{2})^2\pi^2} e^{it(k+\frac{1}{2})\frac{\pi}{a}}, \quad t \in \mathbb{R}.$$

If $a = 1$ and $\tilde{f}_1$ is a positive definite continuation of f_1 to $\mathbb{R}$, then $\tilde{f}_1(2) = -1$ and therefore $\tilde{f}_1$ is periodic with period 4 in view of Corollary 1.4.14.

Using Kreĭn's results it was shown in [38] that for $a < 1$ the relation

$$i \int_0^\infty e^{izt} \overline{\tilde{f}_{a,\tau}(t)} \, dt =$$

$$\frac{\tan az}{z^2} - \frac{1}{z} - \frac{(a-1)^2}{z^2 \cos^2 az(\tau + (a-1)^2 \tan az + (a-1)/z)}, \quad \operatorname{Im} z > 0$$

establishes a bijective correspondence between all continuations $\tilde{f}_{a,\tau}$ of f_a and all functions τ of the form $\tau \equiv \infty$ or

$$\tau(z) = \alpha + \beta z + \int_{-\infty}^{\infty} \frac{tz+1}{t-z} \, d\sigma(t), \ z \in \mathbb{C} \setminus \mathbb{R}$$

where $\alpha \in \mathbb{R}$, $\beta \ge 0$ and σ is nonnegative finite measure.[9]

We consider two special extensions.

- If $\tau = \infty$, then the corresponding extension is the $4a$-periodic extension of f_a.
- If $\tau(z) = (a-1)^2 i$, $\operatorname{Im} z > 0$, and $\mu_{a,\tau}$ is the measure corresponding to $\tilde{f}_{a,\tau}$. then

$$d\mu_{a,\tau}(x) = \frac{(a-1)^2}{\pi((a-1)^2 x^2 + (a-1)x \sin(2ax) + \cos^2(ax))} dx.$$

This implies the relation

$$\frac{2}{\pi} \int_0^\infty \frac{(a-1)^2 \cos(tx)}{(a-1)^2 x^2 + (a-1)x \sin(2ax) + \cos^2(ax)} dx = 1 - |t|$$

where $0 < a < 1$, $-2a < t < 2a$.

We now turn to the multivariate extension problem. Below we show that the extension is not always possible if $d > 1$ (cf. Theorem 4.2.6 and Theorem 4.2.8). For $M \in \mathbb{N}$ write

$$S_M^d := \{i \in \mathbb{Z}^d : |i_k| \le M \text{ for all } k\}.$$

Theorem 4.2.5. *If $G = \mathbb{Z}^d$ and $V = S_M^d$ where d and M are greater than* 1, *then* $T_V^+ \ne T_V^2$.

---

[9] τ is a so-called *Nevanlinna function*: it is holomorphic in the upper half plane and has a nonnegative imaginary part there.

Proof. We consider only the case $d = 2$, the general case can be treated in the same way. Denote by $\mathcal{L}$ the real linear space of all polynomials

$$P(s,t) = \sum_{m,n=0}^{2M} a_{m,n} s^m t^n, \quad a_{m,n} \in \mathbb{R}$$

of two real variables. Since the character group of $\mathbb{Z}^2$ is $\mathbb{T}^2$, the linear space T_V^r consists of functions of the form

$$p(z_1, z_2) = \sum_{m,n=-M}^{M} c_{m,n} z_1^m z_2^n, \quad z_1, z_2 \in \mathbb{T}$$

where $c_{m,n} \in \mathbb{C}$ and $c_{m,n} = \overline{c_{-m,-n}}$. For $p \in T_V^r$ we define the function Lp by

$$(Lp)(s,t) := (1+s^2)^M (1+t^2)^M p\left(\frac{s+\mathrm{i}}{s-\mathrm{i}}, \frac{t+\mathrm{i}}{t-\mathrm{i}}\right), \quad s,t \in \mathbb{R}.$$

Noting that $(r+\mathrm{i})/(r-\mathrm{i}) \in \mathbb{T}$ for all $r \in \mathbb{R}$, we see that $Lp \in \mathcal{L}$. The mapping L is obviously linear and $Lp \neq 0$ if $p \neq 0$. From

$$\dim(\mathcal{L}) = \dim(T_V^r) = (2M+1)^2$$

we conclude that L is a linear isomorphism from T_V^r onto $\mathcal{L}$. Moreover, $p \geq 0$ if and only if $Lp \geq 0$. Let P be the polynomial from Lemma B.6.4. Then $P \in \mathcal{L}$ and setting $p := L^{-1}P$ we have $p \in T_V^+$. We show that $p \notin T_V^2$.

Suppose, on the contrary, that $p \in T_V^2$. Then there exist $q_j = \hat{\mu}_j \in T_V$ with $p = \sum |q_j|^2$ where

$$\operatorname{supp}(\mu_j) \subset V \quad \text{and} \quad \operatorname{supp}(\mu_j) - \operatorname{supp}(\mu_j) \subset V. \tag{1}$$

We may suppose that

$$\operatorname{supp}(\mu_j) \subset S_M^+ := \{(k,l) \in \mathbb{Z}^2 : 0 \leq k, l \leq M\}.$$

Indeed, it follows from (1) that there exists $x_j \in \mathbb{Z}^2$ such that $\operatorname{supp}(\delta_{x_j} * \mu_j) \subset S_M^+$. Moreover, we have

$$\left|(\delta_{x_j} * \mu_j)^\wedge\right|^2 = |\hat{\mu}_j|^2 = |q_j|^2.$$

Since $\operatorname{supp}(\mu_j) \subset S_M^+$, the function

$$Q_j(s,t) := (s-\mathrm{i})^M (t-\mathrm{i})^M q_j\left(\frac{s+\mathrm{i}}{s-\mathrm{i}}, \frac{t+\mathrm{i}}{t-\mathrm{i}}\right)$$

is a complex polynomial in each of the variables s and t. Applying the definition of L we see that $|Q_j|^2 = L|q_j|^2$. Consequently,

$$P = Lp = \sum L|q_j|^2 = \sum |Q_j|^2.$$

Writing $Q_j = g_j + \mathrm{i}h_j$ where g_j and h_j are polynomials with real coefficients, we obtain

$$P = \sum (g_j^2 + h_j^2)$$

in contradiction to the choice of P. Thus, $p \notin T_V^2$. □

Combining Theorem 4.1.10 and Theorem 4.2.5 we obtain the following theorems:

Theorem 4.2.6 (Calderón–Pepinsky). *If d and M are greater than 1, then there exists a positive definite function on S_M^d which cannot be extended to a positive definite function on $\mathbb{Z}^d$.*

Theorem 4.2.7. *Let $M, d \in \mathbb{N}$ be arbitrary. Every positive definite function on S_M^d can be extended to a continuous positive definite function on the set*

$$T_M^d := \{t \in \mathbb{R}^d : |t_j| \le M \text{ for all } j\} \subset \mathbb{R}^d.$$

Proof. Write

$$U := \{t \in \mathbb{R}^d : |t_j| < 1/2 \text{ for all } j\}$$

and choose a continuous function h such that $\operatorname{supp}(h) \subset U$ and $\int_{\mathbb{R}^d} |h|^2 \, \mathrm{d}\lambda = 1$. Let $K := h * \tilde{h}$ and define ϕ by

$$\phi(t) := \sum_{n \in S_M^d} f(n) K(t-n), \quad t \in \mathbb{R}^d.$$

Note that K and ϕ are continuous. Using that $K(0) = 1$ and $K(m) = 0$ if $m \in \mathbb{Z}^d \setminus \{0\}$, we see that ϕ is an extension of f. It remains to prove that ϕ is positive definite, i.e.,

$$\sum_{i,j=1}^{n} \phi(x_i - x_j) \overline{c_i} c_j \ge 0 \tag{1}$$

for every choice of $n \in \mathbb{N}$, $c_i \in \mathbb{C}$ and points $x_i \in T_M^d$ with $x_i - x_j \in T_M^d$. We may suppose that the x_i's have nonnegative coordinates. Indeed, denote by r_k the smallest of the k-th coordinates of the points x_i and write $x_0 := (r_1, \dots, r_d)$ and $y_i := x_i - x_0$. The points y_i lie in T_M^d, have nonnegative coordinates, and $\phi(x_i - x_j) = \phi(y_i - y_j)$.

Put

$$H(y) := \sum_{i=1}^{n} c_i h(y - x_i), \quad y \in \mathbb{R}^d. \tag{2}$$

Setting $f(n) := 0$ outside of S_M^d and using the definition of ϕ we see that the left-hand side of (1) is equal to

$$\sum_{n \in \mathbb{Z}^d} f(n) \int_{\mathbb{R}^d} H(y-n) \overline{H(y)} \, \mathrm{d}y. \tag{3}$$

It follows from

$$\int_{\mathbb{R}^d} H(y-n)\overline{H(y)}\,\mathrm{d}y = \sum_{m\in\mathbb{Z}^d}\int_U H(y-n-m)\overline{H(y-m)}\,\mathrm{d}y$$

that (3) is equal to

$$\int_U \Bigg(\sum_{m,n\in\mathbb{Z}^d} f(m-n)\overline{H(y+m)}H(y+n)\Bigg)\,\mathrm{d}y. \tag{4}$$

Using the definition of H we see that $H(y+n)$, where $y \in U$ and $n \in \mathbb{Z}^d$, can be different from zero only if $n \in S_M^d$ and n has nonnegative coordinates. Thus, (4) is not changed if we restrict m and n to lie in S_M^d and to have nonnegative coordinates. For such m and n we have $m-n \in S_M^d$. Since f is positive definite on S_M^d, the integrand in (4) is nonnegative from which (1) follows. □

The next result follows from Theorems 4.2.6 and 4.2.7.

Theorem 4.2.8 (Rudin). *Let $d \geq 2$ and $V \subset \mathbb{R}^d$ be a symmetric closed square. Then there exists a continuous function on V which cannot be extended to a positive definite function on $\mathbb{R}^d$.*

In Section 4.4 we return to the extension problem and show that if V is a ball in $\mathbb{R}^d$ and f is a continuous radial positive definite function on V, then there exists a radial positive definite function on $\mathbb{R}^d$ that extends f.

4.3 Decomposition of locally defined positive definite functions

Throughout the present section V denotes a symmetric open neighborhood of $0 \in \mathbb{R}^d$. The main result of this section is that every positive definite function f on V can be written as $f = f_c + f_0$ where f_c and f_0 are positive definite, f_c is continuous and f_0 is near to zero in a certain sense.

Definition 4.3.1. Let h be a complex-valued function on V. We say that h *averages to zero on V* if for every neighborhood W of 0 and for any $\epsilon > 0$ there exist $x_1,\dots,x_n \in W$ and positive numbers $p_1,\dots,p_n$ summing to 1 such that

$$\left|\sum_{i=1}^n p_i h(x+x_i)\right| < \epsilon$$

whenever $x \in V$ and $x + x_i \in V$ for all i.

Example 4.3.2. The function $\mathbf{1}_{\{t\}}$, $t \in V$, averages to zero on V. Indeed, if $x_1, \dots, x_n \in V$ are mutually distinct, then for all $x \in V$ the equation $x + x_i = t$ holds for at most one i and hence

$$\left| \sum_{i=1}^{n} p_i \mathbf{1}_{\{t\}}(x + x_i) \right| \leq \frac{1}{n}$$

where $p_i = 1/n$. More generally, functions with finite support average to zero.

To see another example, consider the subgroup $\mathbb{Q}^d \subset \mathbb{R}^d$ and let h be the restriction of $\mathbf{1}_{\mathbb{Q}^d}$ to V. Then for all $n \in \mathbb{N}$ there exist $x_1, \dots, x_n \in V$ such that $x_i - x_j \notin \mathbb{Q}^d$ if $i \neq j$. Choosing an arbitrary $x \in V$, the relation $x + x_i \in \mathbb{Q}^d$ holds for at most one i. As above, we conclude that h averages to zero.

Theorem 4.3.3. *Let h be a bounded Lebesgue measurable function on V. If h averages to zero on V, then h is equal to zero λ-almost everywhere.*

Proof. We may assume that h is real-valued. If h does not vanish λ-almost everywhere, then there exist $x_0 \in V$ and an open symmetric neighborhood W of 0 such that $x_0 + W + W \subset V$ and h does not vanish λ-almost everywhere on $x_0 + W$. We choose a function $g \in \mathrm{C}_{00}^r(\mathbb{R}^d)$ vanishing off $-x_0 + W$ such that

$$\int_{W + x_0} h(y) g(-y)\, dy = 1.$$

Let H be the function on $\mathbb{R}^d$ which is equal to h on V and is equal to zero otherwise. The function $H * g$ is continuous and $H * g(0) = 1$. Let W_0 be an open set such that $0 \in W_0 \subset W$ and $H * g(x) > \frac{1}{2}$ for all $x \in W_0$. If $p_i \geq 0$ and $\sum p_i = 1$, then we have

$$\sum_{i=1}^{n} p_i H * g(x_i) > \frac{1}{2}$$

for all $x_i \in W_0$. Since h averages to zero, for all $\epsilon > 0$ we can choose p_i and x_i such that the inequality

$$\left| \sum_{i=1}^{n} p_i h(y + x_i) \right| < \epsilon$$

holds whenever $x, y + x_i \in V$. Using that $y \in x_0 + W$ implies $y + x_i \in x_i + x_0 + W \subset x_0 + W + W \subset V$, we obtain

$$\begin{aligned}
\frac{1}{2} < \sum_{i=1}^{n} p_i H * g(x_i) &= \int_{\mathbb{R}^d} \left[\sum_{i=1}^{n} p_i H(y + x_i) \right] g(-y)\, dy \\
&\leq \int_{\mathbb{R}^d} |g(-y)|\, dy \cdot \sup_{y \in x_0 + W} \left| \sum_{i=1}^{n} p_i h(y + x_i) \right| \\
&\leq \epsilon \int_{\mathbb{R}^d} |g(-y)|\, dy
\end{aligned}$$

for all $\epsilon > 0$. This contradiction shows that h vanishes λ-almost everywhere. □

The main result of this section is the following theorem:

Theorem 4.3.4 (Sasvári). *Every function $f \in P(V)$ admits a unique decomposition*

$$f = f_c + f_0$$

where $f_c \in P^c(V)$, $f_0 \in P(V)$ and f_0 averages to zero. If f is Lebesgue measurable, then f_0 vanishes λ-almost everywhere.[10]

First we introduce some notation and basic facts that will be used in the proof.

Notation 4.3.5. Let $\mu = \sum_i c_i \delta_{t_i}$ and $\nu = \sum_j d_j \delta_{s_j}$ be measures in $\mathcal{M}^f(V)$ such that $s_j - t_i \in V$ for all i, j. We write[11]

$$(\mu, \nu) := \sum_{i,j} f(s_j - t_i) c_i \overline{d_j}.$$

Since f is positive definite we have $(\mu, \mu) \geq 0$. Moreover, the Cauchy–Schwarz inequality

$$|(\mu, \nu)|^2 \leq (\mu, \mu)(\nu, \nu) \tag{1}$$

holds whenever all expressions are defined. This can be proved in the same way as in positive semidefinite inner product spaces. For $t \in V$ and $\mu \in \mathcal{M}^f(V)$ such that $\delta_t * \mu \in \mathcal{M}^f(V)$ we set

$$U_t \mu := \delta_t * \mu.$$

Then the mapping $\mu \mapsto U_t \mu$ is linear, where it is defined, and we have

$$(U_t \mu, U_t \mu) = (\mu, \mu) \tag{2}$$

$$f(t) = (\delta_0, U_t \delta_0). \tag{3}$$

Let $\{W_n\}_{n=1}^{\infty}$ be a sequence of balls with center zero such that

$$W_1 + W_1 \subset V, \quad W_{n+1} + W_{n+1} \subset W_n \quad \text{and} \quad \bigcap_{n=1}^{\infty} W_n = \{0\}.$$

For all $t \in V$ let $N(t) \in \mathbb{N}$ be the smallest integer such that $t + W_{N(t)} \subset V$. Then $N(0) = 1$ and $\{t + W_n\}_{n=N(t)}^{\infty}$ is a neighborhood basis of t. If $n \geq N(t)$, then $(\cdot, \cdot)$ is a well-defined semidefinite inner product on the linear manifold $\mathcal{M}^f(t + W_n)$. Therefore, $\mathcal{M}^f(t + W_n)$ can be completed to a Hilbert space $H_n(t)$. The element of

---

[10] The case where V is the whole group has been considered in [11] in a more general setting.

[11] Note that if $V \neq \mathbb{R}^d$, then $(\cdot, \cdot)$ is not an inner product in $\mathcal{M}^f(V)$ since it is not defined for all pairs of μ and ν.

$H_n(t)$ corresponding to a $\mu \in \mathcal{M}^f(t + W_n)$ will also be denoted by μ. We assume that the completions are chosen such that[12] $t+W_n \subset s+W_m$ implies $H_n(t) \subset H_m(s)$.

For all $t, s \in V$ such that $t - s \in V$ let $N(t,s) \in \mathbb{N}$ be the smallest integer such that $t + W_{N(t,s)} - (s + W_{N(t,s)}) \subset V$. If $n \geq N(t,s)$, then (μ, ν) is well-defined for all $\mu \in \mathcal{M}^f(t + W_n)$ and $\nu \in \mathcal{M}^f(s + W_n)$. In view of inequality (1), $(\cdot,\cdot)$ is a bounded sequilinear functional on $\mathcal{M}^f(t + W_n) \times \mathcal{M}^f(s + W_n)$ and hence it has a unique extension to a bounded sequilinear functional on $H_n(t) \times H_n(s)$ which we also denote by $(\cdot,\cdot)$. By continuity, the inequality

$$|(h,g)|^2 \leq (h,h)(g,g)$$

holds for all $h \in H_n(t)$ and $g \in H_n(s)$ if $n \geq N(t,s)$.

Let $t \in V$ be fixed. Since $U_t\mu \in \mathcal{M}^f(s - t + W_{N(t,s)})$ for all measures $\mu \in \mathcal{M}^f(s + W_{N(t,s)})$, the isometric linear mapping

$$U_t : \mathcal{M}^f(s + W_{N(t,s)}) \to \mathcal{M}^f(s - t + W_{N(t,s)})$$

can be uniquely extended to an isometric mapping

$$U_t : H_{N(t,s)}(s) \to H_{N(t,s)}(s - t).$$

Doing so for all $s \in V$ we obtain a linear isometric mapping[13] U_t defined on

$$\operatorname{dom}(t) := \bigcup_{s \in V} H_{N(t,s)}(s).$$

Using the properties of U_t on $\mathcal{M}^f(s + W_{N(t)})$ we see that the relations

$$(U_{t+s}h, g) = (U_t U_s h, g) = (U_s h, U_{-t} g) \tag{4}$$

hold for all $t, s \in V$ and for all h, g for which all the expressions above are well-defined.

Set

$$H_0 := \bigcap_{n=1}^{\infty} H_n(0).$$

Then $\delta_0 \in H_0$ and H_0 is a Hilbert space. Since $H_{N(t,0)}(0) \subset \operatorname{dom}(t)$, the mapping U_t is well-defined on H_0.

Next we define two semigroups of operators that will play an important role in the proof of Theorem 4.3.4. If $m > 1$, then $W_m + W_m \subset W_{m-1}$ and hence the relations $t \in W_m$ and $\mu \in \mathcal{M}^f(W_m)$ imply that $U_t \in \mathcal{M}^f(W_1)$. Thus, U_t maps $H_m(0)$ and H_0 into $H_{m-1}(0)$. For $t \in W_m$ we denote by U_t^0 the restriction of U_t to H_0. Then U_t^0

[12] That this is possible can be easily seen, e.g., by applying the completion procedure which uses Cauchy sequences.

[13] We use the word "mapping" to indicate that domain and range of U_t need not be Hilbert spaces.

is an isometric linear operator on H_0 with range in $H_{m-1}(0)$. For $m > 1$ set

$$\mathcal{U}_m := \{U_t^0 : t \in W_m\}$$

and denote by $\mathcal{U}_m^-$ the closure of $\mathcal{U}_m$ in the weak operator topology.[14] Write further

$$S_0 := \bigcap_{m>1} \mathcal{U}_m^-$$

and let S be the closure, in the weak operator topology, of the convex hull of S_0. The set $\mathcal{U}_m$ consists of operators of norm 1 and hence the sets $\mathcal{U}_m^-$ and S_0 are weakly compact. By Theorem D.5.9, the set S is weakly compact as well. Moreover, in view of Theorem D.5.5, the norm of the operators in S and S_0 is at most 1.

We now show several properties of S, S_0 and the mappings U_t.

Lemma 4.3.6. *The operators in S and S_0 map H_0 into H_0.*

Proof. It suffices to prove the statement concerning S_0. If $m' > m > 1$, then we have $W_{m'} + W_{m'} \subset W_m$ and

$$U_t^0 H_0 \subset H_m(0), \quad t \in W_{m'}.$$

Consequently,

$$\mathcal{U}_{m''}^- H_0 \subset H_m(0), \quad m'' \geq m'.$$

We conclude that $S_0 H_0 \subset H_m(0)$ whenever $m > 1$ and therefore $S_0 H_0 \subset H_0$. □

In the sequel we will consider the elements of S and S_0 as operators from H_0 into H_0.

Lemma 4.3.7. *S and S_0 are semigroups.*

Proof. Let $s_1, s_2 \in S_0$ be arbitrary. We show that $s_1 s_2 \in \mathcal{U}_m$ for all $m > 1$. If $m' > m$, then $W_{m'} + W_{m'} \subset W_m$ and $\mathcal{U}_{m'}\mathcal{U}_{m'} \subset \mathcal{U}_m$. Since multiplication of operators is separately continuous in the weak operator topology (cf. Lemma D.5.2), we infer that $\mathcal{U}_{m'}^-\mathcal{U}_{m'}^- \subset \mathcal{U}_m^-$. From the definition of S_0 we see that $s_1 s_2 \in \mathcal{U}_{m'}^-\mathcal{U}_{m'}^-$ and hence $s_1 s_2 \in \mathcal{U}_m^-\mathcal{U}_m^-$ for all $m > 1$. Consequently, $s_1 s_2 \in S_0$ showing that S_0 is a semigroup. Since the convex hull of S_0 is also a semigroup, the same is true for the weak operator closure S. □

Notation 4.3.8. By Theorem D.5.11 the semigroup S contains an orthogonal projection P such that $sP = Ps = P$ for all $s \in S$. We introduce the notation $\mu_c := P\delta_0$ and $\mu_0 := \delta_0 - \mu_c$.

---

[14] We consider the weak operator topology on the set of all bounded linear operators from H_0 to $H_{m-1}(0)$, cf. D.5.1.

Lemma 4.3.9. *Let $m' < m$. The mapping $t \mapsto U_t^0 \mu_c$ from $W_{m'}$ into $H_m(0)$ is weakly continuous at* 0.

Proof. We show that $U_{t_n}^0 \mu_c \to \mu_c$ for an arbitrary sequence $t_n \to 0$, $t_n \in W_m$. The sequence $\{U_{t_n}^0\}$ can have cluster points only in S_0. Consequently, $\{U_{t_n}^0 \mu_c\}$ can only have cluster points in $S_0 \mu_c$. Using the relation

$$sP = P, \quad s \in S_0 \subset S$$

we see that $S_0 \mu_c = S_0 P \delta_0 = \{P\delta_0\} = \{\mu_c\}$ and therefore $U_{t_n} \mu_c \to \mu_c$. □

Lemma 4.3.10. *The function g defined by*

$$g(t) = (U_t^0 \mu_c, h), \quad t \in V$$

is continuous on V for all $h \in H_0$.

Proof. The function g is continuous at $t_0 \in V$ if and only if the function

$$g_0(t) := (U_{t+t_0} \mu_c, h)$$

which is defined in a neighborhood of 0, is continuous at 0. Using (4.3.5.4) we obtain

$$(U_{t+t_0} \mu_c, h) = (U_t \mu_c, U_{-t_0} h). \tag{1}$$

Since $v \mapsto (v, U_{-t_0} h)$ is a continuous linear functional on H_0, the continuity of g_0 at 0 follows at once from equation (1) and from Lemma 4.3.9. □

Lemma 4.3.11. *For all $\epsilon > 0$ and $m \in \mathbb{N}$ there exist $t_1, \dots, t_n \in W_m$ and $p_1, \dots, p_n \geq 0$ summing to 1 such that*

$$\left\| \sum_{i=1}^{n} p_i U_{t_i} \mu_0 \right\| < \epsilon. \tag{1}$$

Proof. Since $0 = P\mu_0 \in S\mu_0$, the lemma follows from the definitions of S and S_0 and from the fact that the weak and norm closures of the convex hull of $S_0 \mu_0$ coincide. □

Lemma 4.3.12. *For all $t_0 \in V$ we have*

$$(U_{t_0} \mu_c, \mu_0) = (U_{t_0} \mu_0, \mu_c) = 0.$$

Proof. In view of Lemma 4.3.10 the function $t \mapsto (U_t \mu_c, \mu_0)$ is continuous on V. Thus, for all $\epsilon > 0$ there exists $N \in \mathbb{N}$ such that

$$|(U_{t_0 - t} \mu_c, \mu_0) - (U_{t_0} \mu_c, \mu_0)| < \epsilon, \quad t \in W_N.$$

Consequently, if $p_i \geq 0,\ p_1 + \cdots + p_n = 1$ and $t_i \in W_N$, then

$$\left|\left(\sum_{i=1}^{n} p_i U_{t_0 - t}\mu_c, \mu_0\right) - (U_{t_0}\mu_c, \mu_0)\right| < \epsilon, \quad t \in W_N. \tag{1}$$

By the previous lemma we can choose p_i and t_i such that inequality (4.3.11.1) holds. We then have

$$\left|\left(\sum_{i=1}^{n} p_i U_{t_0 - t_i}\mu_c, \mu_0\right)\right| = \left|\left(U_{t_0}\mu_c, \sum_{i=1}^{n} p_i U_{t_i}\mu_0\right)\right| \leq \epsilon \cdot \|U_{t_0}\mu_c\|.$$

Comparing this with (1) we obtain $(U_{t_0}\mu_c, \mu_0) = 0$ and hence $(U_{t_0}\mu_0, \mu_c) = (\mu_0, U_{-t_0}\mu_c) = 0$. □

Proof of Theorem 4.3.4. Setting

$$f_c(t) := (\mu_c, U_t\mu_c), \quad f_0(t) := (\mu_0, U_t\mu_0), \quad t \in V$$

the functions f_c and f_0 are positive definite on V. Indeed, write $\varphi(t) := (x, U_t x)$, $t \in V$, where $x \in H_0$ is arbitrary. Using the relation (4.3.5.4) we obtain

$$\sum_{i,j=1}^{n} \varphi(t_i - t_j)c_i\overline{c_j} = \left(x, \sum_{i,j=1}^{n} c_i\overline{c_j}U_{t_i - t_j}x\right) = \left(\sum_{i=1}^{n} c_i U_{t_i}x, \sum_{i=1}^{n} c_i U_{t_i}x\right) \geq 0$$

where $t_j \in V$ and $c_j \in \mathbb{C}$.

By the previous lemma we have

$$f(t) = (\delta_0, U_t\delta_0) = (\mu_c + \mu_0, U_t(\mu_c + \mu_0)) = f_c(t) + f_0(t), \quad t \in V.$$

The continuity of f_c follows from Lemma 4.3.10. To prove that f_0 averages to zero, let $\epsilon > 0$ and $m \in \mathbb{N}$ be arbitrary and choose t_j and p_j as in Lemma 4.3.11. We then have

$$\left|\sum_{i=1}^{n} p_i f_0(t - t_i)\right| = \left|\left(\mu_0, \sum_{i=1}^{n} p_i U_{t - t_i}\mu_0\right)\right| \tag{2}$$

$$= \left|\left(\sum_{i=1}^{n} p_i U_{t_i}\mu_0, U_t\mu_0,\right)\right| \tag{3}$$

$$\leq \|U_t\mu_0\| \cdot \left\|\sum_{i=1}^{n} p_i U_{t_i}\mu_0\right\| \leq \epsilon. \tag{4}$$

To prove the uniqueness, assume that $f = f_c + f_0 = g_c + g_0$ where $g_c \in P^c(V)$, $g_0 \in P(V)$ and g_0 averages to zero. Then the function $h := f_c - g_c = g_0 - f_0$ averages to zero and it is continuous. Theorem 4.3.3 shows that $h = 0$, i.e., we have $f_c = g_c$ and $f_0 = g_0$.

The second statement follows from Theorem 4.3.3. □

Recall that $B^o(r) \subset \mathbb{R}^d$ denotes the open ball with center 0 and radius $r > 0$. A function g defined on $B^o(r)$ is called a *radial function* if $g(t) = g(Ot)$ for all $t \in B^o(r)$ and for all orthogonal matrices $O \in \mathcal{O}(d)$. Taking $O = -I$ we see that radial positive definite functions on $B^o(r)$ are real-valued.

The same argument as in Lemma 3.6.2 shows that $g : B^o(r) \to \mathbb{C}$ is radial if and only if there exists a complex-valued function h on $[0, r)$ such that

$$g(t) = h(\|t\|), \quad t \in B^o(r).$$

Lemma 4.3.13. *If $V = B^o(r)$ and $f \in P(V)$ is radial, then the functions f_c and f_0 in Theorem 4.3.4 are radial as well.*

Proof. Let $e \in B^o(1)$ be arbitrary and define the function h by

$$h(t) := f(te), \quad t \in I := (-r, r).$$

Since f is radial h does not depend on e. By Theorem 4.3.4, $h = h_c + h_0$ where $h_c \in P^c(I)$, $h_0 \in P(I)$ and h_0 averages to zero. On the other hand, we have $h(t) = f_c(te) + f_0(te)$. The uniqueness statement of Theorem 4.3.4 shows that $f_c(te) = h_c(t)$ and $f_0(te) = h_0(t)$ for all t. Thus, f_c and f_0 are radial. □

Lemma 4.3.14. *If $V = B^o_d(r)$, $d \geq 2$, and $f_0 \in P(V)$ is a radial function which vanishes Lebesgue almost everywhere, then $f_0 = p\mathbf{1}_{\{0\}}$ with some nonnegative constant.*[15]

Proof. We may assume that $f_0(0) = 1$. Let k be a positive integer and let $0 < q < r$. Let $t \in \mathbb{R}^d$ be a fixed point of norm q and choose $\delta > 0$ such that $S_\delta - S_\delta \subset B^o(r)$ where

$$S_\delta = \{x \in \mathbb{R}^d : \|x\| = q, \ \|x - t\| \leq \delta\}.$$

Choose independent random vectors $x_1, \ldots, x_k$ on a probability space $(\Omega, \mathcal{A}, P)$ such that $x_j(\omega) \in S_\delta$ for all ω and x_j is uniformly distributed in S_δ for all j. Further, let x_0 be the random variable on Ω which is identically zero. Write $\alpha_0 = 1$ and $\alpha_j = -f_0(t)$, $1 \leq j \leq k$. Since f_0 is positive definite the inequality

$$\sum_{i=0}^{k} \sum_{j=0}^{k} f_0(x_i(\omega) - x_j(\omega))\alpha_i \alpha_j = \\ 1 - kf(t)^2 + 2f(t)^2 \sum_{1 \leq i < j \leq k} f_0(x_i(\omega) - x_j(\omega)) \geq 0$$

holds for all $\omega \in \Omega$. Assume that $d = 2$. It is easy to see that the random variables $\|x_i - x_j\|$, $i \neq j$, are absolutely continuous with respect to the Lebesgue measure on

[15] The example $f = \mathbf{1}_{\mathbb{Q}}$ shows that the lemma is not true if $d = 1$.

$\mathbb{R}$.[16] Since f_0 is radial and vanishes Lebesgue almost everywhere, we conclude that $\mathbb{E}\left(f_0(x_i - x_j)\right) = 0,\ 1 \le i < j \le k$. Thus, the inequality $1 - kf(t)^2 \ge 0$ holds for all k from which $f(t) = 0$ follows. This proves the lemma for $d = 2$.

In the case $d > 2$, we consider the function $g_0(s) := f_0(s, 0, \dots, 0),\ s \in B_2^o(r)$. By the first part of the proof, $g_0(s) = 0$ for all $s \ne 0$. Since f_0 is radial we infer that $f_0(t) = 0$ for all $t \in B^o(r) \setminus \{0\}$. □

Combining Theorem 4.3.4 and the preceding lemmata we obtain the following result.[17]

Theorem 4.3.15. *Let $d \ge 2$ and f be a Lebesgue measurable radial positive definite function on $B^o(r) \subset \mathbb{R}^d$. Then f admits the decomposition*

$$f = f_c + p\mathbf{1}_{\{0\}}$$

where f_c is a continuous radial positive definite function on $B^o(r)$ and p is a non-negative constant.

4.4 Extension of radial positive definite functions

We have seen in Section 4.2 that there exist locally defined positive definite functions which cannot be extended to the whole of $\mathbb{Z}^d$ or $\mathbb{R}^d$. The situation changes if we assume radiality.

Theorem 4.4.1 (Rudin). *Every radial function $\varphi \in P^c(B_d^o(2r))$ can be extended to a radial positive definite function on $\mathbb{R}^d$.*

Proof. Denote by C_r^∞ the set of all radial, infinitely differentiable functions on $\mathbb{R}^d$ whose support lies in $B^o(2r)$ and let P_r be the set of all $f \in C_r^\infty$ which are positive definite. By Theorem 3.10.4, each $f \in P_r$ is the sum of a uniformly convergent series

$$f = \sum_{k=1}^{\infty} f_k * \tilde{f_k} \tag{1}$$

where f_k is infinitely differentiable and vanishes outside $B^o(r)$. Note that $\hat{f}$ is non-negative and radial.

Fix $f \in C_r^\infty$, write $M_f := \|\hat{f}\|_\infty$, and choose $\eta \in (0, 1)$ such that the support of f lies in $B^o(2r - \eta)$. By Lemma 3.2.4 there exists a sequence $\{g_i\}$ of functions $g_i \in P_r$ such that

[16] This argument is due to T. Gneiting.

[17] See also [21] and [50] for decomposition results concerning more general norms.

(i) $\operatorname{supp} g_i \subset B^o(\eta)$

(ii) $\hat{g}_i \to 1$ uniformly on compact sets

(iii) $\int_{\mathbb{R}^d} g_i h \, d\lambda \to h(0)$ for every continuous function h.

Let $\delta > 0$ be arbitrary. There is a compact set $K \subset \mathbb{R}^d$ such that

$$|\hat{f}(y)| < \delta, \quad y \in \mathbb{R}^d \setminus K$$

and there is a positive integer i_0 such that

$$M_f \hat{g}_i(y) < M_f - \delta, \quad y \in K, \, i \geq i_0.$$

Then $\hat{f} < M_f \hat{g}_i + \delta$ on $\mathbb{R}^d$. Multiplying this inequality by $\hat{g}_i$ we see that the Fourier transform of the function

$$h_i := M_f g_i * g_i + \delta g_i - f * g_i$$

is nonnegative. Noting that the support of h_i lies in $B^o(2r)$ we conclude that $h_i \in P_r$. Proposition 4.1.6 together with (1) shows that

$$\int_{B^o(2r)} (f * g_i)\varphi(x) \, dx \leq \int_{B^o(2r)} [\delta g_i(x) + M_f(g_i * g_i)(x)]\varphi(x) \, dx, \quad i \geq i_0.$$

Letting $i \to \infty$ we obtain

$$\int_{B^o(2r)} f(x)\varphi(x) \, dx \leq (\delta + M_f)\varphi(0).$$

Letting $\delta \to 0$ and applying the same argument to $-f$ in place of f we obtain the inequality

$$\left| \int_{B^o(2r)} f(x)\varphi(x) \, dx \right| \leq M_f \varphi(0), \quad f \in C_r^\infty. \tag{2}$$

Denote by E the linear subspace of all $g \in C_0(\mathbb{R}^d)$ which are integrable and $\check{g} \in C_r^\infty$. We endow $C_0(\mathbb{R}^d)$ with the supremum norm and define the linear functional L by

$$Lg := \int_{B^o(2r)} \check{g}(x)\varphi(x) \, dx, \quad g \in E.$$

By inequality (2) we have $|Lg| \leq \varphi(0)\|g\|_\infty$, $g \in E$. The Hahn–Banach Theorem D.4.4 shows that L can be extended to a linear functional $\tilde{L}$ on $C_0(\mathbb{R}^d)$ such that $|\tilde{L}g| \leq \varphi(0)\|g\|_\infty$, $g \in C_0(\mathbb{R}^d)$. Therefore, in view of the Riesz–Markov Theorem E.1.4, there is a complex Radon measure μ on $\mathbb{R}^d$ such that

$$\int_{B^o(2r)} f(x)\varphi(x) \, dx = \int_{\mathbb{R}^d} \hat{f}(y) \, d\mu(y), \quad f \in C_r^\infty \tag{3}$$

and $|\mu|(X) \leq \varphi(0)$. It is not hard to check that μ is radial and real. Applying equation (3) to g_i in place of f and letting $i \to \infty$, we obtain $\varphi(0) = \mu(\mathbb{R}^d) \leq |\mu|(\mathbb{R}^d)$.

Hence μ is nonnegative. Equation (3) is the same as

$$\int_{B^o(2r)} f(x)\varphi(x)\,\mathrm{d}x = \int_{B^o(2r)} f(x) \int_{\mathbb{R}^d} \mathrm{e}^{-\mathrm{i}(x,y)}\,\mathrm{d}\mu(y)\,\mathrm{d}x, \quad f \in \mathrm{C}_r^\infty.$$

Since this holds for all $f \in \mathrm{C}_r^\infty$ we conclude that

$$\varphi(x) = \int_{\mathbb{R}^d} \mathrm{e}^{-\mathrm{i}(x,y)}\,\mathrm{d}\mu(y), \quad x \in B^o(2r).$$

This right-hand side of this equation defines a radial positive definite function on the whole of $\mathbb{R}^d$. This completes the proof of the theorem. □

Combining Theorem 4.4.1 with Theorem 4.3.15 we obtain the following result.

Theorem 4.4.2 (Gneiting–Sasvári). *Every Lebesgue measurable radial function in $P(B_d^o(2r))$ can be extended to a Lebesgue measurable radial positive definite function on the whole of $\mathbb{R}^d$.*

Chapter 5

Selected applications

A chapter on applications could easily fill several hundred pages so that we had to be very restrictive in our choice of the topics. We start with the probably 'most classical' application of characteristic functions: limit theorems. Even here we had to restrict ourselves to present only basic methods. In the remaining sections we treat topics which usually cannot be found in other textbooks in this form.

5.1 Limit theorems

First we prove a version of the *central limit theorem.*

Theorem 5.1.1. *Let μ be a d-dimensional distribution such that the first order moments are all equal to zero while the second order moments are finite. If $X_1, X_2, \dots$ are independent random vectors all having the distribution μ, then the distributions of the random vectors*

$$S_n = \frac{X_1 + X_2 + \dots + X_n}{\sqrt{n}}$$

tend weakly, as $n \to \infty$, to the normal distribution having the same first and second moments as μ.

Proof. Denoting by f the characteristic function of μ, the characteristic function f_n of S_n is given by $f_n(t) = f(t/\sqrt{n})^n$. By Corollary 1.2.7,

$$f_n(t) = \left[1 - \frac{1}{2n} \sum_{i,j=1}^{d} [\mathbb{E}(X_i X_j) + R_{i,j}(t/\sqrt{n})]\, t_i t_j\right]^n, \quad t \in \mathbb{R}^d$$

where $\lim_{t\to 0} R_{i,j}(t) = 0$. Using Lemma B.1.2 we obtain

$$\lim_{n\to\infty} f_n(t) = \mathrm{e}^{-\frac{1}{2}\sum_{i,j=1}^{d} \mathbb{E}(X_i X_j)\, t_i t_j}.$$

The statement now follows from Lévy's continuity Theorem 1.6.3. □

The next application is *Khinchin's weak law of large numbers.*

Theorem 5.1.2. *Let $X_1, X_2, \dots$ be independent identically distributed d-dimensional random vectors and put*

$$S_n = \frac{X_1 + X_2 + \cdots + X_n}{n}.$$

If $\mathbb{E}(X_1) = m$ exists then

$$\lim_{n\to\infty} P\left(\|S_n - m\| \geq \epsilon\right) = 0$$

for all $\epsilon > 0$.

Proof. Denote by f the characteristic function of the random vectors X_j. Then the characteristic function f_n of S_n is given by $f_n(t) = f(t/n)^n$. By Theorem 1.2.6,

$$f_n(t) = \left[1 + \frac{1}{n} \cdot \sum_{j=1}^{d} [\mathrm{i}m_j + R_j(t/n)] \cdot t_j\right]^n, \quad t \in \mathbb{R}^d$$

where $\lim_{t\to 0} R_j(t) = 0$. Lemma B.1.2 shows that

$$\lim_{n\to\infty} f_n(t) = \mathrm{e}^{\mathrm{i}(m,t)}.$$

By Lévy's continuity Theorem 1.6.3, the distributions of the S_n's converge weakly, and hence also in probability, to δ_m. □

The next result is *Poisson's limit theorem.*

Theorem 5.1.3. *Let*

$$\begin{array}{l} X_{1,1} \\ X_{2,1},\ X_{2,2} \\ \vdots \\ X_{n,1},\ X_{n,2}, \dots,\ X_{n,n} \\ \vdots \end{array}$$

be random variables which are independent and identically distributed within each row of the array above and such that

$$P(X_{n,k} = 1) = p_n, \quad P(X_{n,k} = 0) = 1 - p_n$$

where $\lim_{n\to\infty} np_n = \lambda$. Writing

$$S_n = X_{n,1} + X_{n,2} + \cdots + X_{n,n}$$

we have

$$\lim_{n\to\infty} P(S_n = k) = \mathrm{e}^{-\lambda} \cdot \frac{\lambda^k}{k!}, \quad k \in \mathbb{N}_0. \tag{1}$$

Proof. The characteristic function of S_n is given by

$$f_n(t) = \left(1 + p_n\left(\mathrm{e}^{\mathrm{i}t} - 1\right)\right)^n, \quad t \in \mathbb{R}.$$

By our assumption, $p_n = (\lambda + \epsilon_n)/n$ where $\epsilon_n \to 0$. Using this and Lemma B.1.2 we obtain

$$\lim_{n\to\infty} f_n(t) = \mathrm{e}^{\lambda(\mathrm{e}^{\mathrm{i}t}-1)}.$$

Thus, by 1.6.3, the distributions of the S_n's converge weakly to the Poisson distribution with parameter λ (see (1.1.13.d)) from which (1) follows. □

5.2 Sums of independent random vectors and the Jessen–Wintner purity law

In the present section we study infinite series $X_1 + X_2 + \cdots$ where the X_j's are independent random vectors. The investigation of such series leads us to infinite convolutions of distributions and infinite products of characteristic functions. The main results are Kolmogorov's three-series theorem, the equivalence of convergence in law and almost sure convergence, and the Jessen–Wintner purity law (cf. Theorem 5.2.14, Corollary 5.2.12 and Theorem 5.2.15).[1]

If $\{\mu_n\}$ is a sequence of distributions converging weakly to some distribution μ, then we simply write $\mu_n \to \mu$. Recall that the sum $A + B$ of two sets $A, B \subset \mathbb{R}^d$ is defined as the set of those points in $\mathbb{R}^d$ that may be represented in at least one way as $a + b$ where $a \in A$ and $b \in B$. Next we extend this notation to an infinite sequence of sets.

Definition 5.2.1. By the *limit* $\lim B_n$ of a sequence $\{B_n\}$ of subsets of $\mathbb{R}^d$ we understand the set of those points in $\mathbb{R}^d$ that may be represented in at least one way as the limit of a sequence $\{x_n\}$ where $x_n \in B_n$. By the sum

$$B_1 + B_2 + \cdots$$

we mean the set $\lim(B_1 + \cdots + B_n)$.

It is easy to check that $\lim B_n$, which may be empty, is a closed set.

[1] The presentation of the results is based on the paper [31].

Definition 5.2.2. If $\{\mu_n\}_1^\infty$ is a sequence of distributions, we say that the *infinite convolution* $\mu_1 * \mu_2 * \cdots$ *is convergent* if there exists a distribution ν such that

$$\mu_1 * \cdots * \mu_n \to \nu.$$

Lemma 5.2.3. *If $\mu_n \to \mu$ and $\mu_n * \nu_n \to \mu$, then $\nu_n \to \delta_0$.*

Proof. Denote by f_n, f and g_n the characteristic functions of μ_n, μ and ν_n, respectively. For all $t \in \mathbb{R}^d$ we have $f_n(t) \to f(t)$ and $f_n(t)g_n(t) \to f(t)$. This implies that $g_n(t) \to 1$ if $f(t) \neq 0$. Since the set of such t's contains a neighborhood of 0, Corollary 1.4.15 shows that $g_n(t) \to 1$ for all $t \in \mathbb{R}^d$. Thus, $\nu_n \to \delta_0$ by Lévy's continuity Theorem 1.6.3. □

Theorem 5.2.4. *Let μ_n be a distribution with characteristic function f_n, $n \in \mathbb{N}$. The following conditions are equivalent:*

(i) *The infinite convolution $\mu_1 * \mu_2 * \cdots$ is convergent.*

(ii) *The sequence $\{p_n\}_1^\infty$ with $p_n = f_1 \cdots f_n$ converges uniformly on every compact subset of $\mathbb{R}^d$.*

(iii) $\mu_{n+1} * \cdots * \mu_{n+d(n)} \to \delta_0$ *for an arbitrary function $d : \mathbb{N} \to \mathbb{N}$.*

Proof. The equivalence of (i) and (ii) follows immediately from Theorems 1.6.1 and 1.6.3. Let $d : \mathbb{N} \to \mathbb{N}$ be arbitrary. Writing $\rho_n = \mu_{n+1} * \cdots * \mu_{n+d(n)}$ and $\nu_n = \mu_1 * \cdots * \mu_n$ we have $\nu_{n+d(n)} = \nu_n * \rho_n$. Consequently, if $\nu_n \to \nu$ we also have $\nu_n * \rho_n \to \nu$, hence $\rho_n \to \delta_0$ by Lemma 5.2.3. Conversely, assume that $\rho_n \to \delta_0$ for all $d : \mathbb{N} \to \mathbb{N}$ and let $K \subset \mathbb{R}^d$ be compact. Then

$$\|1 - \hat{\rho}_n\| = \|1 - f_{n+1} \cdots f_{n+d(n)}\| \to 0 \tag{1}$$

where $\|\cdot\|$ denotes the supremum norm on K. We show that $\{p_n\}$ is a Cauchy sequence with respect to the supremum metric on K from which (ii) follows. Assume, on the contrary, that there exists $\epsilon > 0$ such that for all $n \in \mathbb{N}$ there exist positive integers $a(n), b(n)$ with $n < a(n) < b(n)$ such that the distance of $p_{a(n)}$ and $p_{b(n)}$ is at least ϵ. We have

$$\begin{aligned}
\epsilon &\leq \left\| p_{a(n)} - p_{b(n)} \right\| \\
&= \left\| p_{a(n)} \left[1 - f_{a(n)+1} \cdots f_{b(n)}\right] \right\| \\
&\leq \left\| 1 - f_{a(n)+1} \cdots f_{b(n)} \right\| \\
&\leq \left\| 1 - f_{n+1} \cdots f_{b(n)} \right\| + \left\| f_{n+1} \cdots f_{b(n)} - f_{a(n)+1} \cdots f_{b(n)} \right\| \\
&= \left\| 1 - f_{n+1} \cdots f_{b(n)} \right\| + \left\| f_{a(n)+1} \cdots f_{b(n)} \cdot \left[1 - f_{n+1} \cdots f_{a(n)}\right] \right\| \\
&\leq \left\| 1 - f_{n+1} \cdots f_{b(n)} \right\| + \left\| 1 - f_{n+1} \cdots f_{a(n)} \right\| \to 0
\end{aligned}$$

where we have used (1). This contradiction completes the proof. □

Using Theorem B.3.6 we obtain the following corollary:

Corollary 5.2.5. *If for each compact set $K \subset \mathbb{R}^d$ there exists a constant $C_K \geq 0$ such that*

$$\sum_{n=1}^{\infty} |1 - f_n(t)| \leq C_K, \quad t \in K,$$

then the infinite convolution $\mu_1 * \mu_2 * \cdots$ is convergent.

5.2.6. If $x \mapsto x_j$ is μ-integrable for each $j = 1, \dots, d$, then we denote by $\mathbb{E}(\mu)$ the *center of gravity* or *barycenter* of μ, i.e., $\mathbb{E}(\mu)$ is the point of $\mathbb{R}^d$ which has the moments

$$M_j(\mu) = \int_{\mathbb{R}^d} x_j \, \mathrm{d}\mu(x)$$

of order one as coordinates. We denote by μ^c the distribution given by

$$\mu^c(B) = \mu(B + \mathbb{E}(\mu)), \; B \in \mathcal{B}_d$$

so that $\mathbb{E}(\mu^c) = 0$. For an arbitrary distribution μ on $\mathbb{R}^d$ with finite moments

$$M_{j,j}(\mu) = \int_{\mathbb{R}^d} x_j^2 \, \mathrm{d}\mu(x), \quad j = 1, \dots, d$$

of order two, we write

$$\operatorname{Var}(\mu) = \int_{\mathbb{R}^d} \|x - \mathbb{E}\,\mu\|^2 \, \mathrm{d}\mu(x) = \int_{\mathbb{R}^d} (x_1 - M_1)^2 + \cdots + (x_d - M_d)^2 \, \mathrm{d}\mu(x)$$

where $M_j = M_j(\mu)$. Note that if μ is the distribution of a random vector X, then $\mathbb{E}(\mu) = \mathbb{E}(X)$, $\operatorname{Var}(\mu) = \mathbb{E}(\|X - \mathbb{E}\,X\|^2)$ and μ^c is the distribution of $X - \mathbb{E}(X)$. This fact can be used to show easily the following simple properties (provided the existence of $\mathbb{E}$ or Var, respectively):

(i) $\mathbb{E}(\mu_1 * \mu_2) = \mathbb{E}(\mu_1) + \mathbb{E}(\mu_2)$;

(ii) $\operatorname{Var}(\mu_1 * \mu_2) = \operatorname{Var}(\mu_1) + \operatorname{Var}(\mu_2)$;

(iii) $(\mu_1 * \mu_2)^c = \mu_1^c * \mu_2^c$;

(iv) $f^c(t) = \mathrm{e}^{-\mathrm{i}\mathbb{E}(\mu)t} \cdot f(t)$, $t \in \mathbb{R}^d$, where f and f^c denote the characteristic functions of μ and μ^c, respectively.

(v) if $\mu_n \to \mu$ and $\liminf_n \operatorname{Var}(\mu_n)$ is finite, then $\operatorname{Var}(\mu)$ is finite, namely $\operatorname{Var}(\mu) \leq \liminf_n \operatorname{Var}(\mu_n)$.[2]

[2] This follows from the relation

$$\int f(x) \, \mathrm{d}\mu(x) \leq \liminf_n \int f(x) \, \mathrm{d}\mu_n(x)$$

which holds for every nonnegative, continuous function on $\mathbb{R}^d$.

Theorem 5.2.7. *If $M_{j,j}(\mu_n)$ is finite for every $n \in \mathbb{N}$ and $j = 1, \dots, d$, then the convergence of the two series*

$$\mathbb{E}(\mu_1) + \mathbb{E}(\mu_2) + \cdots \quad \textit{and} \quad \mathrm{Var}(\mu_1) + \mathrm{Var}(\mu_2) + \cdots$$

implies that $\mu_1 * \mu_2 * \cdots$ converges to some distribution μ. Furthermore, $\mathrm{Var}(\mu)$ *is finite and the series above converge to $\mathbb{E}(\mu)$ and* $\mathrm{Var}(\mu)$, *respectively.*

Proof. Assume first the $\mathbb{E}(\mu_n) = 0$. Since $M_{j,j}(\mu_n) < \infty$, the characteristic function g_n of μ_n possesses continuous partial derivatives of order at least two and those of the first order vanish at 0. Using this, Corollary 1.2.7 shows that

$$g_n(t) - 1 = \frac{1}{2} \cdot \sum_{i,j=1}^{d} [\mathrm{Re}\, g_{n,t_i,t_j}(\theta_n t) + \mathrm{i} \cdot \mathrm{Im}\, g_{n,t_i,t_j}(\eta_n t)]\, t_i t_j$$

where the subscripts t_i, t_j denote partial differentiation. By Theorem 1.2.1,

$$\begin{aligned} |g_{n,t_i,t_j}| &\le \int_{\mathbb{R}^d} |t_i t_j| \,\mathrm{d}\mu_n(t) \\ &\le \Big(\int_{\mathbb{R}^d} t_i^2 \,\mathrm{d}\mu_n(t)\Big)^{\frac{1}{2}} \Big(\int_{\mathbb{R}^d} t_j^2 \,\mathrm{d}\mu_n(t)\Big)^{\frac{1}{2}} \le \mathrm{Var}(\mu_n). \end{aligned}$$

Using this we obtain

$$|g_n(t) - 1| \le \frac{1}{\sqrt{2}} \cdot \mathrm{Var}(\mu_n) \cdot \sum_{i,j=1}^{d} |t_i t_j| \le \frac{d}{\sqrt{2}} \cdot \mathrm{Var}(\mu_n) \cdot \|t\|^2 \tag{1}$$

which proves that $\mu_1 * \mu_2 * \cdots$ converges to some distribution μ (cf. Corollary 5.2.5). From (5.2.6.v) we conclude that $\mathrm{Var}(\mu) \le \sum_{j=1}^{\infty} \mathrm{Var}(\mu_j) < \infty$. Since $\mathrm{Var}(\mu) = \mathrm{Var}(\mu_1 * \cdots * \mu_n) + \mathrm{Var}(\mu_{n+1} * \cdots)$, we have $\mathrm{Var}(\mu) \ge \mathrm{Var}(\mu_1) + \cdots + \mathrm{Var}(\mu_n)$, and hence $\mathrm{Var}(\mu) \ge \sum_{j=1}^{\infty} \mathrm{Var}(\mu_j)$.

Applying (1) to the characteristic function $f_n := g_1 \cdots g_n$ instead of g_n and using (5.2.6.ii) we get

$$|f_n(t) - 1| \le \frac{d}{\sqrt{2}} \cdot \sum_{j=1}^{n} \mathrm{Var}(\mu_j) \cdot \|t\|^2.$$

Taking the limit $n \to \infty$ we see that the partial derivatives of order one of the characteristic function of μ are zero at 0, thus $\mathbb{E}(\mu) = 0$.

The general case where $\mathbb{E}(\mu_n)$ is not supposed to be zero follows from the relation $g_n(t) = \mathrm{e}^{\mathrm{i}\mathbb{E}(\mu_n)t} \cdot g_n^c(t)$. □

Remark 5.2.8. The converse of the previous theorem is false. To see an example, let $X_1, X_2, \dots$ be independent random variables such that $P(X_n = 0) = 1 - p_n$ where $0 \le p_n \le 1$ and $\sum_1^\infty p_n < \infty$. The Borel–Cantelli Lemma F.2.1 shows that

$\sum_1^\infty X_n$ is a finite sum with probability 1. On the other hand, the series $\sum_1^\infty \mathbb{E}(X_n)$ and $\sum_1^\infty \mathrm{Var}(X_n)$ may not converge. We can take for example $p_n = \frac{1}{n^2}$ and choose X_n such that $P(X_n = n^2) = \frac{1}{n^2}$. Then $\mathbb{E}(X_n) = 1$ and $\mathrm{Var}(X_n) = n^2 - 1$.

We have, however, the following theorem.

Theorem 5.2.9. *If for some $R \geq 0$ the supports of all μ_n are contained in the sphere $\{x : \|x\| \leq R\}$, then the convergence of the two series*

$$\mathbb{E}(\mu_1) + \mathbb{E}(\mu_2) + \cdots \quad \textit{and} \quad \mathrm{Var}(\mu_1) + \mathrm{Var}(\mu_2) + \cdots$$

is necessary and sufficient for the convergence of $\mu_1 * \mu_2 * \cdots$.

Proof. The sufficiency follows from Theorem 5.2.7. In order to prove the necessity assume that $\mu_1 * \mu_2 * \cdots$ converges to some distribution μ and let g_n, g_n^c and g be the characteristic functions of μ_n, μ_n^c and μ, respectively. Suppose that the series $\mathrm{Var}(\mu_1) + \mathrm{Var}(\mu_2) + \cdots$ is divergent. Then it follows from $\mathrm{Var}(\mu_n) = \sum_{j=1}^d M_{j,j}(\mu_n^c)$ that for some value of j the series $M_{j,j}(\mu_1^c) + M_{j,j}(\mu_2^c) + \cdots$ is divergent. Without loss of generality assume that $j = 1$. In view of Corollary 1.11.2 there exists an open neighborhood U of $0 \in \mathbb{R}$ such that

$$|g_n(s, 0, \ldots, 0)| = |g_n^c(s, 0, \ldots, 0)| \leq 1 - \frac{1}{8} \cdot s^2 M_{1,1}(\mu_n^c)$$

holds for all $s \in U$ and for all n. The second part of Theorem B.3.4 shows that

$$g(s, 0, \ldots, 0) = \prod_{n=1}^{\infty} g_n(s, 0, \ldots, 0) = 0, \quad s \in U \setminus \{0\}.$$

This, however, contradicts the continuity of g. □

A fundamental theorem of the theory of sums of independent random vectors is that for the sequence of partial sums there is equivalence between convergence in law and convergence almost everywhere. Theorem 5.2.11 is an immediate consequence of this fact. Below we present a different proof given in the paper [31] of B. Jessen and A. Wintner since it contains a nice application of Jessen's martingale Theorem 5.2.10.[3] To formulate this theorem we need some notation (see Section F.3 for the definition of products of probability spaces).

[3] I would like to thank Christian Berg for making Jessen's original work [30] available to me.

For all $n \in \mathbb{N}$ let $(\Omega_n, \mathcal{A}_n, P_n)$ be a probability space and write

$$\begin{aligned}
\tilde{\Omega} &= \Omega_1 \times \Omega_2 \times \cdots \\
\tilde{\mathcal{A}} &= \mathcal{A}_1 \times \mathcal{A}_2 \times \cdots \\
\tilde{P} &= P_1 \times P_2 \times \cdots \\
\tilde{\omega}_n &= (\omega_{n+1}, \omega_{n+2}, \dots), \qquad \omega = (\omega_1, \omega_2, \dots) \in \tilde{\Omega} \\
\tilde{\Omega}_n &= \Omega_{n+1} \times \Omega_{n+2} \times \cdots \\
\tilde{\mathcal{A}}_n &= \mathcal{A}_{n+1} \times \mathcal{A}_{n+2} \times \cdots \\
\tilde{P}_n &= P_{n+1} \times P_{n+2} \times \cdots .
\end{aligned}$$

We will identify $(\omega_1, \omega_2, \dots)$ and $(\omega_1, \dots, \omega_n, \tilde{\omega}_n)$.

Theorem 5.2.10 (Jessen). *Let $f \in L^1(\tilde{P})$ and for each $n \in \mathbb{N}$ define the function f_n by*

$$f_n(\tilde{\omega}) = \int_{\tilde{\Omega}_n} f(\omega_1, \dots, \omega_n, \tilde{\omega}_n)\, \mathrm{d}\tilde{P}_n(\tilde{\omega}_n), \qquad \tilde{\omega} \in \tilde{\Omega}.$$

Then

$$\lim_{n\to\infty} f_n(\tilde{\omega}) = f(\tilde{\omega})$$

holds for $\tilde{P}$-almost all $\tilde{\omega} \in \tilde{\Omega}$.

Theorem 5.2.11. *A necessary and sufficient condition for the convergence of the infinite convolution $\mu_1 * \mu_2 * \cdots$ is that if $X_1, X_2, \dots$ are independent random vectors on a probability space $(\Omega, \mathcal{A}, P)$ having the distributions $\mu_1, \mu_2, \dots$, then the series $X_1 + X_2 + \cdots$ is convergent P-almost everywhere. The distribution of the random vector $S = X_1 + X_2 + \cdots$ is then $\mu = \mu_1 * \mu_2 * \cdots$.*

Proof. By independence, the distribution of the partial sum $S_n = X_1 + \cdots + X_n$ is $\nu_n = \mu_1 * \cdots * \mu_n$. Moreover, by the very definition of the convolution, ν_n is also the distribution of the random vector

$$\tilde{S}_n(\tilde{\omega}) = X_1(\omega_1) + \cdots + X_n(\omega_n), \qquad \tilde{\omega} = (\omega_1, \omega_2, \dots) \in \tilde{\Omega}.$$

If $\{S_n\}$ converges P-almost everywhere to some random vector S, then $S_n \to S$ in law, showing that $\nu_n \to \mu$ where μ is the distribution of S.

Now assume that $\{\mu_n\}$ is convergent and let $d : \mathbb{N} \to \mathbb{N}$ be arbitrary. The distribution of

$$\tilde{R}_n(\tilde{\omega}) = X_{n+1}(\omega_{n+1}) + \cdots + X_{n+d(n)}(\omega_{n+d(n)})$$

is $\rho_n = \mu_{n+1} * \cdots * \mu_{n+d(n)}$. From Theorem 5.2.4 we see that $\rho_n \to \delta_0$ implying that $\tilde{R}_n \to 0$ in probability (see Section F.2). This shows that $\tilde{S}_n$ converges to some random vector $\tilde{S}$ in probability. Setting $\tilde{Q}_n(\tilde{\omega}) = \sum_{k=n+1}^{\infty} X_k(\omega_k)$, $\tilde{Q}_n$ is defined $\tilde{P}$-almost everywhere and

$$f(\tilde{\omega}) := \mathrm{e}^{\mathrm{i}(\tilde{S}(\tilde{\omega}), y)} = \mathrm{e}^{\mathrm{i}(\tilde{S}_n(\tilde{\omega}), y)} \cdot \mathrm{e}^{\mathrm{i}(\tilde{Q}_n(\tilde{\omega}), y)}$$

for every $y \in \mathbb{R}^d$. The function f is measurable and bounded, hence integrable, so that we may apply Theorem 5.2.10 with $(\Omega_n, \mathcal{A}_n, P_n) = (\Omega, \mathcal{A}, P)$ for all n. We find

$$f_n(\tilde{\omega}) = \mathrm{e}^{\mathrm{i}(\tilde{S}_n(\tilde{\omega}),y)} \cdot z_n(y)$$

with some $z_n(y) \in \mathbb{C}$ satisfying $\lim_n z_n(y) = 1$. Hence, $\mathrm{e}^{\mathrm{i}(\tilde{S}_n(\tilde{\omega}),y)}$ converges $\tilde{P}$-almost everywhere for every y. Lemma B.1.17 shows that $\tilde{S}_n$ is convergent $\tilde{P}$-almost everywhere. Since the random variables

$$\omega \mapsto (X_1(\omega), X_2(\omega), \dots) \quad \text{and} \quad \tilde{\omega} \mapsto (X_1(\omega_1), X_2(\omega_2), \dots)$$

defined on Ω and $\tilde{\Omega}$, respectively, have the same distribution (by independence), we conclude that S_n is convergent P-almost everywhere. □

The set of points where an infinite series of random vectors converges is a tail event. Combining the previous theorem and the zero–one law of Kolmogorov (cf. Theorem F.2.5) we obtain the following corollary.

Corollary 5.2.12. *An infinite series $X_1 + X_2 + \cdots$ of independent random vectors is convergent in law if and only if it is convergent almost surely. Moreover, the series is always either almost surely convergent or almost surely divergent.*

5.2.13. The following notation will be used in the next theorem. Let $\{X_n\}$ be a sequence of d-dimensional random vectors and, as in Lemma 1.11.3, define the random vectors $X_{n,R}$, where $n \in \mathbb{N}$ and $R \in (0, \infty)$, by

$$X_{n,R}(\omega) = \begin{cases} X_n(\omega) & \text{if } \|X_n(\omega)\| \le R \\ 0 & \text{if } \|X_n(\omega)\| > R \end{cases} \qquad \omega \in \Omega.$$

Denoting by μ_n and $\mu_{n,R}$ the corresponding distributions we have

$$\mu_{n,R} = \nu_{n,R} + (1 - \mu_n(B_R)) \cdot \delta_0$$

where $B_R = \{x \in \mathbb{R}^d : \|x\| \le R\}$ and $\nu_{n,R}(A) = \mu_n(B_R \cap A)$, $A \in \mathcal{B}_d$.

Theorem 5.2.14 (Kolmogorov's three-series theorem). *A necessary and sufficient condition for the convergence of the infinite convolution $\mu_1 * \mu_2 * \cdots$ is the convergence of the three series*

$$(1 - \mu_1(B_R)) + (1 - \mu_2(B_R)) + \cdots \tag{1}$$

$$\mathbb{E}(\mu_{1,R}) + \mathbb{E}(\mu_{2,R}) + \cdots \quad \textit{and} \quad \operatorname{Var}(\mu_{1,R}) + \operatorname{Var}(\mu_{2,R}) + \cdots \tag{2}$$

for a fixed $R > 0$ (or for all $R > 0$).

Proof. Let $X_1, X_2, \dots$ be independent random vectors having the distributions $\mu_1, \mu_2, \dots$. By Theorem 5.2.11 the convergence of $\mu_1 * \mu_2 * \cdots$ is necessary and sufficient for the almost sure convergence of $X_1 + X_2 + \cdots$.

If $X_1 + X_2 + \cdots$ is almost surely convergent, then $X_n \to 0$ almost surely, hence the probability that the event $[\|X_n\| > R]$ happens for infinitely many n, is equal to 0. By Theorem F.2.1, we then have

$$\sum_{n=1}^{\infty} P(X_n \neq X_{n,R}) = \sum_{n=1}^{\infty} (1 - \mu_n(B_R)) < \infty$$

so that $X_{1,R} + X_{2,R} + \cdots$ is almost surely convergent in view of Lemma F.2.2. From Theorem 5.2.9 we see that the two series in (2) are convergent.

Assume now that the three series in (1) and (2) are convergent. Theorem 5.2.9 then shows that $X_{1,R} + X_{2,R} + \cdots$ is almost surely convergent. Applying again Lemma F.2.2 we see that $X_1 + X_2 + \cdots$ is almost surely convergent. □

Theorem 5.2.15 (Jessen–Wintner purity law). *If $\mu = \mu_1 * \mu_2 * \cdots$ is a convergent infinite convolution of distributions μ_n each of which is discrete, then the following three cases are possible: μ is*

(i) *discrete;*

(ii) *singular and continuous;*

(iii) *absolutely continuous.*

Proof. Let $X_1, X_2, \ldots$ be a sequence of independent random vectors on a probability space $(\Omega, \mathcal{A}, P)$ such that μ_n is the distribution of X_n. By Theorem 5.2.11, the series $X_1(\omega) + X_2(\omega) + \cdots$ converges with probability 1. We may suppose that each X_n takes values in a countable set and that $S = X_1 + X_2 + \cdots$ converges everywhere.

Let G denote the smallest group in $\mathbb{R}^d$ containing all possible values of the X_n's. Then G is countable. Let $E \subset \mathbb{R}^d$ be an arbitrary Borel set and write

$$A := \{\omega \in \Omega : S(\omega) \in E + G\}.$$

Since

$$X_1(\omega) + \cdots + X_n(\omega) \in G, \quad n \in \mathbb{N},\ \omega \in \Omega$$

and G is a group, we see that

$$A = \{\omega : X_n(\omega) + X_{n+1}(\omega) + \cdots \in E + G\}, \quad n \in \mathbb{N}.$$

Thus, A is a tail event (cf. F.2.4) with respect to $X_1, X_2, \ldots$. By Theorem F.2.5, $P(A) = 1$ whenever $P(A) > 0$. This means that $\mu(E + G) = 1$ whenever $\mu(E + G) > 0$ and a fortiori when $\mu(E) > 0$ (since $E \subset E + G$).

If μ is not continuous, then there is a one-point set E with $\mu(E) > 0$. The set $E + G$ is then countable, its μ-measure is 1, showing that μ is discrete.

Assume now that μ is continuous. If there is a Borel set E such that $\lambda(E) = 0$ and $\mu(E) > 0$, then $\lambda(E + G) = 0$ (since G is countable) and $\mu(E + G) = 1$, i.e., μ is singular. If there is no such set E then, by the theorem of Radon and Nikodym, μ is absolutely continuous. □

5.3 Ergodic theorems for stationary fields

We start with a simple lemma. As an immediate corollary, we obtain a limit theorem for stationary sequences by applying the isometric operator I_Z from Remark 2.9.12.

Lemma 5.3.1. *The sequence $\{s_n\}_{n=1}^{\infty}$ of complex-valued functions where*

$$s_n(x) = \frac{1}{n}\sum_{k=1}^{n} e^{ikx}, \quad x \in \mathbb{R}, \quad n \in \mathbb{N}$$

is uniformly bounded and converges pointwise to $\mathbf{1}_{\{0\}}$ (see Figure 5.1).

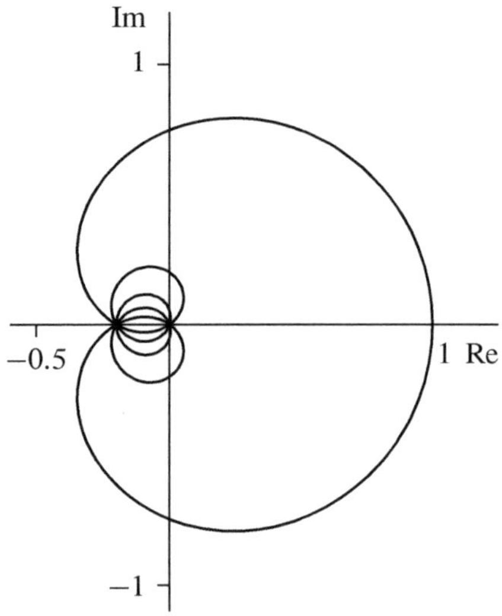

Figure 5.1. The function s_5 from Lemma 5.3.1.

Proof. We have $s_n(0) = 1$ and $|s_n(x)| \leq 1$ for all x. If $x \neq 0$ then

$$|s_n(x)|^2 = \frac{1}{n^2} \cdot \left| \frac{e^{ix} - e^{i(n+1)x}}{1 - e^{ix}} \right|^2 \tag{1}$$

$$\leq \frac{1}{n^2} \cdot \frac{4}{|1 - e^{ix}|^2} \to 0, \quad n \to \infty. \quad \square \tag{2}$$

Theorem 5.3.2. *Let Z be a stationary time series on $\mathbb{Z}$ with representing measure ζ. Then we have*

$$\lim_{n\to\infty} \frac{1}{n} \sum_{k=1}^{n} Z(k) = \zeta(\{0\}).$$

Proof. In view of Lemma 5.3.1 the sequence $\{s_n\}$ converges in $L^2(\sigma)$ to $\mathbf{1}_{\{0\}}$. The theorem follows by applying the isometric operator I_Z (cf. Remark 2.9.12) to s_n and noting that $I_Z(\mathbf{1}_{\{0\}}) = \zeta(\{0\})$ (cf. equation (2.7.1.3)). □

Next we prove the continuous version of the previous result.

Theorem 5.3.3. *If Z is a continuous stationary process on $\mathbb{R}$ with representing measure ζ, then*

$$\lim_{T\to\infty} \frac{1}{T} \int_0^T Z(t)\,\mathrm{d}t = \zeta(\{0\}).$$

Proof. This theorem can be proved in the same way as the previous one by using the relations

$$\frac{1}{T}\int_0^T \mathrm{e}^{\mathrm{i}tx}\,\mathrm{d}t = \frac{\mathrm{e}^{\mathrm{i}Tx}-1}{\mathrm{i}Tx}, \qquad x \neq 0,\ T > 0 \tag{1}$$

and

$$I_Z\left(\frac{1}{T}\int_0^T \mathrm{e}^{\mathrm{i}tx}\,\mathrm{d}t\right) = \frac{1}{T}\int_0^T Z(t)\,\mathrm{d}t$$

(see equation (2.9.13.1)). □

Remark 5.3.4. Let Z be as in Theorem 5.3.2 or in Theorem 5.3.3 and denote by m the mean of Z. We say that Z is *ergodic* if

$$\lim_{n\to\infty} \frac{1}{n}\sum_{k=1}^{n} Z(k) = m$$

or

$$\lim_{T\to\infty} \frac{1}{T}\int_0^T Z(t)\,\mathrm{d}t = m$$

respectively. By the theorems above, Z is ergodic if and only if $\zeta(\{0\}) = m$. If Z is ergodic and σ denotes the spectral measure of Z, then

$$\sigma(\{0\}) = \int 1_{\{0\}}\,\mathrm{d}\sigma = \mathbb{E}\left(|\zeta(\{0\})|^2\right) = |m|^2.$$

Consequently, if $m = 0$, then Z is ergodic if and only if its spectral measure has no atom at the origin.

Remark 5.3.5. Let Z be a stationary time series and write $S_n = Z_1 + \cdots + Z_n$, $n \in \mathbb{N}$. We already know that the sequence $\{S_n/n\}$ converges to some $Y \in H(Z)$. Denote by (U_n) the canonical unitary representation of Z in $H(Z)$. We then have

$$U_{-k}\left(\frac{S_n}{n}\right) = \frac{Z_{1+k} + \cdots + Z_{n+k}}{n} = \frac{S_{n+k} - S_k}{n} = \frac{S_{n+k}}{n+k}\cdot\frac{n+k}{n} - \frac{S_k}{n}$$

for all $k \in \mathbb{Z}$ such that $n + k \neq 0$. Taking the limit $n \to \infty$ and replacing k by $-k$ we obtain

$$U_k Y = Y, \quad k \in \mathbb{Z}$$

i.e., Y is a so-called *fixed point* of (U_n). Similar computations can be carried out for continuous stationary processes and integrals instead of sums. This is the motivation for the following investigation of fixed points of unitary representations.

Notation 5.3.6. Let (U_t) be a unitary representation of $T = \mathbb{R}^d$ or $T = \mathbb{Z}^d$ in a Hilbert space H. For $g \in H$ we denote by $H(g)$ the closure of the linear manifold

$$\operatorname{span}\{U_t g : t \in T\}.$$

It is easy to check that $H(g)$ is a (U_t)-invariant subspace of H. The orthogonal projection from $H(Z)$ onto $H(g)$ will be denoted by P_g. It follows from Lemma 2.10.16 that P_g commutes with all operators U_t:

$$U_t P_g = P_g U_t, \quad t \in T,\ g \in H. \tag{1}$$

An element $h \in H$ is called a *fixed point* of (U_t) if

$$U_t h = h, \quad t \in T.$$

We denote by H_0 the set of all fixed points. This set is a (U_t)-invariant subspace, the orthogonal projection onto H_0 will be denoted by P^0. We have

$$P^0 U_t = U_t P^0 = P^0. \tag{2}$$

The first equation follows from the (U_t)-invariance of H_0, the second one from $P^0 H = H_0$.

Lemma 5.3.7. *For all $g \in H$ we have*

(i) $P_g P^0 = P^0 P_g$

(ii) $H(g) \cap H_0 = \mathbb{C} \cdot P^0 g$.

Proof. (i) Equations (5.3.6.1) and (5.3.6.2) show that $U_t P_g P^0 = P_g P^0$. Thus, for all $h \in H$ the vector $P_g P^0 h$ is a fixed point, i.e., $P_g P^0 h \in H_0$. Consequently,

$$P^0 P_g P^0 h = P_g P^0 h, \quad h \in H$$

and hence

$$P^0 P_g P^0 = P_g P^0. \tag{1}$$

Since P_g and P^0 are orthogonal projections, we see that

$$(P^0 P_g h, h') = (h, P_g P^0 h') = (h, P^0 P_g P^0 h') = (P^0 P_g P^0 h, h')$$

for all $h, h' \in H$. This relation and (1) imply that $P^0 P_g = P^0 P_g P^0 = P_g P^0$.

(ii) By (i) we have

$$P^0 g = P^0 P_g g = P_g P^0 g \in H(g)$$

and hence $\mathbb{C} \cdot P^0 g \subset H(g) \cap H_0$. If $h_0 \in H_0 \cap H(g)$, then

$$h_0 = \lim_{n\to\infty} \sum_{j=1}^{k_n} c_{j,n} U_{t_{j,n}} g$$

with some $k_n \in \mathbb{N}$, $c_{j,n} \in \mathbb{C}$ and $t_{j,n} \in T$. Using equation (5.3.6.1) we see that

$$h_0 = P^0 h_0 = \lim_{n\to\infty} \sum_{j=1}^{k_n} c_{j,n} P^0 U_{t_{j,n}} g = \left(\lim_{n\to\infty} \sum_{j=1}^{k_n} c_{j,n} \right) \cdot P^0 g$$

i.e., h_0 is proportional to $P^0 g$. □

Theorem 5.3.8. *Let Z be a continuous stationary field on $T = \mathbb{R}^d$ or $T = \mathbb{Z}^d$ with representing measure ζ, spectral measure σ and correlation function C. Denote by (U_t), (V_t) and (W_t) the canonical unitary representations in $H(Z)$, $L^2(\sigma)$ and $H(C)$, respectively. Then we have:*

(i) *The set of all fixed points of (V_t) is $\mathbb{C} \cdot \mathbf{1}_{\{0\}}$.*

(ii) *The set of all fixed points of (W_t) is $\mathbb{C} \cdot \mathbf{1}_T \cap H(C)$.*

(iii) *The set H_0 of all fixed points of (U_t) is $\mathbb{C} \cdot \zeta(\{0\})$ and*

$$\zeta(\{0\}) = P^0 Z(s)$$

for all $s \in T$.

Proof. We consider only the case $\mathbb{R}^d$, the case $\mathbb{Z}^d$ can be treated in the same way.

(i) If h is a fixed point of (V_t), then

$$h(x) = V_t h(x) = \mathrm{e}^{-\mathrm{i}(t,x)} h(x), \quad x, t \in \mathbb{R}^d.$$

We conclude that $h(x) = 0$ if $x \neq 0$, i.e., $h \in \mathbb{C} \cdot \mathbf{1}_{\{0\}}$. On the other hand, $\mathbf{1}_{\{0\}}$ is a fixed point of (V_t) so that (i) holds.

(ii) A function $h \in H(C)$ is a fixed point for the translation operators W_t if and only if $h(x - t) = h(x)$ for all $x, t \in T$. This equation is satisfied if and only if h is constant.

(iii) It is easy to see that the isometric operator I_Z from Remark 2.10.15 maps the set of fixed points of (V_t) onto H_0. From (i) we conclude that

$$H_0 = \mathbb{C} \cdot I_Z(\mathbf{1}_{\{0\}}) = \mathbb{C} \cdot \zeta(\{0\}).$$

If $\sigma(\{0\}) = 0$, then $\zeta(\{0\}) = 0$. The equation above shows that $H_0 = \{0\}$ and hence $P^0 = 0$. Thus, the equation in (iii) holds whenever $\sigma(\{0\}) = 0$.
If $\sigma(\{0\}) \neq 0$, then $\zeta(\{0\}) \neq 0$, the subspace H_0 is one-dimensional and $P^0 \neq 0$. We obviously have $P^0 Z(s) \in H_0$. By equation (5.3.6.2),

$$P^0 Z(s) = P^0 U_{-s} Z(0) = U_{-s} P^0 Z(0) = P^0 Z(0)$$

and hence $P^0 Z(s)$ does not depend on s. From the definition of $H(Z)$ and from the fact that $P^0 \neq 0$ we conclude that $P^0 Z(s) \neq 0$. Since H_0 is one-dimensional, $P^0 Z(s) = P^0 Z(0) = c I_Z(\mathbf{1}_{\{0\}})$ with some $c \in \mathbb{C} \setminus \{0\}$. As P^0 is an orthogonal projection we have

$$(P^0 Z(0), P^0 Z(0)) = (Z(0), P^0 Z(0)).$$

Using the fact that I_Z is isometric we obtain

$$(P^0 Z(0), P^0 Z(0)) = (c\mathbf{1}_{\{0\}}, c\mathbf{1}_{\{0\}})$$

and

$$(Z(0), P^0 Z(0)) = (\mathbf{1}, c\mathbf{1}_{\{0\}}) = (\mathbf{1}_{\{0\}}, c\mathbf{1}_{\{0\}}).$$

Thus, $c = 1$. □

Corollary 5.3.9. $H_0 = \{0\}$ *if and only if* $\sigma(\{0\}) = 0$.

We are now able to prove a generalization of Theorem 5.3.2 and Theorem 5.3.3. Note that the expressions

$$\frac{1}{n} \sum_{k=1}^{n} Z(k) \quad \text{and} \quad \frac{1}{T} \int_0^T Z(t)\,\mathrm{d}t$$

can be written as $\int Z\,\mathrm{d}\mu$ where

$$\mu = \frac{1}{n} \sum_{k=1}^{n} \delta_k \quad \text{or} \quad \mu = \frac{1}{T} \lambda|_{[0,T]}.$$

The Fourier–Stieltjes transforms of these measures converge to $\mathbf{1}_{\{0\}}$ if $n \to \infty$ and $T \to \infty$, respectively (cf. Lemma 5.3.1 and equation (5.3.3.1)).

Theorem 5.3.10 (ergodic theorem). *Let Z be a continuous stationary field on the set $T = \mathbb{R}^d$ or $T = \mathbb{Z}^d$ and let $\{\mu_n\}$ be a sequence of probability measures such that*

$$\lim_{n\to\infty} \check{\mu}_n(t) = 0, \quad t \in T \setminus \{0\}.$$

Then

$$\lim_{n\to\infty} \int_T Z(t)\,\mathrm{d}\mu_n(t) = \zeta(\{0\}) = P^0 Z(s), \quad s \in T$$

where ζ denotes the representing measure of Z.

Proof. Since $|\check{\mu}_n| \leq 1$, the sequence $\{\check{\mu}_n\}$ converges in $L^2(\sigma)$ to $\mathbf{1}_{\{0\}}$. By (2.9.13.1) and (2.9.14.1) we have $\int Z \, \mathrm{d}\mu_n = I_Z(\check{\mu}_n)$. Using this and Theorem 5.3.8 we obtain

$$\lim_{n\to\infty} \int_T Z(t)\,\mathrm{d}\mu_n(t) = I_Z\left(\mathbf{1}_{\{0\}}\right) = \zeta(\{0\}) = P^0 Z(s). \qquad \square$$

Example 5.3.11. (i) Let μ_n be the Gaussian distribution on $\mathbb{R}^d$ with density

$$x \mapsto \frac{1}{(2\pi n)^{d/2}} \cdot \mathrm{e}^{-\frac{\|x\|^2}{2n}}$$

and characteristic function $t \mapsto \check{\mu}_n(t) = \mathrm{e}^{-\frac{n}{2}\|t\|^2}$. By the previous theorem we have

$$\lim_{n\to\infty} \frac{1}{(2\pi n)^{d/2}} \cdot \int_{\mathbb{R}^d} Z(t) \cdot \mathrm{e}^{-\frac{\|t\|^2}{2n}}\,\mathrm{d}t = \zeta(\{0\}).$$

(ii) Let μ_n be the uniform distribution on the cube $[0,n]^d \subset \mathbb{R}^d$. The characteristic function is then

$$t \mapsto \prod_{j=1}^{d} \frac{1 - \mathrm{e}^{2\pi\mathrm{i}t_j n}}{2\pi\mathrm{i}t_j n}.$$

The sequence $\{\mu_n\}$ satisfies the condition of the previous theorem.

(iii) Examples of sequences of distributions on $\mathbb{Z}^d$ satisfying the condition of Theorem 5.3.10 can be given by considering the uniform distribution on $\{1,\ldots,n\}^d$ or on $\{-n,\ldots,n\}^d$.

5.4 Filtration of discrete stationary fields

First we prove a more detailed version of Theorem 2.9.19 where we replace μ by a square integrable function.

Theorem 5.4.1. *Let Z be a stationary field on $\mathbb{Z}^d$ with representing measure ζ_Z and spectral measure σ_Z. Further, let $a \in L^2(\mathbb{Z}^d)$ be such that the series*

$$\hat{a}(\cdot) := \sum_{n\in\mathbb{Z}^d} a(n)\mathrm{e}^{-\mathrm{i}(\cdot,n)} \tag{1}$$

converges in $L^2(\sigma_Z)$. Then the series

$$a * Z(t) := \sum_{n\in\mathbb{Z}^d} a(n) Z(t-n), \quad t \in \mathbb{Z}^d$$

converges in $H(Z)$. The field $Y = a * Z$ is stationary with representing measure ζ_Y and spectral measure σ_Y given by

$$\mathrm{d}\zeta_Y = \hat{a}\,\mathrm{d}\zeta_Z \quad \textit{and} \quad \mathrm{d}\sigma_Y = |\hat{a}|^2\,\mathrm{d}\sigma_Z.$$

If $\sigma_Z(\{x : \hat{a}(x) = 0\}) = 0$, then

$$Z(t) = \int_{[0,2\pi)^d} \mathrm{e}^{\mathrm{i}(t,x)} \cdot \frac{1}{\hat{a}(x)}\,\mathrm{d}\zeta_Y(x), \quad t \in \mathbb{Z}^d$$

i.e., the convolution $Z \mapsto a * Z$ is invertible.

Proof. Using the spectral representation of Z we obtain

$$\sum_{j:\,|j|\le n} a(j)Z(t-j) = \int_{[0,2\pi)^d} \mathrm{e}^{\mathrm{i}(t,x)} \sum_{j:\,|j|\le n} a(j)\mathrm{e}^{-\mathrm{i}(j,x)}\,\mathrm{d}\zeta_Z(x).$$

The sum on the right-hand side converges in $L^2(\sigma_Z)$. Using the isometric operator I_Z we see that the sum on the left-hand side converges in $H(Z)$. Taking the limit $n \to \infty$ we obtain the spectral representation of Y. The rest follows from Lemma 2.6.11. □

Remark 5.4.2. If σ_Z has a constant density, then the series (5.4.1.1) converges in $L^2(\sigma)$ for all $a \in L^2(\mathbb{Z}^d)$, since the functions $\mathrm{e}^{-\mathrm{i}(\cdot,n)}$ have the same norm and build a (complete) orthogonal system in this Hilbert space.

The next theorem is important in signal processing.

Theorem 5.4.3. *A stationary field Y on $\mathbb{Z}^d$ admits the representation*

$$Y(t) = a * X(t) = \sum_{n\in\mathbb{Z}^d} a(n)X(t-n), \quad t \in \mathbb{Z}^d$$

with a white noise $X \sim WN(0,1)$ on $\mathbb{Z}^d$ and $a \in L^2(\mathbb{Z}^d)$ if and only if the spectral measure of Y is absolutely continuous.

Proof. Assume that Y has the representation above. By Example 2.9.5 and Theorem 5.4.1, the spectral measure σ of Y is given by

$$\mathrm{d}\sigma = \frac{1}{(2\pi)^d} \cdot |\hat{a}|^2\,\mathrm{d}\lambda_d$$

i.e., σ is absolutely continuous.

Now assume that σ is absolutely continuous and let C be the correlation function of Y. By Theorem 1.8.16 there exists a function $g \in L^2(\mathbb{Z}^d)$ such that

$$C(t) = g * \tilde{g}(t) = \sum_{n\in\mathbb{Z}^d} g(t+n)\overline{g(n)}, \quad t \in \mathbb{Z}^d$$

and $d\sigma(x) = |\hat{g}(x)|^2 \, d\lambda_d(x)$. Let

$$Y(t) = \int_{[0,2\pi)^d} e^{i(t,x)} \, d\zeta(x)$$

be the spectral representation of Y where ζ is a random orthogonal measure with structure measure σ. We define the random orthogonal measure ζ_h as in Lemma 2.6.11 by

$$\zeta_h(A) := \int_A h \, d\zeta, \quad A \in \mathcal{B}\left([0, 2\pi)^d\right)$$

where

$$h = \frac{1}{(2\pi)^{d/2} \hat{g}} \in L^2(\sigma)$$

(since $d\sigma(x) = |\hat{g}(x)|^2 \, d\lambda_d(x)$ we may ignore the zeros of $\hat{g}$). In view of

$$(\zeta_h(A), \zeta_h(B)) = \int_{[0,2\pi)^d} \mathbf{1}_A \mathbf{1}_B |h|^2 |\hat{g}|^2 \, d\lambda_d = \frac{1}{(2\pi)^d} \int_{A \cap B} 1 \, d\lambda_d$$

the structure measure of ζ_h is $\lambda_d/(2\pi)^d$. Consequently,

$$X(t) = \int_{[0,2\pi)^d} e^{i(t,x)} \, d\zeta_h(x), \quad t \in \mathbb{Z}^d$$

defines a white noise $X \sim WN(0,1)$ (see Example 2.9.5). By Lemma 2.6.11 we have

$$\begin{aligned}
Y(t) &= \int e^{i(t,x)} \, d\zeta(x) = \int e^{i(t,x)} (2\pi)^{d/2} \hat{g}(x) \, d\zeta_h(x) \\
&= \int e^{i(t,x)} (2\pi)^{d/2} \sum_{n=-\infty}^{\infty} e^{-i(x,n)} g(n) \, d\zeta_h(x) \\
&= (2\pi)^{d/2} \sum_{n=-\infty}^{\infty} g(n) \int e^{i(t-n,x)} \, d\zeta_h(x) \\
&= (2\pi)^{d/2} \sum_{n=-\infty}^{\infty} g(n) X(t-n)
\end{aligned}$$

completing the proof. □

Appendix A

Basic notation

A.1 Standard notation

The list below contains some standard notation that is used throughout the book.

$\mathbb{N}$	$=$	$\{1, 2, 3, \dots\}$
$\mathbb{N}_0$	$=$	$\{0, 1, 2, \dots\}$
$\mathbb{Z}$	$=$	$\{0, \pm 1, \pm 2, \dots\}$
$\mathbb{Q}$		rational numbers
$\mathbb{R}$		real numbers
$\mathbb{C}$		complex numbers
$\mathbb{T}$		complex numbers with modulus one
$\mathrm{C}^\infty(\mathbb{R}^d)$		the set of infinitely differentiable functions on $\mathbb{R}^d$
$\mathrm{C}^\infty_{00}(\mathbb{R}^d)$		the set of infinitely differentiable functions on $\mathbb{R}^d$ with compact support
$\mathcal{L}^p(\mathbb{R}^d)$		the set of complex-valued Lebesgue measurable functions f on $\mathbb{R}^d$ for which $\lvert f\rvert^p$ is Lebesgue integrable
$L^p(\mathbb{R}^d)$		L^p-space with respect to Lebesgue measure on $\mathbb{R}^d$
$L^p(\mathbb{Z}^d)$		L^p-space with respect to the counting measure on $\mathbb{Z}^d$
$\mathcal{L}^p_r, L^p_r$		the corresponding spaces of *real-valued* functions
x_+	$=$	$\max(0, x),\ x \in \mathbb{R}$
$\exp(z)$	$=$	$\mathrm{e}^z,\ z \in \mathbb{C}$
$\mathbf{1}_A$		indicator function of a set A
δ_x		one-point or Dirac measure concentrated at x
$\lVert \cdot \rVert_p$		L^p-norm
$\tilde{f}(x)$	$=$	$\overline{f(-x)}$
span		linear span
$f(x+0)$		right-hand limit
$f(x-0)$		left-hand limit
$\mathbb{E}(X)$		expectation of the random variable or random vector X
$\operatorname{cov}(X)$		covariance of the random variable or random vector X

Let S be a topological space. By $\mathrm{C}(S)$ we denote the set of continuous complex-valued functions on S. The symbol $\mathrm{C}_{00}(S)$ denotes the set of continuous functions with compact support while $\mathrm{C}_0(S)$ is the set of continuous functions vanishing at infinity.[1]

[1] Recall that a continuous function f on S vanishes at infinity if and only if for all $\epsilon > 0$ there exists a compact set $K \subset S$ such that $|f(x)| < \epsilon,\ x \in S \setminus K$.

A.2 Multidimensional notation

If $x \in \mathbb{R}^d$ ($\mathbb{C}^d$, $\mathbb{N}_0^d$, etc.), then x_i, $1 \le i \le d$, denotes the i-th coordinate of x and we write x as $x = (x_1, \dots, x_d)$. In expressions involving matrix operations, e.g., in Ax where A is a matrix, we consider x as a column vector. The zero element of $\mathbb{R}^d$ will also be denoted by 0. The *standard basis* of $\mathbb{R}^d$ or $\mathbb{Z}^d$ consists of elements $e_j \in \mathbb{R}^d$, $1 \le j \le d$, such that the j-th coordinate of e_j is equal to 1 and the other coordinates are zero.

For $\alpha, \beta \in \mathbb{R}^d$ we write $\beta \le \alpha$ if $\beta_j \le \alpha_j$ for all j. The relation $\alpha < \beta$ holds if and only if $\alpha \le \beta$ and $\alpha \ne \beta$.

Let $x \in \mathbb{R}^d$ or $\mathbb{C}^d$ and let $\alpha, \beta \in \mathbb{N}_0^d$. Then we write

$$\begin{aligned} x^\alpha &= x_1^{\alpha_1} \cdot x_2^{\alpha_2} \cdot \ldots \cdot x_d^{\alpha_d} \\ \|x\| &= \sqrt{|x_1|^2 + \cdots + |x_d|^2} \\ |x| &= |x_1| + \cdots + |x_d| \\ \alpha! &= \alpha_1! \cdot \ldots \cdot \alpha_d! \\ \binom{\alpha}{\beta} &= \frac{\alpha!}{(\alpha - \beta)! \cdot \beta!} = \binom{\alpha_1}{\beta_1} \cdot \ldots \cdot \binom{\alpha_d}{\beta_d}, \quad \beta \le \alpha \\ D^\alpha &= \frac{\partial^{|\alpha|}}{\partial x_1^{\alpha_1} \cdots \partial x_d^{\alpha_d}} \end{aligned}$$

When $\alpha = (0, \dots, 0)$ then $x^\alpha = 1$. Note that $D^\alpha D^\beta g = D^{\alpha+\beta} g$ where g denotes an arbitrary real- or complex-valued function on $\mathbb{R}^d$ for which the partial derivative $D^{\alpha+\beta} g$ exists. By definition, $D^\alpha g = g$ when $\alpha = (0, \dots, 0)$. We also use the notation

$$g_{x_i} := \frac{\partial g}{\partial x_i} \quad \text{or} \quad g_{x_i, x_j} := \frac{\partial g}{\partial x_i \partial x_j}.$$

By $\mathrm{d}x$, $\mathrm{d}\lambda(x)$ or $\mathrm{d}\lambda_d(x)$ we denote integration with respect to the Lebesgue measure $\lambda = \lambda_d$ on $\mathbb{R}^d$, while (x, y) is the inner product of $x, y \in \mathbb{R}^d$ or $\mathbb{C}^d$.

For $r \ge 0$ the open and closed balls with radius r and center $0 \in \mathbb{R}^d$ are denoted by $B^o(r) = B_d^o(r)$ and $B^c(r) = B_d^c(r)$, respectively. That is,

$$B_d^o(r) = \{t \in \mathbb{R}^d : \|t\| < r\}$$

and

$$B_d^c(r) = \{t \in \mathbb{R}^d : \|t\| \le r\}.$$

Appendix B
Basic analysis

B.1 Miscellaneous results from classical analysis

Lemma B.1.1. *The inequality*

$$\sum_{k=1}^{n} \frac{|z^k - w^k|}{k!} \leq |z - w| \cdot e^{\max(|z|,|w|)}$$

holds for all $z, w \in \mathbb{C}$ *and* $n \in \mathbb{N}$.

Proof. Using the identity

$$z^k - w^k = (z - w) \cdot \sum_{s=0}^{k-1} z^s w^{k-1-s}, \quad k \in \mathbb{N}$$

we obtain

$$\begin{aligned}
\left| \sum_{k=1}^{n} \frac{z^k - w^k}{k!} \right| &\leq |z - w| \cdot \sum_{k=1}^{n} \frac{1}{k!} \sum_{s=0}^{k-1} |z|^s |w|^{k-1-s} \\
&\leq |z - w| \cdot \sum_{k=1}^{n} \frac{1}{k!} \cdot k \cdot \max(|z|, |w|)^{k-1} \\
&= |z - w| \cdot \sum_{k=0}^{n-1} \frac{1}{k!} \cdot \max(|z|, |w|)^{k} \\
&\leq |z - w| \cdot e^{\max(|z|,|w|)}.
\end{aligned}$$

□

Lemma B.1.2. *Let* $\{z_n\}_1^\infty$ *be a sequence of complex numbers tending to some complex number* z. *Then*

$$\lim_{n \to \infty} \left(1 + \frac{z_n}{n}\right)^n = e^z.$$

Proof. Setting

$$d_n = \sum_{k=0}^{n} \frac{z^k}{k!} - \left(1 + \frac{z_n}{n}\right)^n, \quad n \in \mathbb{N}$$

it suffices to prove that $\lim_{n\to\infty} d_n = 0$. We have

$$\begin{aligned}|d_n| &= \left|\sum_{k=1}^{n}\left[\frac{z^k}{k!} - \frac{n(n-1)\cdots(n-k+1)}{k!}\cdot\frac{z_n^k}{n^k}\right]\right| \\ &\le \sum_{k=1}^{n}\frac{1}{k!}\cdot\left|z^k - z_n^k\cdot\prod_{j=1}^{k-1}\left(1-\frac{j}{n}\right)\right| \\ &\le \sum_{k=1}^{n}\frac{\left|z^k - z_n^k\right|}{k!} + \sum_{k=1}^{n}\frac{|z_n|^k}{k!}\cdot\left[1-\prod_{j=1}^{k-1}\left(1-\frac{j}{n}\right)\right].\end{aligned}$$

By Lemma B.1.1, the first sum in the last display tends to zero if $n \to \infty$. Denote by s_n the second sum. Applying Bernoulli's inequality[1] we obtain

$$|s_n| \le \sum_{k=1}^{n}\frac{|z_n|^k}{k!}\cdot\frac{(k-1)k}{2n} \le \frac{|z_n|^2}{2n}\cdot \mathrm{e}^{|z_n|} \to 0 \quad (n\to\infty).$$ □

Proposition B.1.3 (binomial series). *For all $\alpha > 0$ and $s \in (-1,1)$ we have*

$$\frac{1}{(1+s)^\alpha} = \sum_{k=0}^{\infty}(-1)^k\frac{\Gamma(k+\alpha)}{\Gamma(k+1)\Gamma(\alpha)}\cdot s^k \tag{1}$$

where the series is absolutely convergent.

Proof. Denote by a_k the coefficient of s^k. Using the relation

$$\Gamma(x+1) = x\Gamma(x) \tag{2}$$

(cf. Lemma C.4.2) we see that $\lim_{k\to\infty}|a_{k+1}/a_k| = 1$. Thus, the radius of convergence of the power series above is equal to 1. Applying again (2), the relation $\Gamma(k+1) = k!$, and using induction on k it is easy to check that (1) is the Taylor series expansion of $s \mapsto \frac{1}{(1+s)^\alpha}$ at $s = 0$. This completes the proof. □

Theorem B.1.4 (Fejér–Riesz). *Let p be a function of the form*

$$p(z) = \sum_{n=-N}^{N} c_n z^n, \quad c_n \in \mathbb{C},\ z \in \mathbb{C}\setminus\{0\}.$$

---

[1] Bernoulli's inequality states that

$$(1+x_1)(1+x_2)\cdots(1+x_n) \ge 1 + x_1 + x_2 + \cdots + x_n$$

where $x_j \in (-1,0]$ for all j or $x_j \in [0,\infty)$ for all j. This inequality can easily be proved by induction on n.

If $p(z) \geq 0$ for all $z \in \mathbb{T}$, then there exists a polynomial q of the form

$$q(z) = \sum_{n=0}^{N} b_n z^n, \quad z, b_n \in \mathbb{C}$$

such that

$$p(z) = q(z)\overline{q(1/\overline{z})}, \quad z \in \mathbb{C} \setminus \{0\}.$$

In particular, $p(z) = |q(z)|^2$ holds for all $z \in \mathbb{T}$.

Proof. Without loss of generality assume that $c_N \neq 0$. Since p is real-valued on $\mathbb{T}$ we have

$$0 = p(z) - \overline{p(z)} = \sum_{n=-N}^{N} (c_n - \overline{c}_{-n})z^n, \quad z \in \mathbb{T}$$

showing that $c_{-n} = \overline{c}_n$, $n = 0, 1, \ldots, N$. Using this we see that the polynomial g given by

$$g(z) := z^N p(z) = \overline{c}_N + \cdots + c_0 z^N + \cdots + c_N z^{2N}, \quad z \in \mathbb{C}$$

satisfies the equation

$$g(z) = z^{2N}\overline{g(1/\overline{z})}, \quad z \in \mathbb{C} \setminus \{0\}. \tag{1}$$

Denote by $z_1, \ldots, z_{2N}$ the zeros of g counted according to their multiplicities. Since $c_N \neq 0$, all these zeros are different from 0. It follows from (1) that $1/\overline{z}_j$ is a zero of g having the same multiplicity as z_j.

Next we show that if $z_j = e^{it_j} \in \mathbb{T}$, then z_j has even multiplicity. The function $\varphi(t) := p(e^{it})$, $t \in \mathbb{R}$, is analytic and nonnegative on $\mathbb{R}$. Moreover, t_j is a zero of φ. It suffices to prove that the multiplicity of t_j is even but this is a consequence of the fact that the function φ, being nonnegative, has a local minimum at t_j. It follows from the statements concerning the zeros of g that we can, if necessary, rearrange these zeros z_j such that $z_1, \ldots, z_N, 1/\overline{z}_1, \ldots, 1/\overline{z}_N$ represent all the $2N$ zeros of g. We then have

$$g(z) = c_N \prod_{j=1}^{N} (z - z_j) \prod_{j=1}^{N} (z - 1/\overline{z}_j)$$

and hence

$$\begin{aligned} p(z) = z^{-N} g(z) &= c_N \prod_{j=1}^{N} (z - z_j) \prod_{j=1}^{N} (1 - 1/z\overline{z}_j) \\ &= c \prod_{j=1}^{N} (z - z_j) \prod_{j=1}^{N} (1/z - \overline{z}_j) \end{aligned} \tag{2}$$

where $c = c_N \left(\prod_{j=1}^{N}(-\overline{z}_j)\right)^{-1}$. Since p is nonnegative on $\mathbb{T}$ the constant c is positive. Setting $q(z) := \sqrt{c}\prod_{j=1}^{N}(z - z_j)$ and applying (2) we obtain

$$p(z) = q(z)\overline{q(1/\overline{z})}, \quad z \in \mathbb{C} \setminus \{0\}. \qquad \square$$

Theorem B.1.5. *The inequality*

$$\left|\mathrm{e}^{\mathrm{i}z} - 1 - \mathrm{i}z - \cdots - \frac{(\mathrm{i}z)^n}{n!}\right| \le \begin{cases} \dfrac{|z|^{n+1}}{(n+1)!}, & \text{if} \quad \operatorname{Im} z \ge 0, \\[2ex] \dfrac{|z|^{n+1}}{(n+1)!} \cdot \mathrm{e}^{|z|}, & \text{if} \quad \operatorname{Im} z \le 0. \end{cases}$$

holds for all $n \in \mathbb{N}_0$ and $z \in \mathbb{C}$.

Proof. Assume first that $n = 0$ and let $z = a + ib,\ a, b \in \mathbb{R}$, be arbitrary. Using the identity

$$\mathrm{e}^{\mathrm{i}z} - 1 = \mathrm{i}z \int_0^1 \mathrm{e}^{\mathrm{i}tz}\,\mathrm{d}t$$

we obtain

$$\begin{aligned} \left|\mathrm{e}^{\mathrm{i}z} - 1\right| &= |z| \left|\int_0^1 \mathrm{e}^{\mathrm{i}tz}\,\mathrm{d}t\right| \le |z| \int_0^1 \left|\mathrm{e}^{\mathrm{i}tz}\right| \mathrm{d}t \\ &= |z| \int_0^1 \left|\mathrm{e}^{\mathrm{i}ta - tb}\right| \mathrm{d}t = |z| \int_0^1 \mathrm{e}^{-tb}\,\mathrm{d}t \le |z| \begin{cases} 1, & \text{if} \quad b \ge 0, \\ \mathrm{e}^{|z|}, & \text{if} \quad b \le 0. \end{cases} \end{aligned}$$

Setting

$$r_n(z) := \mathrm{e}^{\mathrm{i}z} - 1 - \mathrm{i}z - \cdots - \frac{(\mathrm{i}z)^n}{n!}, \quad n \in \mathbb{N}_0$$

it is not hard to check that

$$r_{n+1}(z) = \mathrm{i}z \int_0^1 r_n(tz)\,\mathrm{d}t.$$

Using this equation, the theorem follows by induction on n. $\square$

B.1.6. Here we collect some simple but useful inequalities for trigonometric functions. Integrating the inequalities

$$-1 \le \cos x \le 1 \quad \text{and} \quad -1 \le \sin x \le 1, \quad x \in \mathbb{R}$$

several times we find

$$|\sin x| < |x|, \quad x \ne 0 \tag{1}$$

$$1 - \frac{1}{2}x^2 < \cos x < 1 - \frac{1}{2}x^2 + \frac{1}{24}x^4, \quad x \ne 0 \tag{2}$$

$$x - \frac{1}{6}x^3 < \sin x < x - \frac{1}{6}x^3 + \frac{1}{120}x^5, \quad x > 0. \tag{3}$$

The inequality

$$\sin x \le x \cdot \frac{\sin y}{y}, \quad 0 < y \le \frac{\pi}{2}, \ x \ge y \tag{4}$$

can be proved easily, e.g., by using geometrical arguments. Setting here $y = 1$ and applying (3) we conclude that

$$\frac{\sin x}{x} \le \frac{\sin 1}{1} < \frac{101}{120}, \quad |x| \ge 1. \tag{5}$$

Next we show that

$$h(t) := n \cos^n t - \cos nt \le n - 1, \quad t \in \mathbb{R},\ n \in \mathbb{N}. \tag{6}$$

The function h is periodic, so that it attains its maximum at some $y \in \mathbb{R}$. Since

$$h'(y) = -n^2 \cos^{n-1} y \sin y + n \sin ny$$

either $\sin y = 0$ or

$$\sin y \ne 0 \quad \text{and} \quad n \cos^{n-1} y = \frac{\sin ny}{\sin y}. \tag{7}$$

It is easy to check that $\sin y = 0$ implies (6). If (7) holds, then

$$\begin{aligned} h(y) &= \frac{\cos y \sin ny}{\sin y} - \cos ny \\ &= \frac{\cos y \sin ny - \sin y \cos ny}{\sin y} = \frac{\sin(n-1)y}{\sin y}. \end{aligned}$$

A simple induction argument shows that

$$\frac{\sin(n-1)y}{\sin y} \le n - 1, \quad n \in \mathbb{N}$$

and therefore $h(y) \le n - 1$.

The function

$$\mathrm{Si}(T) := \int_0^T \frac{\sin t}{t}\, \mathrm{d}t, \quad T \in \mathbb{R} \tag{8}$$

is frequently used (see Figure B.1). We will need its limit at infinity.

Theorem B.1.7. *We have*

$$\lim_{T \to \infty} \mathrm{Si}(T) = \frac{\pi}{2}.$$

Proof. We have

$$\mathrm{Si}(T) = \int_0^T \sin t \int_0^\infty \mathrm{e}^{-ty}\, \mathrm{d}y\, \mathrm{d}t = \int_0^\infty \int_0^T \sin t\, \mathrm{e}^{-ty}\, \mathrm{d}t\, \mathrm{d}y.$$

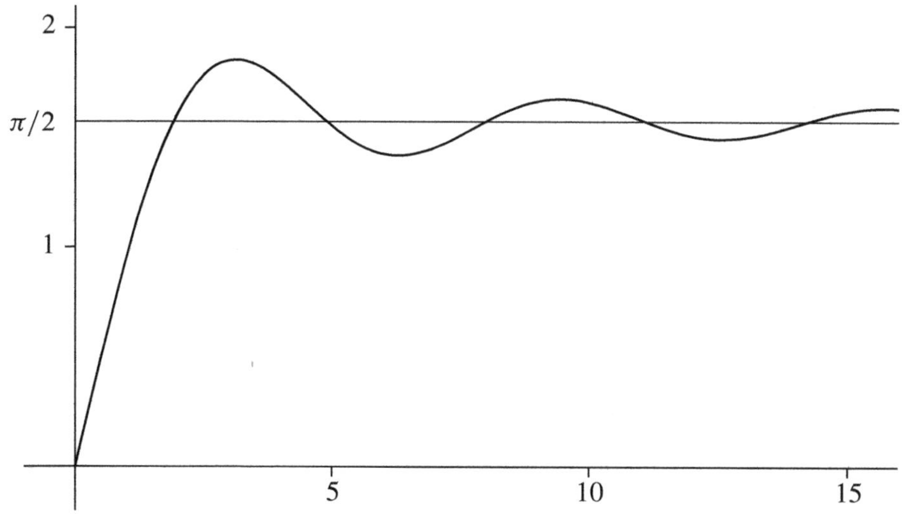

Figure B.1. The function Si from equation (B.1.6.8).

Applying integration by parts twice gives

$$\mathrm{Si}(T) = \int_0^\infty \left[\frac{1}{1+y^2} - \mathrm{e}^{-Ty}\,\frac{y\sin T + \cos T}{1+y^2}\right] \mathrm{d}y$$

and therefore

$$\lim_{T\to\infty} \mathrm{Si}(T) = \int_0^\infty \frac{1}{1+y^2}\,\mathrm{d}y = \arctan y\big|_0^\infty = \frac{\pi}{2}. \qquad \square$$

From Theorem B.1.7 we easily obtain the following corollary.

Corollary B.1.8.

$$\frac{1}{\pi}\lim_{T\to\infty}\int_{-T}^{T} \frac{\sin xt}{t}\,\mathrm{d}t = \operatorname{sign} x, \qquad x \in \mathbb{R}.$$

Theorem B.1.9. *If* $-\infty < a \le b < \infty$ *then*

$$\left|\int_a^b \frac{\sin t}{t}\,\mathrm{d}t\right| < 2\pi.$$

Proof. Define the real numbers c_n, $n = 0, 1, 2, \ldots$, by

$$c_n := \int_{n\pi}^{(n+1)\pi} \frac{\sin t}{t}\,\mathrm{d}t = (-1)^n \int_0^\pi \frac{\sin t}{n\pi + t}\,\mathrm{d}t.$$

Then $|c_n| \geq |c_{n+1}|$. From this we conclude that the inequalities

$$0 \leq \int_0^T \frac{\sin t}{t}\, \mathrm{d}t \leq c_0 < \pi$$

hold for all $T \geq 0$, from which the theorem follows. □

Lemma B.1.10. *We have*

$$\int_0^\infty \mathrm{e}^{-x^2}\, \mathrm{d}x = \frac{\sqrt{\pi}}{2}.$$

Proof. Denote by I the integral above. Using Fubini's theorem and the substitution $y = xs$ we obtain

$$\begin{aligned}
I^2 &= \Big(\int_0^\infty \mathrm{e}^{-x^2}\, \mathrm{d}x\Big) \cdot \Big(\int_0^\infty \mathrm{e}^{-y^2}\, \mathrm{d}y\Big) \\
&= \int_0^\infty \Big(\int_0^\infty \mathrm{e}^{-(x^2+y^2)}\, \mathrm{d}y\Big)\, \mathrm{d}x \\
&= \int_0^\infty \Big(\int_0^\infty x \cdot \mathrm{e}^{-x^2(1+s^2)}\, \mathrm{d}s\Big)\, \mathrm{d}x \\
&= \int_0^\infty \frac{1}{-2(1+s^2)} \cdot \mathrm{e}^{-x^2(1+s^2)}\Big|_{x=0}^{\infty}\, \mathrm{d}s \\
&= \frac{1}{2}\int_0^\infty \frac{1}{1+s^2}\, \mathrm{d}s = \frac{1}{2}\arctan s\Big|_0^\infty = \frac{\pi}{4}.
\end{aligned}$$

□

Using an obvious substitution we obtain the following corollary.

Corollary B.1.11. *For every $a, \sigma \in \mathbb{R}$, $\sigma > 0$, the function*

$$\varphi_{a,\sigma}(x) = \frac{1}{\sqrt{2\pi} \cdot \sigma} \cdot \mathrm{e}^{-\frac{x^2}{2\sigma^2}}, \quad x \in \mathbb{R}$$

is a probability density (see Figure B.2).

Lemma B.1.12. *For all $x > 0$ and $a > -x$ we have*

$$\int_0^\infty \frac{\mathrm{e}^{-xt} - \mathrm{e}^{-(x+a)t}}{t}\, \mathrm{d}t = \log\Big(1 + \frac{a}{x}\Big).$$

Proof. Denote by $f(x)$ and $g(x)$ the left- and right-hand side of the equation above. We have

$$\lim_{x\to\infty} f(x) = \lim_{x\to\infty} g(x) = 0$$

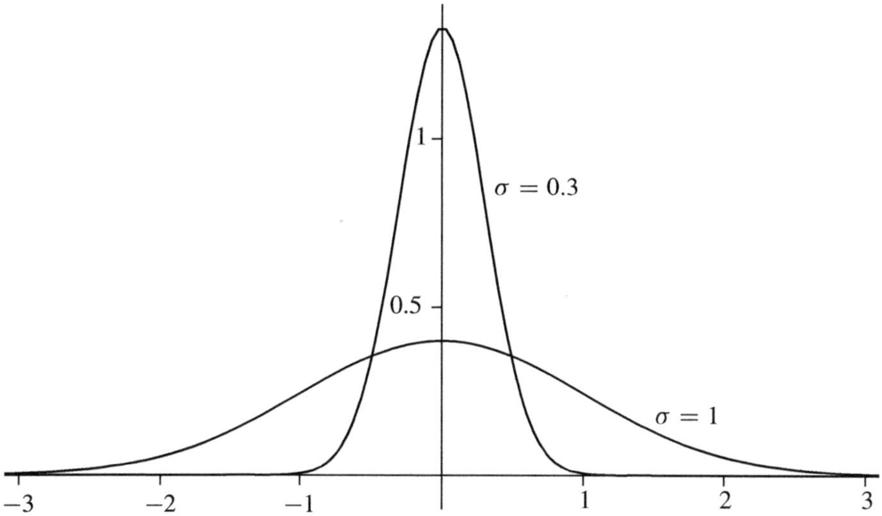

Figure B.2. The function $\varphi_{0,\sigma}$ from Corollary B.1.11.

and $f'(x) = g'(x)$.[2] Consequently, $f(x) = g(x)$ for all $x > 0$. □

B.1.13. The numbers

$$\binom{m}{k_1, k_2, \ldots, k_n} := \frac{m!}{k_1! k_2! \cdots k_n!}$$

are called *multinomial coefficients*. They satisfy the *multinomial formula*

$$(x_1 + \cdots + x_n)^m = \sum_{|k|=m} \binom{m}{k_1, k_2, \ldots, k_n} x^k, \quad n \in \mathbb{N},\ m \in \mathbb{N}_0$$

where $k \in \mathbb{N}_0^n$.[3] This formula can be proved by induction on n, applying the binomial formula to the identity

$$(x_1 + x_2 + \cdots + x_n + x_{n+1})^m = (x_1 + x_2 + \cdots + [x_n + x_{n+1}])^m.$$

It follows from the multinomial formula that the multinomial coefficients have a combinatorial interpretation as the number of partitions[4] of a set with m elements in n blocks, with k_1 elements in the first block, k_2 elements in the second block, and so on.

[2] Differentiating under the integral sign can be avoided by replacing $\frac{1}{t}$ by $\int_0^\infty e^{-st}\, ds$ and then using Fubini's theorem.

[3] Expressions of the form 0^0 are defined to be equal to 1.

[4] A *partition* of a set S is a collection of nonempty pairwise disjoint subsets of S whose union is S. The subsets are also called *blocks*.

The multinomial coefficients also occur in the identity

$$\frac{d^m}{dt^m}\prod_{i=1}^{n} f_i(t) = \sum_{|k|=m}\binom{m}{k_1,\dots,k_n}\prod_{i=1}^{n} f_i^{(k_i)}(t) \tag{1}$$

called the *generalized Leibniz rule*. This identity can be proved by induction on n showing first that

$$(fg)^{(m)} = \sum_{k=0}^{m}\binom{m}{k} f^{(k)} g^{(m-k)}.$$

Next we prove *Faà di Bruno's formula* for the higher chain rule for differentiation.[5]

Theorem B.1.14 (Faà di Bruno's formula). *If f and g are real functions with a sufficient number of derivatives, then*

$$\frac{d^m}{dt^m} g(f(t)) =$$

$$\sum \frac{m!}{b_1!b_2!\cdots b_m!} g^{(|b|)}(f(t))\left(\frac{f'(t)}{1!}\right)^{b_1}\left(\frac{f''(t)}{2!}\right)^{b_2}\cdots\left(\frac{f^{(m)}(t)}{m!}\right)^{b_m} \tag{1}$$

where the sum is over all different solutions in nonnegative integers $b_1,\dots,b_m$ of $b_1 + 2b_2 + \cdots + mb_m = m$, and $|b| = b_1 + \cdots + b_m$.

Proof. We prove that

$$\frac{d^m}{dt^m} g(f(t)) = \sum g^{(k)}(f(t))\,(f'(t))^{b_1}(f''(t))^{b_2}\cdots(f^{(m)(t)})^{b_m} \tag{2}$$

where the sum is over all partitions of $S_m = \{1, 2, \dots, m\}$, and, for each partition, k is its number of blocks and b_i is the number of blocks with exactly i elements. The number of partitions of S_m into b_1 blocks with 1 element, b_2 blocks with 2 elements, etc. would be (cf. B.1.13) the multinomial coefficient

$$\binom{m}{\underbrace{1,\dots,1}_{b_1\ 1's},\underbrace{2,\dots,2}_{b_2\ 2's},\underbrace{3,\dots,3}_{b_3\ 3's},\dots}$$

except that this makes artificial distinctions among the i-blocks for each i. Therefore, we have to divide this number by $b_1!\,b_2!\cdots b_m!$. Thus, (1) follows from equation (2).

We prove (2) by induction on m. The case $m = 1$ being simple, assume that the statement is true for some m. Every partition of S_{m+1} can be obtained in a unique way by adjoining $m + 1$ to a partition of S_m.

[5] Our proof is taken from the survey paper [32] which contains interesting remarks on the history of this formula.

If we add $\{m+1\}$ as a new singleton block, we increase the number of blocks of size 1 by one, and the total number of blocks by one. This corresponds to applying d/dt to $g^{(k)}(f(t))$ to get $g^{(k+1)}(f(t))f'(t)$.

If we add $m+1$ to an existing block of size i, then the number of such blocks decreases by one, the number of blocks of size $i+1$ increases by one, and the total number of blocks remains the same. This corresponds to applying d/dt to $(f^{(i)}(t))^{b_i}$, to get $b_i(f^{(i)}(t))^{b_i-1}f^{(i+1)}(t)$. □

Lemma B.1.15. *Let $a_k \geq 0$, $k \in \mathbb{N}$, and denote by R the radius of convergence of the power series*

$$S(z) = \sum_{k=1}^{\infty} a_k z^k, \quad z \in \mathbb{C}.$$

If $0 \leq R < \infty$, then $S(R+\epsilon) = \infty$ for all $\epsilon > 0$.

Proof. Indeed, if $S(R+\epsilon)$ were finite for some positive ϵ, then we would have

$$|S(z)| \leq \sum_{k=1}^{\infty} a_k (R+\epsilon)^k < \infty$$

whenever $|z| \leq R+\epsilon$, i.e., the radius of convergence would be larger than R. □

Lemma B.1.16. *Let $\{a_k\}_1^{\infty}$ be a sequence of nonnegative real numbers. There exists a sequence $\{p_k\}_1^{\infty}$ of positive real numbers such that*

$$\sum_{k=1}^{\infty} p_k a_k^n < \infty$$

for all $n \in \mathbb{N}$.

Proof. Setting $b_k := \max(a_k, k)$ we have $a_k \leq b_k$ and $b_k \to \infty$. Let $p_k = \mathrm{e}^{-b_k}$. Then for each $n \in \mathbb{N}$ there exists $N(n) \in \mathbb{N}$ such that

$$p_k b_k^n \leq \mathrm{e}^{-b_k/2} \leq \mathrm{e}^{-k/2}$$

if $k \geq N(n)$. Thus,

$$\sum_{k=1}^{\infty} p_k a_k^n \leq \sum_{k=1}^{\infty} p_k b_k^n < \infty$$

for all $n \in \mathbb{N}$. □

Lemma B.1.17. *Let $x \in \mathbb{R}^d$ and let $\{x_n\}$ be a sequence in $\mathbb{R}^d$ such that*

$$\lim_{n\to\infty} \mathrm{e}^{\mathrm{i}(t,x_n)} = \mathrm{e}^{\mathrm{i}(t,x)}$$

for all $t \in \mathbb{R}^d$. Then $\lim_{n\to\infty} x_n = x$.[6]

[6] Note that for the divergent sequence $\{x_n\}$ in $\mathbb{R}$ with $x_n = 2\pi n!$ the relation above holds for all rational t.

Proof. We may assume that $x = 0$ and $d = 1$ so that

$$\lim_{n\to\infty} \mathrm{e}^{\mathrm{i}tx_n} = 1 \tag{1}$$

for all t. Assume first that the sequence $\{x_n\}$ is bounded. If s is an accumulation point of this sequence then, by equation (1), $\mathrm{e}^{\mathrm{i}ts} = 1$ for all t and hence $s = 0$. From this we conclude that $\{x_n\}$ tends to 0.

Next assume that $\{x_n\}$ is unbounded and choose a subsequence $\{y_n\}$ tending to ∞ or to $-\infty$. Using equation (1) we see that

$$1 = \int_0^1 \lim_{n\to\infty} \mathrm{e}^{\mathrm{i}ty_n}\,\mathrm{d}t = \lim_{n\to\infty} \int_0^1 \mathrm{e}^{\mathrm{i}ty_n}\,\mathrm{d}t = \lim_{n\to\infty} \frac{\mathrm{e}^{\mathrm{i}y_n} - 1}{\mathrm{i}y_n} = 0.$$

Thus, the sequence $\{x_n\}$ cannot be unbounded. □

Lemma B.1.18. *For all* $t, a \geq 0$ *we have*

$$\lim_{\lambda\to\infty} \mathrm{e}^{-\lambda t} \sum_{0\leq k\leq \lambda a} \frac{(\lambda t)^k}{k!} = \mathbf{1}_{[0,a]}.$$

Proof. Let X be a Poisson random variable with parameter λt. Then

$$P(X \leq \lambda a) = \mathrm{e}^{-\lambda t} \sum_{0\leq k\leq \lambda a} \frac{(\lambda t)^k}{k!}.$$

Thus, we have to prove that

$$\lim_{\lambda\to\infty} P(X \leq \lambda a) = \mathbf{1}_{[0,a]}.$$

We have $\mathbb{E}(X) = \mathrm{Var}(X) = \lambda t$ (cf. Example 1.2.4). If $t > a$, then Chebyshev's inequality (cf. (F.1.6)) shows that

$$P(X \leq \lambda a) \leq P(|X - \lambda t| \geq \lambda(t-a)) \leq \frac{t}{\lambda(t-a)^2} \to 0, \quad \lambda \to \infty.$$

If $t \leq a$, we find in the same way

$$\begin{aligned} P(X \leq \lambda a) &= 1 - P(X - \lambda t > \lambda(a-t)) \\ &\geq 1 - P(|X - \lambda t| > \lambda(a-t)) \to 1, \quad \lambda \to \infty. \end{aligned}$$

□

Example B.1.19. Let h be the 2-periodic function on $\mathbb{R}$ with

$$h(x) = 1 - |x|, \quad |x| \leq 1.$$

We show that the function f defined by (see Figure B.3)

$$f(x) = \frac{1}{4} \sum_{n=0}^{\infty} \left(\frac{3}{4}\right)^n h(4^n x)$$

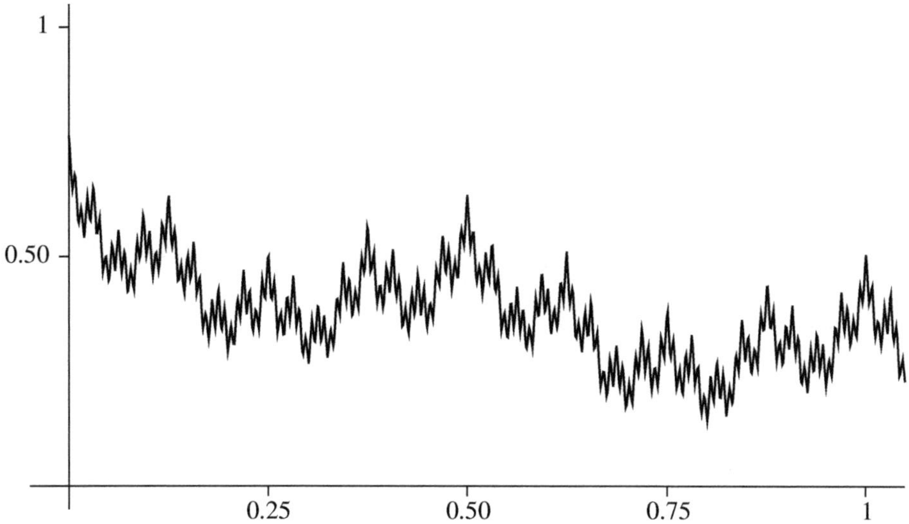

Figure B.3. The 5th partial sum of the series from Example B.1.19.

is nowhere differentiable.[7] Let $x \in \mathbb{R}$, $m \in \mathbb{N}$, and

$$\varepsilon_m := \pm\frac{1}{2 \cdot 4^m}$$

where the sign is chosen in such a way that there is no integer lying between $4^m x$ and $4^m(x + \varepsilon_m)$. Write

$$d_{nm} := \frac{h(4^n(x + \varepsilon_m)) - h(4^n x)}{\varepsilon_m}.$$

Using the obvious inequality $|h(s) - h(t)| \le |s - t|$, we see that $|d_{nm}| \le 4^n$ for all n, m. Since there is no integer between $4^m x$ and $4^m(x + \varepsilon_m)$, we conclude that $|d_{mm}| = 4^m$. If $n > m$, then $4^n \varepsilon_m$ is an even integer and hence $d_{nm} = 0$. Therefore,

$$4 \cdot \left|\frac{f(x + \varepsilon_m) - f(x)}{\varepsilon_m}\right| = \left|\sum_{n=0}^{m}\left(\frac{3}{4}\right)^n d_{nm}\right| \ge 3^m - \sum_{n=0}^{m-1} 3^n = \frac{1}{2}(3^m + 1).$$

Since $\lim_{m\to\infty} \varepsilon_m = 0$ we see that f is not differentiable at x.

[7] Note that h is equal to the function $\tilde{f}_{1/2}$ from Remark 4.2.4. Therefore h, and hence also f, are positive definite. Bochner's Theorem 1.7.3 shows that they are characteristic functions.

B.2 Uniform convergence of continuous functions

The first theorem of this section is due to S. N. Bernstein.

Theorem B.2.1 (Bernstein). *Let f be a continuous real-valued function on* $[0,1]$. *Then*

$$f(x) = \lim_{n\to\infty} \sum_{k=0}^{n} f\Big(\frac{k}{n}\Big)\binom{n}{k} x^k (1-x)^{n-k}, \qquad x \in [0,1]$$

where the convergence is uniform on $[0,1]$.

Proof. For each $x \in [0,1]$ and $n \in \mathbb{N}$ let Y_n^x be a random variable on a probability space $(\Omega, \mathcal{A}, P)$ such that nY_n^x has binomial distribution with parameters n and x. We then have $\mathbb{E}(Y_n^x) = x$, $\operatorname{Var}(Y_n^x) = \frac{x(1-x)}{n}$ and

$$\mathbb{E}(f(Y_n^x)) = \sum_{k=0}^{n} f\Big(\frac{k}{n}\Big)\binom{n}{k} x^k (1-x)^{(n-k)}.$$

Therefore, it suffices to show that

$$\lim_{n\to\infty} \mathbb{E}(f(Y_n^x) - f(x)) = 0, \qquad x \in [0,1]$$

uniformly on $[0,1]$. By Chebyshev's inequality (see (F.1.6))

$$P\left(|Y_n^x - x| > n^{-1/4}\right) \le n^{1/2}\operatorname{Var}(Y_n^x) = \frac{x(1-x)}{\sqrt{n}} \le \frac{1}{\sqrt{n}}. \tag{1}$$

Choose $K \in \mathbb{R}$ such that $|f| \le K$ and denote by A the random event on the left-hand side of (1). Then $P(A) \le \frac{1}{\sqrt{n}}$ and

$$\begin{aligned}
\mathbb{E}(|f(Y_n^x) - f(x)|) &= \int_\Omega |f(Y_n^x) - f(x)|\,\mathrm{d}P \\
&= \int_A |f(Y_n^x) - f(x)|\,\mathrm{d}P + \int_{\Omega\setminus A} |f(Y_n^x) - f(x)|\,\mathrm{d}P \\
&\le \frac{2K}{\sqrt{n}} + \sup_{|y-z|\le n^{-1/4}} |f(y) - f(z)|.
\end{aligned}$$

The theorem follows now from the fact that f is uniformly continuous. □

Corollary B.2.2 (Weierstraß). *Every continuous, real-valued (complex-valued) function on a finite interval $[a,b] \subset \mathbb{R}$ can be uniformly approximated by real (complex, respectively) polynomials.*

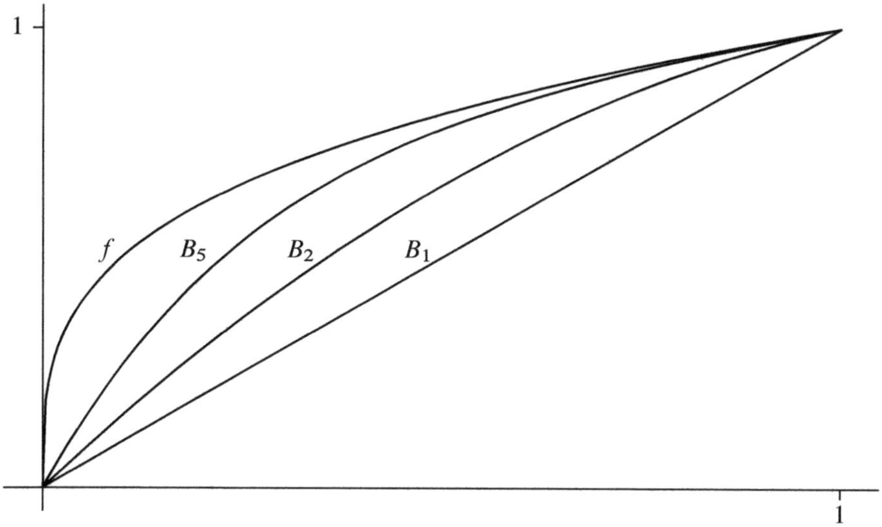

Figure B.4. The Bernstein polynomials B_1, B_2 and B_5 from Remark B.2.3 for the function $f : x \mapsto x^{0.3}$.

Remark B.2.3. Let f be a function on $[0, 1]$. The polynomials $B_n(f)$, $n \in \mathbb{N}_0$, defined by

$$B_n(f)(x) := \sum_{k=0}^{n} f\left(\tfrac{k}{n}\right) \binom{n}{k} x^k (1-x)^{n-k}, \quad x \in [0, 1]$$

are called *Bernstein polynomials* (see Figure B.4).

Theorem B.2.4 (Stone–Weierstraß). *Suppose X is a compact Hausdorff space and $\mathcal{A}$ is a linear subspace of $\mathrm{C}(X)$ such that*

(i) *if $f, g \in \mathcal{A}$, then $\bar{f}$ and $f \cdot g$ belong to $\mathcal{A}$, as well;*

(ii) *$\mathcal{A}$ contains the constant functions;*

(iii) *for all $x, y \in X$, $x \neq y$, there exists a function $f \in \mathcal{A}$ such that $f(x) \neq f(y)$.*

Then every function in $C(X)$ can be uniformly approximated by functions in $\mathcal{A}$.

Proof. Denote by $\overline{\mathcal{A}}$ the closure of $\mathcal{A}$ in the topology of uniform convergence. We have to show that $\overline{\mathcal{A}} = \mathrm{C}(X)$.

It is easy to check that $\overline{\mathcal{A}}$ is a linear subspace and $\bar{f}$, $f \cdot g$, $\operatorname{Im} f$, $\operatorname{Re} f \in \overline{\mathcal{A}}$ whenever $f, g \in \overline{\mathcal{A}}$. From this we conclude that if p is a polynomial, then $p(f) \in \overline{\mathcal{A}}$ for all $f \in \overline{\mathcal{A}}$.

Now let $g \in \overline{\mathcal{A}}$ be a real-valued function. We show that $|g| \in \overline{\mathcal{A}}$. We may suppose that $-1 \leq g \leq 1$. By Corollary B.2.2, for each $n \in \mathbb{N}$ there exists a real polynomial such that

$$\big||t| - p_n(t)\big| \le \frac{1}{n}, \ -1 \le t \le 1.$$

We already know that $p_n(g)$ belongs to $\overline{\mathcal{A}}$. The inequalities above show that the sequence $\{p_n(g)\}$ converges uniformly to $|g|$, and hence $|g| \in \overline{\mathcal{A}}$.

If $f, g \in \overline{\mathcal{A}}$ are real-valued, then $\max(f, g)$ and $\min(f, g)$ belong to $\overline{\mathcal{A}}$, since

$$\max(f, g) = \frac{f + g + |f - g|}{2}, \quad \min(f, g) = \frac{f + g - |f - g|}{2}.$$

To finish the proof it suffices to show that every real-valued function $h \in C(X)$ belongs to $\overline{\mathcal{A}}$. It follows at once from (iii) that for all $x, y \in X$, $x \ne y$, we can choose a real-valued function $f_{xy} \in \mathcal{A}$ such that $f_{xy}(x) = h(x)$ and $f_{xy}(y) = h(y)$. Let $\epsilon > 0$ be arbitrary and define the sets U_{xy} and V_{xy} by

$$U_{xy} = \{t \in X : \ f_{xy}(t) < h(t) + \epsilon\},$$

$$V_{xy} = \{t \in X : \ f_{xy}(t) > h(t) - \epsilon\}.$$

These sets are open and $x, y \in U_{xy} \cap V_{xy}$. Therefore, $\{U_{xy}\}_{x \in X}$ is an open covering of X for each y. By compactness of X, there exist $n = n(y) \in \mathbb{N}$ and $x_i = x_i(y) \in X$, $i = 1, \dots, n$, such that

$$X = \bigcup_{i=1}^{n} U_{x_i y}.$$

By what we have already proved, $f_y := \min(f_{x_1 y}, \dots, f_{x_n y}) \in \overline{\mathcal{A}}$. Moreover,

$$\begin{aligned} f_y(t) &< h(t) + \epsilon, \quad && t \in X, \\ f_y(t) &> h(t) - \epsilon, \quad && t \in V_y := \cap_{i=1}^{n} V_{x_i y}. \end{aligned}$$

The sets V_y are open and $y \in V_y$. Using again the compactness of X we see that there exist $m \in \mathbb{N}$ and $y_1, \dots, y_m \in X$ such that $\cup_1^m V_{y_i} = X$. Choosing $f := \max(f_{y_1}, \dots, f_{y_m})$ we have $f \in \overline{\mathcal{A}}$ and

$$h(t) - \epsilon < f(t) < h(t) + \epsilon, \quad t \in X$$

completing the proof. □

Theorem B.2.5 (Dini). *Let $\{f_n\}$ be a sequence of continuous real-valued functions on a compact space X converging pointwise to a continuous function g. If this sequence is monotonically increasing or decreasing, then the convergence is uniform.*

Proof. We assume that the sequence is increasing. For every $\epsilon > 0$ and $x \in X$ there exists an index $n(x) = n(x, \epsilon)$ such that

$$0 \le g(x) - f_m(x) \le \frac{\epsilon}{3}, \quad m \ge n(x).$$

Let $V(x)$ be an open neighborhood of x such that the inequalities

$$|g(x) - g(y)| \leq \frac{\epsilon}{3}$$
$$|f_{n(x)}(x) - f_{n(x)}(y)| \leq \frac{\epsilon}{3}$$

hold for all $y \in V(x)$. For all $y \in V(x)$ we then have

$$0 \leq g(y) - f_{n(x)}(y) \leq \epsilon.$$

By compactness we can choose finitely many points $x_i \in X$ such that the union of the sets $V(x_i)$ is equal to X. Let n_0 be the largest of the numbers $n(x_i)$. Since each $y \in X$ belongs to some $V(x_i)$ we have

$$g(y) - f_n(y) \leq g(y) - f_{n(x_i)}(y) \leq \epsilon, \quad n \geq n_0.$$

The proof is complete. □

Definition B.2.6. Let X be a topological space. A set F of complex-valued functions on X is called *equicontinuous* if for all $x \in X$ and $\epsilon > 0$ there exists a neighborhood V_x of x such that the inequality

$$|f(x) - f(y)| < \epsilon$$

holds for all $f \in F$ and for all $y \in V_x$.

Theorem B.2.7 (Ascoli). *Let $\{f_\alpha\}$ be a net of equicontinuous complex-valued functions on a topological space X converging pointwise to a function f. Then this net converges uniformly on compact sets and f is continuous.*

Proof. We show first that f is continuous. For all $x, y \in X$ and for all α we have

$$|f(x) - f(y)| \leq |f(x) - f_\alpha(x)| + |f_\alpha(x) - f_\alpha(y)| + |f_\alpha(y) - f(y)|. \quad (1)$$

Let $x \in X$ and $\epsilon > 0$ be arbitrary and let V_x be a neighborhood of x such that

$$|f_\alpha(x) - f_\alpha(y)| < \epsilon, \quad y \in V_x.$$

For each $y \in V_x$ we choose an index $\alpha(y) = \alpha$ such that $|f(x) - f_\alpha(x)| < \epsilon$ and $|f(y) - f_\alpha(y)| < \epsilon$. It follows from inequality (1) that $|f(x) - f(y)| < 3\epsilon$, $y \in V_x$. Thus, f is continuous in x.

Obviously the set $\{f_\alpha, f\}$ is also equicontinuous. Let $K \subset X$ be compact. For all $x \in K$ and for all $\epsilon > 0$ there exists an open neighborhood V_x of x such that the inequalities

$$|f(x) - f(y)| < \epsilon \text{ and } |f_\alpha(x) - f_\alpha(y)| < \epsilon$$

hold for all α and for all $y \in V_x$. Since K is compact there exist $x_1, \dots, x_n$ such that $K \subset \bigcup_{i=1}^n V_{x_i}$. Choose β such that

$$|f_\alpha(x_i) - f(x_i)| < \epsilon, \quad \alpha \geq \beta,\ i = 1, \dots, n.$$

Now let $y \in K$ be arbitrary. Then $y \in V_{x_i}$ for some i. Using this we see that

$$|f_\alpha(y) - f(y)| \leq |f_\alpha(y) - f_\alpha(x_i)| + |f_\alpha(x_i) - f(x_i)| + |f(x_i) - f(y)| < 3\epsilon$$

holds whenever $\alpha \geq \beta$. Thus, $f_\alpha \to f$ uniformly on K. □

Theorem B.2.8. *Let F be a set of complex-valued functions on a set $X \neq \emptyset$ and consider on F the topology of pointwise convergence. If*

$$\sup\{|f(x)| : f \in F\} < \infty$$

for all $x \in X$, then the closure of F is compact.

Proof. Let $C(x)$ be the closure of the set $\{f(x) : f \in F\} \subset \mathbb{C}$. Then $C(x)$ is bounded and hence compact. Now F is a subset of the product space

$$C = \prod_{x \in X} C(x)$$

and pointwise convergence is the same as convergence in the product topology. By Tychonoff's theorem, C is compact. □

Combining Theorem B.2.7 and Theorem B.2.8 we obtain the following result.

Theorem B.2.9 (Arzelà–Ascoli). *Let $\{f_\alpha\}$ be an equicontinuous net of complex-valued functions on a topological space X such that*

$$\sup\{|f(x)| : f \in F\} < \infty$$

for all $x \in X$. Then this net contains a subnet converging uniformly on compact sets to a continuous function.

A proof of the next theorem can be found for example in [47].

Theorem B.2.10. *Let $\{f_n\}$ be a sequence of differentiable real-valued functions on a finite interval $I = [a, b] \subset \mathbb{R}$ such that*

(i) *for some $x_0 \in I$ the sequence $\{f_n(x_0)\}$ is convergent;*

(ii) *the sequence $\{f_n'\}$ is uniformly convergent on I.*

Then the sequence $\{f_n\}$ converges uniformly on I to a differentiable function f and

$$f'(x) = \lim_{n \to \infty} f_n'(x), \quad x \in I.$$

B.3 Infinite products

Definition B.3.1. Given a sequence $\{z_j\}_{j=1}^{\infty}$ of complex numbers let $p_n = z_1 z_2 \cdots z_n$, $n \geq 1$, be the *n-th partial product*. If there is a $p \in \mathbb{C}$ such that $\lim_n p_n = p$, then we write[8]

$$p = \prod_{n=1}^{\infty} z_n.$$

Lemma B.3.2. *For a sequence $\{z_j\}_1^{\infty}$ of complex numbers the following conditions are equivalent:*

(i) *The n-th partial products p_n converge to a nonzero limit as $n \to \infty$.*

(ii) *$p_n \neq 0$ for all n and $\epsilon > 0$ there is an $N \in \mathbb{N}$ such that $|p_n/p_m - 1| < \epsilon$ for all $n, m \geq N$.*

(iii) *There is a $\delta > 0$ such that $|p_n| > \delta$ for all n, and for each $\epsilon > 0$ there is an $N \in \mathbb{N}$ such that and $|p_n - p_m| < \epsilon$ for all $n, m \geq N$.*

Proof. The equivalence of (i) and (iii) is obvious. If (iii) holds then

$$\left| \frac{p_n}{p_m} - 1 \right| < \frac{\epsilon}{|p_m|} < \frac{\epsilon}{\delta}, \quad n, m \geq N$$

from which (ii) follows. Assume finally that (ii) is true and let $0 < \epsilon < 1$. Then

$$\left| \left| \frac{p_n}{p_m} \right| - 1 \right| < \left| \frac{p_n}{p_m} - 1 \right| < \epsilon, \quad n, m \geq N.$$

Replacing m by N and writing $A = |p_N|$ we obtain

$$-\epsilon < \frac{|p_n|}{A} - 1 < \epsilon$$

or

$$(1 - \epsilon)A < |p_n| < (1 + \epsilon)A, \quad n \geq N.$$

The lower estimate shows that the condition $|p_n| > \delta$ in (iii) is satisfied while the upper estimate shows that the sequence $\{p_n\}$ is bounded. Multiplying the inequality $|p_n/p_m - 1| < \epsilon$ by $|p_m|$ and using boundedness the condition $|p_n - p_m| < \epsilon$ in (iii) is satisfied as well. □

Since $p_n/p_{n-1} = z_n$, $n > 1$, we obtain the following corollary.

Corollary B.3.3. *If* $\prod_{n=1}^{\infty} z_n = p \neq 0$ *then* $\lim_{n\to\infty} z_n = 1$.

8 Note that we do not introduce the terminology *convergent infinite product* which is treated differently in the literature.

Theorem B.3.4. *Let $\{a_n\}_1^\infty$ be a sequence of nonnegative real numbers. Then*

$$\prod_{n=1}^{\infty}(1+a_n)<\infty$$

if and only if $\sum a_n<\infty$. If $a_n<1$ for all n, then

$$\prod_{n=1}^{\infty}(1-a_n)>0$$

if and only if $\sum a_n<\infty$.

Proof. Writing as shorthands $s_n=a_1+\cdots+a_n$, $p_n=(1+a_1)\cdots(1+a_n)$ and $q_n=(1-a_1)\cdots(1-a_n)$, the sequences $\{s_n\}$ and $\{p_n\}$ are increasing while $\{q_n\}$ is decreasing so that their (possibly infinite) limits exist. Moreover,

$$s_n<p_n\le e^{s_n}. \tag{1}$$

Indeed, the lower estimate for p_n is a consequence of

$$p_n=(1+a_1)\cdots(1+a_n)=1+a_1+\cdots+a_n+a_1a_2+\cdots$$

while the upper one follows by applying the inequalities $1+a_k\le e^{a_k}$, $k=1,\dots,n$. From (1) we conclude that $\{s_n\}$ and $\{p_n\}$ are at the same time bounded or unbounded. This is shown by the first statement. If $a_k<1$, then $0<1-a_k\le e^{-a_k}$ hence $0<q_n\le e^{-s_n}$. Thus, $\lim q_n=0$ if $\lim s_n=\infty$. The case $\lim s_n<\infty$ follows from the next theorem.[9] □

Theorem B.3.5. *Let $\{u_n\}_1^\infty$ be a sequence of complex numbers different from -1 such that*

$$\sum_{n=1}^{\infty}|u_n|<\infty.$$

Then there exists $p\in\mathbb{C}$ such that

$$\prod_{n=1}^{\infty}(1+u_n)=p\neq 0.$$

Proof. Write

$$a_n=|u_n|,\quad q_n=\prod_{j=1}^{n}(1+a_j),\quad p_n=\prod_{j=1}^{n}(1+u_j)$$

[9] Note that Theorem B.3.5 uses only the first part of Theorem B.3.4.

and let $\epsilon > 0$ arbitrary. It follows from the first part of Theorem B.3.4 that $\lim q_n < \infty$. By Lemma B.3.2, there exists $N \in \mathbb{N}$ such that

$$\frac{q_{n+d}}{q_n} - 1 < \epsilon, \quad n \geq N,\ d \in \mathbb{N}_0.$$

We have

$$\begin{aligned}
\left|\frac{p_{n+d}}{p_n} - 1\right| &= |(1 + u_{n+1})(1 + u_{n+2}) \cdots (1 + u_{n+d}) - 1| \\
&= |u_{n+1} + \cdots + u_{n+d} + u_{n+1}u_{n+2} + \cdots| \\
&\leq a_{n+1} + \cdots + a_{n+d} + a_{n+1}a_{n+2} + \cdots \\
&= \frac{q_{n+d}}{q_n} - 1 < \epsilon, \quad n \geq N,\ d \in \mathbb{N}_0.
\end{aligned}$$

Thus, the theorem follows from Lemma B.3.2. □

Theorem B.3.6. *Let $\{u_n\}_1^\infty$ be a sequence of complex-valued functions on some set $D \neq \emptyset$ such that for some $K > 0$*

$$\sum_{n=1}^{\infty} |u_n(z)| \leq K, \quad z \in D.$$

Then the sequence

$$p_n(z) = (1 + u_1(z)) \cdots (1 + u_n(z))$$

converges uniformly on D. The limit $u(z)$ is equal to zero if and only if $1 + u_n(z) = 0$ for some n.

Proof. We have $p_{n+1}(z) - p_n(z) = u_{n+1}(z)p_n(z)$ and hence for $m > n$

$$p_m(z) - p_n(z) = p_m(z) - p_{m-1}(z) + \cdots + p_{n+1}(z) - p_n(z) = \sum_{j=n}^{m-1} u_{j+1}(z)p_j(z).$$

Using this and the inequality $1 + x \leq \mathrm{e}^x$, $x \in \mathbb{R}$, we obtain

$$\begin{aligned}
|p_m(z) - p_n(z)| &= \left|\sum_{j=n}^{m-1} u_{j+1}(z)p_j(z)\right| \\
&\leq \sum_{j=n}^{m-1} |u_{j+1}(z)| \cdot \mathrm{e}^{|u_1(z)|} \cdots \mathrm{e}^{|u_j(z)|} \\
&\leq \sum_{j=n}^{m-1} |u_{j+1}(z)| \cdot \mathrm{e}^K.
\end{aligned}$$

By our assumption the right-hand side tends to 0 uniformly as $m, n \to \infty$, i.e., the sequence $\{p_n\}$ is uniformly convergent. The last statement follows from Theorem B.3.5. □

B.4 Convex functions

For proofs of the results presented in this section we refer to [61].

Definition B.4.1. A real-valued function f on an interval $I \subset \mathbb{R}$ is said to be *convex* if the inequality

$$f(\lambda x_1 + (1-\lambda)x_2) \le \lambda f(x_1) + (1-\lambda) f(x_2)$$

holds for all $x_1, x_2 \in I$ and $\lambda \in [0, 1]$.

Geometrically, a convex function f has the property that for any two points $x_1 < x_2$ in I the chord joining the points $(x_1, f(x_1))$ and $(x_2, f(x_2))$ is above the graph of f.

Proposition B.4.2. *Let f be a convex function on an interval $I \subset \mathbb{R}$ and $u, v, w \in I$ be such that $u < v < w$. Then we have*

$$\frac{f(v) - f(u)}{v - u} < \frac{f(w) - f(u)}{w - u} < \frac{f(w) - f(v)}{w - v}.$$

The inequalities in Proposition B.4.2 play a basic role in the proof of the next theorem.

Theorem B.4.3. *Let f be a convex function on an open interval $I \subset \mathbb{R}$. Then we have*

(i) *f is continuous on I;*

(ii) *f is both left- and right-differentiable at every $x_0 \in I$, i.e.,*

$$(D_l f)(x_0) := \lim_{x \uparrow x_0} \frac{f(x) - f(x_0)}{x - x_0} \in \mathbb{R}$$
$$(D_r f)(x_0) := \lim_{x \downarrow x_0} \frac{f(x) - f(x_0)}{x - x_0} \in \mathbb{R};$$

(iii) *D_l and D_r are increasing functions, $D_l \le D_r$ and*

$$(D_r f)(x_1) \le \frac{f(x_2) - f(x_1)}{x_2 - x_1} \le (D_l f)(x_2)$$

for all $x_1, x_2 \in I$ such that $x_1 < x_2$;

(iv) *f is differentiable everywhere except at countably many points in I and f' is an increasing function on the subset of I on which it exists.*

Theorem B.4.4. *Let f be a convex function on an open interval $I \subset \mathbb{R}$. Then either f is monotone on I or else there exists $x_0 \in I$ such that f is decreasing on $I \cap (-\infty, x_0]$ and increasing on $I \cap [x_0, \infty)$.*

Theorem B.4.7 below states that convex functions are absolutely continuous. Before formulating it we recall the definition of absolute continuity.

Definition B.4.5. A complex-valued function F, defined on an interval $[a; b] \subset \mathbb{R}$, is called *absolutely continuous* if there corresponds to every $\epsilon > 0$ a $\delta > 0$ such that

$$\sum_{i=1}^{n} |F(b_i) - F(a_i)| < \epsilon$$

for any $n \in \mathbb{N}$ and any disjoint collection of intervals $[a_i, b_i] \subset [a, b]$ satisfying

$$\sum_{i=1}^{n} |(b_i - a_i)| < \delta.$$

Taking $n = 1$ in the definition above, we see that such an F is continuous. Moreover, F is of bounded variation.[10] Indeed, let $\delta > 0$ correspond to $\epsilon = 1$. If $a = x_0 < \cdots < x_m = b$, by inserting further subdivision points if necessary, we can collect the intervals (x_{i-1}, x_i) into at most $N := \lfloor (b-a)/\delta \rfloor + 1$ groups such that the sum of the interval lengths within a group is less than δ. It follows that the total variation of F is at most N.

The next theorem is usually referred to as the *fundamental theorem of calculus for Lebesgue integrals.*

Theorem B.4.6 (Lebesgue). *If $-\infty < a < b < \infty$ and $F : [a, b] \to \mathbb{C}$, then the following statements are equivalent:*

(i) *F is absolutely continuous;*

(ii) $F(x) - F(a) = \int_a^x f(t)\,\mathrm{d}t, \ x \in [a, b]$
for some Lebesgue integrable function f on $[a, b]$;

(iii) *F is Lebesgue almost everywhere differentiable, F' is Lebesgue integrable and* $F(x) - F(a) = \int_a^x F'(t)\,\mathrm{d}t, \ x \in [a, b]$.

Theorem B.4.7. *Let f be a convex function on an interval $I \subset \mathbb{R}$.*

(i) *If I is open, then f is absolutely continuous on every finite closed interval contained in I.*

(ii) *If I is a closed finite interval and f is continuous at the endpoints of I, then f is absolutely continuous on I.*

---

[10] See page 288 for the definition of bounded variation.

Theorem B.4.8. *Let f be a real-valued function on an open interval $I \subset \mathbb{R}$. Suppose*

(i) *f is absolutely continuous on every finite closed interval contained in I, and*

(ii) *f' is an increasing function on the subset of I on which it exists.*

Then f is a convex function on I.

B.5 The Riemann–Stieltjes integral

At a few places in the book we need the Riemann–Stieltjes integral. D. V. Widder's classic book [60] contains a short introduction to this topic.

Let α and f be real-valued functions defined on some finite interval $[a,b] \subset \mathbb{R}$. Denote by $\Delta = \Delta(x_0, x_1, \ldots, x_n)$ a subdivision of this interval by the points $x_0, x_1, \ldots, x_n$, where

$$a = x_0 < x_1 < \cdots < x_n = b.$$

Further, let $\delta = \delta(\Delta)$ be the largest of the numbers $x_{i+1} - x_i$, $i = 0, \ldots, n-1$.

Definition B.5.1. If the limit

$$\lim_{\delta \to 0} \sum_{i=0}^{n-1} f(s_i)[\alpha(x_{i+1}) - \alpha(x_i)]$$

where $x_i \le s_i \le x_{i+1}$ exists independently of the manner of subdivision and of the choice of the numbers s_i, then the limit is called the *Riemann–Stieltjes integral* of f with respect to α from a to b and is denoted by

$$\int_a^b f(x)\,\mathrm{d}\alpha(x).$$

The Riemann–Stieltjes integral reduces to the Riemann integral if $\alpha(x) = x$.

Theorem B.5.2. *If f is continuous and α is of bounded variation in $[a,b]$, then the Riemann–Stieltjes integral of f with respect to α from a to b exists.*

Theorem B.5.3 (integration by parts). *If the Riemann–Stieltjes integral of f with respect to α from a to b exists, then so does the Riemann–Stieltjes integral of α with respect to f, and*

$$\int_a^b f(x)\,\mathrm{d}\alpha(x) = f(b)\alpha(b) - f(a)\alpha(a) - \int_a^b \alpha(x)\,\mathrm{d}f(x).$$

Another useful property of the Riemann–Stieltjes integral is given next.

Theorem B.5.4. *If f and φ are continuous and α is of bounded variation in $[a,b]$, and if*

$$\beta(x) = \int_c^x \varphi(t)\,\mathrm{d}\alpha(t), \quad x \in [a,b]$$

where $c \in [a,b]$ is fixed, then

$$\int_a^b f(x)\,\mathrm{d}\beta(x) = \int_a^b f(x)\varphi(x)\,\mathrm{d}\alpha(x).$$

Definition B.5.5 (improper Riemann–Stieltjes integral). Let α and f be real-valued functions defined on the interval $[a,\infty) \subset \mathbb{R}$. Then

$$\int_a^\infty f(x)\,\mathrm{d}\alpha(x)$$

is said to be *convergent* and equal to A provided $\int_a^b f(x)\,\mathrm{d}\alpha(x)$ exists for all $b \geq a$ and

$$A = \lim_{b\to\infty} \int_a^b f(x)\,\mathrm{d}\alpha(x)$$

is finite.

B.5.6. Using the notation from the previous definition and applying Theorem B.5.3 we obtain the formula

$$\int_a^\infty f(x)\,\mathrm{d}\alpha(x) = \lim_{b\to\infty}[f(b)\alpha(b)] - f(a)\alpha(a) - \int_a^\infty \alpha(x)\,\mathrm{d}f(x) \tag{1}$$

provided the improper integrals and the limit exist.

B.6 Multivariate calculus

Theorem B.6.1 (Taylor's theorem for multivariate functions). *Let f be a real- or complex-valued function defined on an open convex set $S \subset \mathbb{R}^d$ such that for some positive integer n all partial derivatives $D^\alpha f$, $\alpha \in \mathbb{N}_0^d$, $|\alpha| \leq n$, exist and are continuous. Then for all $t, a \in S$ we have*

$$f(t) = \sum_{|\alpha|\leq n} \frac{D^\alpha f(a)}{\alpha!} \cdot (t-a)^\alpha + \sum_{|\alpha|=n} R_\alpha(t) \cdot (t-a)^\alpha$$

where

$$R_\alpha(t) = \frac{n}{\alpha!} \cdot \int_0^1 \left(D^\alpha f(a + x(t-a)) - D^\alpha f(a)\right) \cdot (1-x)^{n-1}\,\mathrm{d}x.$$

The remainder terms R_α satisfy the inequality

$$|R_\alpha(t)| \le \frac{1}{\alpha!} \cdot \sup_{y \in L} |D^\alpha f(y) - D^\alpha f(a)|$$

where L denotes the line segment connecting a and t.

Proof. Since S is open and convex we can choose an open interval $I \supset [0,1]$ such that $a + s(t-a) \in S$ for all $s \in I$. The function g of one real variable defined by

$$g(s) = f(a + s(t-a)), \quad s \in I \tag{1}$$

is n-times continuously differentiable. Thus, for $n > 0$ the one-dimensional Taylor formula

$$g(s) = \sum_{k=0}^{n} \frac{g^{(k)}(0)}{k!} \cdot s^k + \frac{1}{(n-1)!} \cdot \int_0^s \big(g^{(n)}(x) - g^{(n)}(0)\big) \cdot (s-x)^{n-1}\,\mathrm{d}x, \quad s \in I$$

holds. Consequently,

$$f(t) = g(1) = \sum_{k=0}^{n} \frac{g^{(k)}(0)}{k!} + \frac{1}{(n-1)!} \cdot \int_0^1 \big(g^{(n)}(x) - g^{(n)}(0)\big) \cdot (1-x)^{n-1}\,\mathrm{d}x. \tag{2}$$

Using (1), the chain rule, and induction on k we get

$$\begin{aligned} g^{(k)}(s) &= \sum_{i_1=1}^{d} \cdots \sum_{i_k=1}^{d} \frac{\partial^k f}{\partial t_{i_1} \cdots \partial t_{i_k}} (a + s(t-a)) \cdot (t_{i_1} - a_{i_1}) \cdot \ldots \cdot (t_{i_k} - a_{i_k}) \\ &= \sum_{|\alpha|=k} \frac{k!}{\alpha!} \cdot D^\alpha f(a + s(t-a)) \cdot (t-a)^\alpha. \end{aligned}$$

Inserting this into (2) we obtain the first statement of the theorem.

The remainders are easily seen to satisfy the given upper estimate. □

Theorem B.6.2. *Under the conditions of Theorem B.6.1 there exist $\theta, \eta \in (0,1)$ depending on t such that*

$$f(t) = \sum_{|\alpha| \le n-1} \frac{D^\alpha f(a)}{\alpha!} \cdot (t-a)^\alpha + \sum_{|\alpha|=n} Q_\alpha(t) \cdot (t-a)^\alpha$$

where

$$Q_\alpha(t) = \frac{1}{\alpha!} \cdot [\operatorname{Re} D^\alpha f(a + \theta \cdot (t-a)) + \mathrm{i} \cdot \operatorname{Im} D^\alpha f(a + \eta \cdot (t-a))].$$

Proof. The proof is almost the same as that of Theorem B.6.1, but now we apply the one-dimensional Taylor formula

$$h(s) = \sum_{k=0}^{n-1} \frac{h^{(k)}(0)}{k!} \cdot s^k + \frac{1}{n!} \cdot h^{(n)}(\theta s) \cdot s^n$$

where h is real-valued, for the real and imaginary part of g. □

Lemma B.6.3. *Let f be a complex-valued function on $(a,b) \times (a,b) \subset \mathbb{R}^2$ having continuous partial derivatives of the second order. Then*

$$D^{(1,1)} f(x,y) = f_{xy}(x,y) = \lim_{\epsilon,h \to 0} \frac{\Delta^2 f(x,y,\epsilon,h)}{\epsilon h}, \quad t,s \in (a,b)$$

where

$$\Delta^2 f(x,y,\epsilon,h) = f(x+\epsilon, y+h) - f(x+\epsilon, y) - f(x, y+h) + f(x,y).$$

Proof. We may assume that f is real-valued. Let $x, y \in (a,b)$ and choose $\delta > 0$ such that $x + \epsilon, y + h \in (a,b)$ whenever $\epsilon, h \in I_\delta := (-\delta, \delta)$. Setting

$$g(t) := f(x+t, y+h) - f(x+t, y) - f_{xy}(x,y)th, \quad t \in I_\delta$$

the function g is continuously differentiable on I_δ. For $\epsilon \in I_\delta$ we have

$$g(\epsilon) - g(0) = \Delta^2 f(x,y,\epsilon,h) - f_{xy}(x,y)\epsilon h.$$

Applying the mean value theorem to g we obtain

$$\begin{aligned} |g(\epsilon) - g(0)| &\le |\epsilon| \cdot \sup_{s \in I_\delta} |g'(s)| \\ &= |\epsilon| \cdot \sup_{s \in I_\delta} |f_x(x+s, y+h) - f_x(x+s, y) - f_{xy}(x,y)h|. \end{aligned}$$

Now we apply the mean value theorem to the continuously differentiable function $r(h) = f_x(x+s, y+h) - f_{xy}(x,y)h$:

$$\begin{aligned} |r(h) - r(0)| &= |f_x(x+s, y+h) - f_x(x+s, y) - f_{xy}(x,y)h| \\ &\le |h| \cdot \sup_{q \in I_\delta} |r'(q)| \\ &= |h| \cdot \sup_{q \in I_\delta} |f_{xy}(x+s, y+q) - f_{xy}(x,y)|. \end{aligned}$$

Combining the two upper estimates we see that

$$\left| \frac{\Delta^2 f(x,y,\epsilon,h)}{\epsilon h} - f_{xy}(x,y) \right| \le \sup_{s,q \in I_\delta} |f_{xy}(x+s, y+q) - f_{xy}(x,y)|, \quad \epsilon, h \in I_\delta.$$

Since f_{xy} is continuous, the right-hand side tends to zero as $\delta \to 0$. □

Lemma B.6.4. *Let $a, b \in (0, \infty)$ be such that $b > (a/3)^3$. Then the polynomial*

$$P(s,t) = s^2t^2(s^2 + t^2 - a) + b, \quad s,t \in \mathbb{R}$$

is strictly positive but it is not a sum of squares of polynomials with real coefficients.

Proof. It is easy to show that P attains its minimum at $s = \pm\sqrt{a/3}$, $t = \pm\sqrt{a/3}$ and the minimum is equal to $b - (a/3)^3 > 0$.

Assume that

$$P = P_1^2 + \cdots + P_n^2 \tag{1}$$

where P_j is a polynomial with real coefficients. It follows from $P(s,0) = P(0,t) = b$ that $P_j(s,0)$ and $P_j(0,t)$ are constant for all j. Hence,

$$P_j(s,t) = a_j + stQ_j(s,t) \tag{2}$$

where Q_j is a polynomial and $a_j \in \mathbb{R}$. From (1) we see that the degree of Q_j is at most one. Putting (2) into (1) and comparing the coefficients we obtain

$$s^2 + t^2 - a = \sum_{j=1}^{n} Q_j^2(s,t).$$

This, however, is a contradiction because the right-hand side is always nonnegative whereas the left-hand side is negative if $s^2 + t^2 < a$. □

In the rest of this section we prove some simple results about multivariate polynomials.

Lemma B.6.5. *Let P be a polynomial of degree m on $\mathbb{R}^d$. Then for all $t, s \in \mathbb{R}^d$ the function*

$$Q_{t,s}(r) = P(t + r(s-t)), \quad r \in \mathbb{R}$$

is a polynomial of degree at most m. Moreover, every nonempty open set $S \subset \mathbb{R}^d$ contains elements t, s such that the degree of $Q_{t,s}$ is equal to m.

Proof. The first statement follows from the fact that $Q_{t,s}$ is a linear combination of terms of the form

$$\prod_{j=1}^{d} \left(t_j + r(s_j - t_j)\right)^{\alpha_j}$$

where $\alpha_j \in \mathbb{N}_0$ and $\alpha_1 + \cdots + \alpha_d \leq m$. We now show the second statement. If $m = 0$, then there is nothing to show; otherwise we have

$$P(t) = \sum_{|\alpha|=m} a_\alpha t^\alpha + R(t)$$

where not all of the coefficients a_α are equal to zero and R is a polynomial of degree less than m. We choose $t \in S$ such that the sum above is different from zero. Further, let $\delta \neq 0$ be a real number such that $s = (1+\delta)t \in S$. Then

$$Q_{t,s}(r) = P\big((1+r\delta)t\big) = (1+r\delta)^m \sum_{|\alpha|=m} a_\alpha t^\alpha + R\big(t + r(s-t)\big)$$

Applying the first statement to R we see that the degree of $Q_{t,s}$ is equal to m. □

Theorem B.6.6. *Let f be a real- or complex-valued function defined on an open convex set $S \subset \mathbb{R}^d$. Assume that for some nonnegative integer m and for all $t, s \in S$ the function*

$$r \mapsto f(t + r(s-t)), \quad r \in [0,1]$$

is a polynomial of degree at most m. Then f is a polynomial of degree at most m.

Proof. To simplify the notation we assume that $d = 2$ and write the function f as $f(x,y)$, $(x,y) \in S$. We choose mutually different a_j, mutually different b_j, $0 \leq j \leq m$, such that $(a_i, b_j) \in S$ for all i and j and denote by A_i and B_i the corresponding *Lagrange polynomials*:

$$A_i(x) = \prod_{k=1,\, k\neq i}^{m} \frac{x - a_k}{a_i - a_k}, \quad B_i(y) = \prod_{k=1,\, k\neq i}^{m} \frac{y - b_k}{b_i - b_k}.$$

The function

$$P(x,y) = \sum_{i=1}^{m} [A_i(x) f(a_i, y) + B_i(y) f(x, b_i)] - \sum_{i,j=1}^{m} A_i(x) B_j(y) f(a_i, b_j)$$

is a polynomial of degree at most m in each variable. Moreover, we have $P(a_k, b_l) = f(a_k, b_l)$ for all $0 \leq k, l \leq m$. Since f and P are polynomials of degree at most m of the first variable, we conclude the $P(x, b_l) = f(x, b_l)$ whenever $(x, b_l) \in S$. Knowing this, the same argument shows that $P(x,y) = f(x,y)$ for all $(x,y) \in S$, i.e., f is a polynomial. That the degree of f is at most m follows from Lemma B.6.5. □

Lemma B.6.7. *Let P be a polynomial on $\mathbb{R}^d$ with complex coefficients such that for all $t \in \mathbb{R}^d$ the polynomial $s \mapsto P(st)$, $s \in \mathbb{R}$, of one variable is of degree at most k. Then the degree of P is at most k.*

Proof. We write P in the form $P = P_0 + \cdots + P_n$ where n is the degree of P and

$$P_j(t) = \sum_{|\alpha|=j} c_\alpha t^\alpha, \quad c_\alpha \in \mathbb{C},\ t \in \mathbb{R}^d.$$

Choose t_0 such that $P_n(t_0) \neq 0$. Using the relation $P_j(st_0) = s^j P_j(t_0)$ we see that the degree of $s \mapsto P(st_0)$ is equal to n and hence $n \leq k$. □

B.7 The Lebesgue integral on $\mathbb{R}^d$

We refer to [18] for basic properties of the multidimensional Lebesgue integral.

B.7.1. Assume that the real-valued functions $h_1, \dots, h_d$ are defined on the open set $U \subset \mathbb{R}^d$ and let $h(x) = (h_1(x), \dots, h_d(x)),\ x \in U$. Suppose further that all partial derivatives $\frac{\partial h_i}{\partial x_j}$, $i, j = 1, \dots, d$ exist. Then the matrix-valued function

$$Jh(x) = \left(\frac{\partial h_i}{\partial x_j}(x)\right)_{i,j=1}^{d} = \begin{bmatrix} \frac{\partial h_1}{\partial x_1}(x) & \cdots & \frac{\partial h_1}{\partial x_d}(x) \\ \vdots & & \vdots \\ \frac{\partial h_d}{\partial x_1}(x) & \cdots & \frac{\partial h_d}{\partial x_d}(x) \end{bmatrix}, \quad x \in U$$

is called the *Jacobian matrix* of h. The real-valued function $x \mapsto \det(Jh(x))$ is usually called the *Jacobian* of h.

Theorem B.7.2. *Let U be an open subset of $\mathbb{R}^d$ and $h : U \to \mathbb{R}^d$ be an injective function with continuous partial derivatives, the Jacobian of which is nonzero for every x in U. A real- or complex-valued function g is Lebesgue integrable over $h(U)$ if and only if $x \mapsto g(h(x)) \cdot \det(Jh(x)),\ x \in U$, is Lebesgue integrable over U. In this case*[11]

$$\int_{h(U)} g(y)\,dy = \int_U g(h(x)) \cdot |\det(Jh(x))|\,dx.$$

Corollary B.7.3. *Let h be as in the previous theorem and write $\varphi = h^{-1}$. If X is a d-dimensional random vector having density p_X, then the random vector $Y = h(X)$ has density*

$$p_Y(x) = p_X(\varphi(x)) \cdot |\det(J\varphi(x))|, \quad x \in \mathbb{R}^d.$$

If $h(x) = Ax$ with a nonsingular matrix A, then

$$p_Y(x) = \frac{1}{|\det A|} \cdot p_X(A^{-1}x), \quad x \in \mathbb{R}^d.$$

[11] Note that the assumptions imply that $h(U)$ is open.

B.7.4. Let $S^{d-1} = \{t \in \mathbb{R}^d : \|t\| = 1\}$. The map F defined by

$$F(t) := (\|t\|, t/\|t\|), \quad t \in \mathbb{R}^d \setminus \{0\}$$

is a continuous bijection from $\mathbb{R}^d \setminus \{0\}$ to $(0,\infty) \times S^{d-1}$ whose inverse is

$$F^{-1}(r,s) = rs, \quad (r,s) \in (0,\infty) \times S^{d-1}.$$

Theorem B.7.5. *There is a unique Borel measure $\sigma = \sigma_d$ on S^{d-1} such that if f is Borel measurable on $\mathbb{R}^d$ and $f \geq 0$ or $f \in L^1(\lambda)$, then*

$$\int_{\mathbb{R}^d} f(t)\,\mathrm{d}t = \int_0^\infty \int_{S^{d-1}} f(rs) r^{d-1}\,\mathrm{d}\sigma(s)\,\mathrm{d}r. \tag{1}$$

Moreover, σ is rotation-invariant.

Proof. For $E \in \mathcal{B}(S^{d-1})$ let $\sigma(E) := d\lambda(E_1)$, where

$$E_a = F^{-1}((0,a] \times E) = \{rs : 0 < r \leq a, \ s \in E\}, \quad a > 0.$$

Since the map $E \mapsto E_1$ takes Borel sets to Borel sets and commutes with unions, intersections and complements, it is clear that σ is a Borel measure on S^{d-1}. The rotation-invariance of λ implies that σ is rotation-invariant as well. Since E_a is the image of E_1 under the map $t \mapsto at$, Theorem B.7.2 shows that $\lambda(E_a) = a^d \lambda(E_1)$.

We denote by λ_* the Borel measure on $(0,\infty) \times S^{d-1}$ induced by the mapping F, i.e., $\lambda_*(B) = \lambda(F^{-1}(B))$. Further, define the measure ρ on $(0,\infty)$ by $\rho(A) := \int_A r^{d-1}\,\mathrm{d}r$. We show that $\lambda_* = \rho \times \sigma$ from which equation (1) follows. We have

$$\begin{aligned}\lambda_*((a,b] \times E) &= \lambda(E_b \setminus E_a) = \frac{b^d - a^d}{d}\sigma(E) \\ &= \int_a^b r^{d-1}\,\mathrm{d}r \cdot \sigma(E) = \rho((a,b])\sigma(E), \quad E \in \mathcal{B}(S^{d-1}).\end{aligned}$$

Using this, a standard uniqueness argument shows that for fixed $E \in \mathcal{B}(S^{d-1})$ the measures λ_* and $\rho \times \sigma$ coincide on the σ-algebra

$$\{A \times E : A \in \mathcal{B}((0,\infty))\}.$$

Since $E \in \mathcal{B}(S^{d-1})$ is arbitrary we conclude that $\lambda_* = \rho \times \sigma$ on all Borel sets of $(0,\infty) \times S^{d-1}$. □

Remark B.7.6. By considering the completion of the measure σ, which we denote also by σ, equation (B.7.5.1) can be extended to Lebesgue measurable functions. Note that the completion is rotation-invariant, as well.

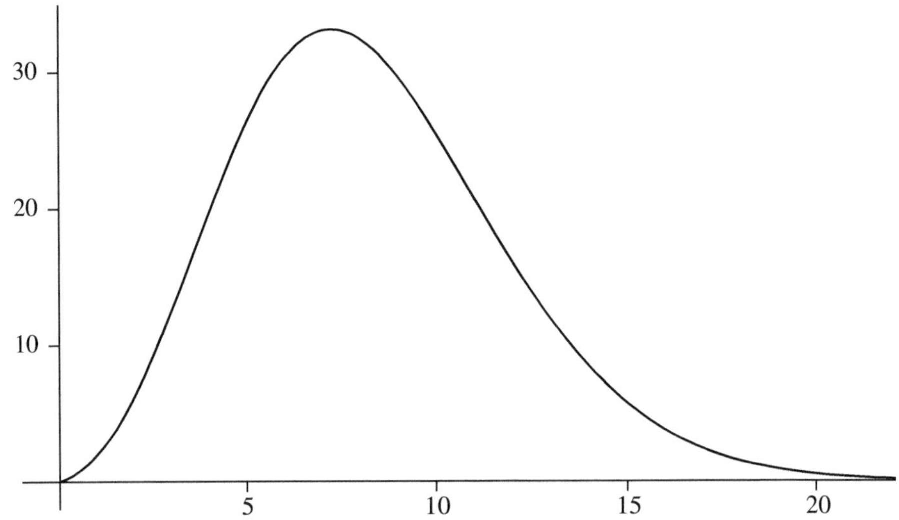

Figure B.5. The function $d \mapsto 2\pi^{d/2}/\Gamma(d/2)$, $d \in (0, \infty)$, from Proposition B.7.8.

Corollary B.7.7. *If f is a measurable function on $\mathbb{R}^d$, nonnegative or integrable, such that $f(t) = g(\|t\|)$, $t \in \mathbb{R}^d$, for some function g on $(0, \infty)$, then*

$$\int_{\mathbb{R}^d} f(t)\,\mathrm{d}t = \sigma(S^{d-1}) \int_0^\infty g(r) r^{d-1}\,\mathrm{d}r.$$

Proposition B.7.8. *We have*[12]

$$\sigma(S^{d-1}) = \frac{2\pi^{d/2}}{\Gamma(d/2)}\,.$$

Proof. Using Lemma B.1.10, Corollary B.7.7, and the substitution $s = r^2$ we obtain

$$\begin{aligned}\pi^{d/2} &= \prod_{j=1}^{d} \int_{-\infty}^{\infty} \mathrm{e}^{-t_j^2}\,\mathrm{d}t_j = \int_{\mathbb{R}^d} \mathrm{e}^{-\|t\|^2}\,\mathrm{d}t \\ &= \sigma(S^{d-1}) \int_0^\infty r^{d-1}\mathrm{e}^{-r^2}\,\mathrm{d}r = \frac{\sigma(S^{d-1})}{2} \int_0^\infty s^{(d-2)/2}\mathrm{e}^{-s}\,\mathrm{d}s \\ &= \frac{\sigma(S^{d-1})}{2}\,\Gamma(d/2)\,.\end{aligned}$$

□

[12] See Figure B.5 for a plot of this expression.

Proposition B.7.9. *For $R \geq 0$ we have*

$$\lambda(\{t \in \mathbb{R}^d : \|t\| \leq R\}) = \frac{R^d}{d} \cdot \sigma(S^{d-1}) = \frac{R^d \pi^{d/2}}{\Gamma(d/2+1)}.$$

Proof. By Corollary B.7.7 and the previous proposition

$$\begin{aligned}\lambda(\{t \in \mathbb{R}^d : \|t\| \leq R\}) &= \sigma(S^{d-1}) \int_0^R r^{d-1}\,\mathrm{d}r \\ &= \frac{R^d}{d} \cdot \sigma(S^{d-1}) = \frac{R^d \pi^{d/2}}{\Gamma(d/2+1)}.\end{aligned}$$ □

Spherical coordinates B.7.10. Recall that the *angle* ψ between $x, y \in \mathbb{R}^d \setminus \{0\}$ is defined by

$$\cos\psi = \frac{(x,y)}{\|x\| \cdot \|y\|}, \qquad 0 \leq \psi \leq \pi.$$

If $x = 0$ or $y = 0$, then we set $\psi := 0$.

Let $e_1, \dots, e_d$ be the standard orthonormal basis in $\mathbb{R}^d$, $d \geq 2$. For $x \in \mathbb{R}^d$ write $r = \|x\|$. Let u_j be the unit vector in the direction of the orthogonal projection of x onto the space spanned by $e_j, \dots, e_d$, $j = 2, \dots, d-1$. If the orthogonal projection is 0, then we set $u_j := 0$. Denote by ψ_{j-1} the angle between u_j and e_j. Below we show that

$$x = \sum_{j=1}^{d-2} r \left(\prod_{k=1}^{j-1} \sin\psi_k \right) \cos\psi_j \cdot e_j + r \left(\prod_{k=1}^{d-2} \sin\psi_k \right) \cdot u_{d-1}. \tag{1}$$

Since u_{d-1} is a linear combination of e_{d-1} and e_d, there is a unique $\theta \in [0, 2\pi)$ such that $u_{d-1} = \cos\theta \cdot e_{d-1} + \sin\theta \cdot e_d$. From equation (1) we obtain

$$x_1 = r\cos\psi_1 \tag{2}$$

$$x_j = r\cos\psi_j \prod_{k=1}^{j-1} \sin\psi_k, \quad j = 2, \dots, d-2 \tag{3}$$

$$x_{d-1} = r\cos\theta \prod_{k=1}^{d-2} \sin\psi_k, \tag{4}$$

$$x_d = r\sin\theta \prod_{k=1}^{d-2} \sin\psi_k \tag{5}$$

where $r \in [0, \infty)$, $\psi_j \in [0, \pi]$, $1 \leq j \leq d-2$, and $\theta \in [0, 2\pi)$.

The numbers $r, \psi_1, \dots, \psi_{d-1}$ and θ are called the *d-dimensional spherical coordinates* of x.[13]

[13] This derivation of the spherical coordinates is taken from [8].

Proof of (1). Since $x_1 = (x, e_1) = r \cos \psi_1$,

$$x = r \cos \psi_1 \cdot e_1 + \sum_{j=2}^{d} x_j \cdot e_j.$$

We have

$$r^2 = \|x\|^2 = r^2 \cos^2 \psi_1 + \sum_{j=2}^{d} x_j^2 \quad \text{and hence} \quad \sum_{j=2}^{d} x_j^2 = r^2 \sin^2 \psi_1.$$

Define a_j by $x_j = a_j r \sin \psi_1$ if $\sin \psi_1 \neq 0$ and let $a_j = 0,\ j = 2, \dots, d$, otherwise. Then $\sum_{j=2}^{d} a_j e_j$ has unit length and

$$x = r \cos \psi_1 \cdot e_1 + r \sin \psi_1 \sum_{j=2}^{d} a_j \cdot e_j.$$

Since $\sin \psi_1 \geq 0$ we see that $u_2 = \sum_{j=2}^{d} a_j \cdot e_j$. We have $\cos \psi_2 = (u_2, e_2) = a_2$ and

$$u_2 = \cos \psi_2 \cdot e_2 + \sum_{j=3}^{d} a_j \cdot e_j.$$

Moreover,

$$1 = \|u_2\|^2 = \cos^2 \psi_2 + \sum_{j=3}^{d} a_j^2 \quad \text{and hence} \quad \sum_{j=3}^{d} a_j^2 = \sin^2 \psi_2.$$

Define b_j by $a_j = b_j \sin \psi_2$ if $\sin \psi_2 \neq 0$ and let $b_j = 0,\ j = 3, \dots, d$, otherwise. Then $\sum_{j=3}^{d} b_j e_j$ has unit length and

$$u_2 = \cos \psi_2 \cdot e_2 + \sin \psi_2 \cdot \sum_{j=3}^{d} b_j e_j$$

i. e., $u_3 = \sum_{j=3}^{d} b_j e_j$. Thus,

$$x = r \cos \psi_1 \cdot e_1 + r \sin \psi_1 \cos \psi_2 \cdot e_2 + r \sin \psi_1 \sin \psi_2 \cdot u_3.$$

Continuing this process we obtain the equation (1). □

Remark B.7.11. Note that the Jacobian of the transformation (B.7.10.1) is given by

$$J(r, \psi_1, \dots, \psi_{d-2}, \theta) = r^{d-1} \prod_{k=1}^{d-2} \sin^k \psi_{d-1-k}.$$

Let $d \geq 2$. Setting $r = 1$ in equation (B.7.10.1) we see that S^{d-1} can be parameterized by angles $\psi_j \in [0, \pi]$, $1 \leq j \leq d-2$, and $\theta \in [0, 2\pi)$, which are related to the Cartesian coordinates $x_1, \ldots, x_d$ by

$$x_1 = \cos\psi_1, \quad x_2 = \sin\psi_1 \cos\psi_2, \quad x_3 = \sin\psi_1 \sin\psi_2 \cos\psi_2, \ldots,$$
$$x_{d-1} = \sin\psi_1 \cdots \sin\psi_{d-2} \cos\theta, \quad x_d = \sin\psi_1 \cdots \sin\psi_{d-2} \sin\theta.$$

Using this parameterization the surface measure σ_d on S^{d-1} can be given explicitly by

$$\mathrm{d}\sigma_d(\psi_1, \ldots, \psi_{d-2}, \theta) = \sin^{d-2}\psi_1 \sin^{d-3}\psi_2 \cdots \sin\psi_{d-2}\,\mathrm{d}\psi_1 \cdots \mathrm{d}\psi_{d-2}\,\mathrm{d}\theta.$$

Making the substitution $t = \cos\psi_1$ on the right-hand side we obtain the recursive relation

$$\mathrm{d}\sigma_d(\psi_1, \ldots, \psi_{d-2}, \theta) = (1-t^2)^{(d-3)/2}\,\mathrm{d}\sigma_{d-1}(\psi_2, \ldots, \psi_{d-2}, \theta)\,\mathrm{d}t \qquad (1)$$

for $d \geq 3$.

Appendix C

Advanced analysis

C.1 Functions of a complex variable

We refer to [46] and [56] for proofs of the classical results not presented in the present section.

Throughout this section the symbol D denotes a nonempty open subset of $\mathbb{C}$. If $r \in (0, \infty]$ and $z_0 \in \mathbb{C}$, then we write

$$D(z_0, r) = \{z \in \mathbb{C} : |z - z_0| < r\}.$$

Definition C.1.1. A complex-valued function f on D is said to be *differentiable* at $z_0 \in D$ if the limit

$$\lim_{z \to z_0} \frac{f(z) - f(z_0)}{z - z_0}$$

exists. This limit is then denoted by $f'(z_0)$ and is called the *derivative* of f.

Higher order derivatives are defined recursively in the same way as for real-valued functions. If f is differentiable at all points of D, then f is said to be *holomorphic* on D. The set of all holomorphic functions on D will be denoted by $H(D)$. Functions that are holomorphic on $\mathbb{C}$ are also called *entire*.

Unlike the real case, holomorphic functions are infinitely often differentiable.

Theorem C.1.2. *If $f \in H(D)$ then $f' \in H(D)$.*

Power series and Taylor expansion C.1.3.
A *power series* is a series of the form

$$\sum_{k=0}^{\infty} c_k \cdot (z - z_0)^k$$

where z, z_0 and c_k are complex numbers. The *radius of convergence* r of this series is a nonnegative number or ∞ given by

$$r = \frac{1}{\limsup_{k \to \infty} \sqrt[k]{|c_k|}}.$$

If $|z - z_0| < r$, then the series converges absolutely; if $|z - z_0| > r$, then the series is not convergent. If $r > 0$, then the function

$$g(z) = \sum_{k=0}^{\infty} c_k \cdot (z - z_0)^k, \quad z \in D(z_0, r)$$

is holomorphic on $D(z_0, r)$.

If f is holomorphic on D and $D(z_0, q) \subset D$ for some $q > 0$, then the convergence radius of the power series

$$\sum_{k=0}^{\infty} \frac{f^{(k)}(z_0)}{k!} \cdot (z - z_0)^k$$

is at least q and the series coincides with f on $D(z_0, q)$. This series is called the *Taylor expansion* of f at z_0.

Line integrals C.1.4. A *curve* in the complex plane is a continuous mapping γ of a finite interval $[a, b] \subset \mathbb{R}$ into $\mathbb{C}$. If $\gamma(a) = \gamma(b)$, then γ is said to be *closed.*

A curve γ can be written in the form

$$\gamma(t) = x(t) + \mathrm{i}y(t), \quad t \in [a, b]$$

where x and y are continuous real-valued functions on $[a, b]$. If x and y are piecewise continuously differentiable, then γ is called a *path.* We define $\gamma'(t)$ by

$$\gamma'(t) = x'(t) + \mathrm{i}y'(t)$$

for all t where the derivatives $x'(t)$ and $y'(t)$ exist.

Let $f : D \to \mathbb{C}$ be a continuous function and $\gamma : [a, b] \to D$, $a, b \in \mathbb{R}$, be a path in D. The *line integral*

$$\int_\gamma f(z)\,\mathrm{d}z$$

of f along γ is defined by

$$\int_\gamma f(z)\,\mathrm{d}z = \int_a^b f(\gamma(t)) \cdot \gamma'(t)\,\mathrm{d}t.$$

We use the notation

$$\int_{[z_0,z_1]} f(z)\,\mathrm{d}z := \int_\gamma f(z)\,\mathrm{d}z = (z_1 - z_0) \int_0^1 f(z_0 + (z_1 - z_0)t)\,\mathrm{d}t, \qquad z_0, z_1 \in \mathbb{C}$$

where γ is the path given by

$$\gamma(t) = z_0 + (z_1 - z_0)t, \quad 0 \le t \le 1.$$

For $z_0, z_1, z_2 \in \mathbb{C}$ let $\Delta = \Delta(z_0, z_1, z_2)$ denote the triangle with vertices at z_0, z_1 and z_2 and define

$$\int_{\partial\Delta} f(z)\,\mathrm{d}z := \int_{[z_0,z_1]} f(z)\,\mathrm{d}z + \int_{[z_1,z_2]} f(z)\,\mathrm{d}z + \int_{[z_2,z_0]} f(z)\,\mathrm{d}z.$$

Theorem C.1.5 (Cauchy). *Suppose D is a convex open set and $f \in H(D)$. Then*

$$\int_\gamma f(z)\,\mathrm{d}z = 0$$

for every closed path in D. If $z_0 \in D$, then the function

$$F(z) := \int_{[z_0,z]} f(w)\,\mathrm{d}w, \quad z \in D$$

belongs to $H(D)$ and $F' = f$.

Theorem C.1.6 (Morera). *Let $f : D \to \mathbb{C}$ be a continuous function such that*

$$\int_{\partial\Delta} f(z)\,\mathrm{d}z = 0$$

for every triangle $\Delta \subset D$. Then $f \in H(D)$.

Simply connected sets C.1.7. Suppose γ_0 and γ_1 are closed curves in $X \subset \mathbb{C}$, both with parameter interval [0, 1]. We say that these curves are *X-homotopic* if there is a continuous mapping H from $[0,1] \times [0,1]$ into X such that

$$H(s,0) = \gamma_0(s), \quad H(s,1) = \gamma_1(s), \quad H(0,t) = H(1,t), \quad t,s \in [0,1].$$

If X is connected and every closed curve in X is X-homotopic to a constant mapping, then X is said to be *simply connected.* Intuitively, this means that every closed curve can be continuously deformed to a single point within X.

Theorem C.1.8. *Suppose that D is connected and let $\{z_n\}$ be a sequence in D having at least one accumulation point in D. If f and g are holomorphic functions in D and $f(z_n) = g(z_n)$ for all n, then $f(z) = g(z)$ for all $z \in D$.*

Next we prove a result on the zeros of certain holomorphic functions.

Theorem C.1.9 (Eneström–Kakeya[1]). *Let $a_0 \geq a_1 \geq a_2 \geq \cdots \geq 0$ and $a_0 \neq 0$. Then*

$$f(z) := \sum_{j=0}^{\infty} a_j z^j \neq 0, \quad |z| < 1.$$

Proof. We may assume that $a_0 = 1$. Setting

$$g(z) := \sum_{j=1}^{\infty} (a_{j-1} - a_j) z^j, \quad |z| < 1$$

[1] The original Eneström–Kakeya theorem is on the zeros of polynomials. The present formulation and its simple proof is due to T. Koornwinder.

we have $(1-z)f(z) = 1 - g(z)$. Since

$$\sum_{j=1}^{\infty}(a_{j-1} - a_j) = 1 - \lim_{j\to\infty} a_j \le 1$$

and $a_{j-1} - a_j \ge 0$ we see that

$$|g(z)| \le \sum_{j=1}^{\infty}(a_{j-1} - a_j)|z|^k \le \sum_{j=1}^{\infty}(a_{j-1} - a_j)|z| \le |z|.$$

So

$$|(1-z)f(z)| \ge 1 - |g(z)| \ge 1 - |z| > 0, \quad |z| < 1. \qquad \square$$

Proposition C.1.10. *For $n \in \mathbb{N}_0$ we have*

$$\int_{[z,1]} (1-t^2)^n \, \mathrm{d}t = (1-z)^{n+1} Q_n(z)$$

where Q_n is a polynomial of degree n, the zeros of which lie in the closed disc $\overline{D(z_0, r)}$ where[2]

$$z_0 = -\frac{1}{2}\left(n + 1 + \frac{1}{n+1}\right) \quad \textit{and} \quad r = \frac{1}{2}\left(n + 1 - \frac{1}{n+1}\right).$$

Proof. We consider the polynomial

$$P_j(z) = \int_{[z,1]} (1-t)^{n+j}(1+t)^{n-j} \, \mathrm{d}t, \quad 0 \le j \le n, \ z \in \mathbb{C}.$$

Then

$$P_n(z) = \frac{1}{2n+1}(1-z)^{2n+1} \tag{1}$$

and integration by parts gives

$$P_j(z) = \frac{1}{n+j+1}(1-z)^{n+j+1}(1+z)^{n-j} + \frac{n-j}{n+j+1} P_{j+1}(z) \tag{2}$$

for all $j < n$. Setting

$$c_j := \frac{1}{n+1} \cdot \frac{n}{n+2} \cdot \frac{n-1}{n+3} \cdots \frac{n-j+1}{n+j+1}$$

from (1) and (2) we obtain

$$P_0(z) = \sum_{j=0}^{n} c_j (1-z)^{n+1+j}(1+z)^{n-j}$$

---

2 This proposition is taken from [39] where the authors state that the zeros of Q_n lie in the open disc $D(z_0, r)$. However, the real part of two zeros of Q_1 is equal to $-\frac{1}{2}$.

and therefore

$$Q_n(z) = (1+z)^n \sum_{j=0}^{n} c_j \left(\frac{1-z}{1+z}\right)^j. \tag{3}$$

Using the notation

$$a_j := \left(\frac{n+2}{n}\right)^j c_j, \quad H(w) := \sum_{j=0}^{n} a_j w^j$$

equation (3) yields

$$Q_n(z) = (1+z)^n H\left(\frac{n}{n+2} \cdot \frac{1-z}{1+z}\right).$$

Since

$$0 < \frac{a_{j+1}}{a_j} = \frac{n-j}{n+j+2} \cdot \frac{n+2}{n} \leq 1$$

the Enström–Kakeya Theorem C.1.9 shows that all of the zeros of H lie in the set $S := \mathbb{C} \setminus D(0,1)$. The proposition follows from the fact that the image of $S \cup \{\infty\}$ by the reciprocal map of

$$z \mapsto w = \frac{n}{n+2}\frac{1-z}{1+z}$$

is $\overline{D(z_0, r)}$. □

Lemma C.1.11. *Let f be a holomorphic function on a convex open set $D \subset \mathbb{C}$ and assume that f has no zeros. Then there exists a holomorphic function g on D such that $f = e^g$. If $z_0 \in D$ and $f(z_0) = 1$, then g can be chosen such that $g(0) = 0$.*[3]

Proof. It suffices to prove the second statement. Since f has no zeros, the function $h = f'/f$ is holomorphic. The second part of Theorem C.1.5 shows that

$$g(z) = \int_{[z_0,z]} h(w)\,dw, \quad z \in D$$

is holomorphic, as well. Since $g' = h$, we have

$$\frac{d}{dz}\left[f(z)e^{-g(z)}\right] = e^{-g(z)}\left[f'(z) - f(z) \cdot \frac{f'(z)}{f(z)}\right] = 0.$$

Thus, fe^{-g} is constant and therefore $f(z)e^{-g(z)} = f(z_0)e^{-g(z_0)} = 1$. □

Lemma C.1.12. *Let g be a holomorphic function on $D(0,R)$ given by*

$$g(z) = \sum_{k=0}^{\infty} c_k z^k, \quad |z| < R.$$

[3] See also Section C.8.

Then

$$c_n r^n = \frac{1}{\pi}\int_0^{2\pi} \left[\operatorname{Re} g\left(r\mathrm{e}^{\mathrm{i}t}\right)\right] \cdot \mathrm{e}^{-\mathrm{i}nt}\,\mathrm{d}t, \quad r \in \mathbb{R},\ |r| < R,\ n \in \mathbb{N}.$$

Proof. Writing $c_k = a_k + \mathrm{i}b_k$, $a_k, b_k \in \mathbb{R}$ and using that the series is absolutely convergent for $|z| \le r$ we obtain

$$\int_0^{2\pi} \left[\operatorname{Re} g\left(r\mathrm{e}^{\mathrm{i}t}\right)\right] \cdot \mathrm{e}^{-\mathrm{i}nt}\,\mathrm{d}t =$$

$$\sum_{k=0}^{\infty} r^k \int_0^{2\pi} [a_k \cos kt - b_k \sin kt] \cdot [\cos nt - \mathrm{i}\sin nt]\,\mathrm{d}t = \pi c_n r^n. \qquad \square$$

Corollary C.1.13. *Let g be as in the previous lemma. Then the inequality*

$$|c_n| \le \frac{2}{r^n} \cdot \sup_{|z|=r} |\operatorname{Re} g(z)|, \quad 0 < r < R$$

holds for all $n \in \mathbb{N}$.

Corollary C.1.14. *Let g be an entire function such that for some $k \in \mathbb{N}_0$ and constant $C > 0$, $R > 0$ the inequality*

$$|\operatorname{Re} g(z)| \le C|z|^k, \quad |z| \ge R$$

holds. Then g is a polynomial of degree at most k.

Proof. By Corollary C.1.13, the inequality $|c_n| \le 2Cr^{k-n}$ holds for all $r \ge R$ and $n \ge 1$. Thus, $c_n = 0$ if $n > k$. $\square$

Theorem C.1.15 (Jensen's formula). *Suppose $0 < r < R$, and let f be a holomorphic function on $D(0, R)$ such that $f(0) \neq 0$. Denote by $z_1, \ldots, z_N$ the zeros of f in the closed disc $\overline{D(0, r)}$, listed according to their multiplicities. Then*

$$|f(0)| \cdot \prod_{n=1}^{N} \frac{r}{|z_n|} = \exp\left[\frac{1}{2\pi}\int_{-\pi}^{\pi} \log|f(r\mathrm{e}^{\mathrm{i}\theta})|\,\mathrm{d}\theta\right].$$

Corollary C.1.16. *Let f be an entire function such that $f(0) = 1$ and denote by $n(r)$, $0 < r < \infty$, the number of zeros of f in the closed disc $\overline{D(0, r)}$. Then*

$$n(r) \le \frac{\log M(2r)}{\log 2}$$

where

$$M(r) = \sup_{\theta \in [0, 2\pi)} |f(r\mathrm{e}^{\mathrm{i}\theta})|.$$

Proof. Let $\{z_k\}$ be the sequence of zeros of f, listed according to their multiplicities and arranged so that the sequence $\{|z_k|\}$ is increasing. By Jensen's formula and inequality

$$M(2r) \geq \exp\left[\frac{1}{2\pi}\int_{-\pi}^{\pi} \log|f(2r\mathrm{e}^{\mathrm{i}\theta})|\,\mathrm{d}\theta\right]$$
$$= \prod_{k=1}^{n(2r)} \frac{2r}{|z_k|} \geq \prod_{k=1}^{n(r)} \frac{2r}{|z_k|} \geq 2^{n(r)}.$$

Taking logarithms we obtain the desired inequality. □

Definition C.1.17. An entire function f is said to be of *finite order* if there exist positive numbers A and C such that

$$|f(z)| \leq C\mathrm{e}^{|z|^A}, \quad z \in \mathbb{C}.$$

The infimum ρ of all numbers A for which this is true is called the *order* of f.

A polynomial is of order zero. The functions e^z, $\cos z$ and $\sin z$ are of order 1 while e^{z^k}, $k \in \mathbb{N}$, is of order k. The function $\mathrm{e}^{\mathrm{e}^z}$ is of infinite order.

Lemma C.1.18. *If f is an entire function of order ρ, $f(0) \neq 0$, and $r_1, r_2, \dots$ are the moduli of the zeros of f, then the series*

$$\sum_{n\geq 1} \frac{1}{r_n^{\alpha}}$$

is convergent if $\alpha > \rho$.

Theorem C.1.19 (Hadamard). *If f is an entire function of order ρ, with zeros $z_1, z_2, \dots$ such that $f(0) \neq 0$, then*

$$f(z) = \mathrm{e}^{Q(z)} \prod_{n\geq 1}\left(1 - \frac{z}{z_n}\right)\exp\left(\frac{z}{z_n} + \frac{1}{2}\left(\frac{z}{z_n}\right)^2 + \dots + \frac{1}{p}\left(\frac{z}{z_n}\right)^p\right)$$

where Q is a polynomial of degree not greater than ρ and p is the smallest integer for which

$$\sum_{n\geq 1} \frac{1}{|z_n|^{p+1}}$$

is convergent.

Remark C.1.20. Since the functions cos and sin are of order 1, the well-known product formulae

$$\sin z = z \prod_{n=1}^{\infty} \left(1 - \frac{z^2}{n^2 \pi^2}\right)$$

$$\cos z = \prod_{n=1}^{\infty} \left(1 - \frac{z^2}{(n - 1/2)^2 \pi^2}\right)$$

are consequences of Hadamard's theorem.[4]

Theorem C.1.21 (Phragmen–Lindelöf). *Let $\alpha < \beta \leq \alpha + \pi$ and let f be a function, holomorphic in the sector*

$$S = \{z \in \mathbb{C} : \alpha < \arg z < \beta\}$$

and continuous on its boundary. If $|f(z)| \leq M$ for all z on the boundary of S and

$$|f(z)| \leq C e^{|z|^A}, \quad z \in S$$

where $0 \leq C$ and $0 \leq A < \frac{\pi}{\beta - \alpha}$, then $|f(z)| \leq M$ for all $z \in S$.

C.2 Almost periodic functions

Definition C.2.1. A complex-valued function g on $\mathbb{R}$ is called *almost periodic* if for every $\varepsilon > 0$ there exist finitely many subsets $U_1, \dots, U_n$ of $\mathbb{R}$ with

$$\bigcup_{j-1}^{n} U_j = \mathbb{R}$$

such that the inequality

$$|g(s + x) - g(s + y)| < \varepsilon$$

holds for every $s \in \mathbb{R}$ whenever $x, y \in U_j$ with some $j = 1, \dots, n$. We call $\{U_1, \dots, U_n\}$ an *ε-partition* for g.

Next we prove a few basic facts about almost periodic functions.[5]

Lemma C.2.2. *Every continuous periodic function g is almost periodic.*

---

[4] See, e.g., § 3.23 in [56] for another proof of these product representations.

[5] See Maak [41] for more information on almost periodic functions.

Proof. Assume, without loss of generality, that g is periodic with period 1. Since g is uniformly continuous on $[0, 1]$ for every $\varepsilon > 0$, there exists $n \in \mathbb{N}$ such that $|g(t) - g(s)| < \varepsilon$ whenever $t, s \in [0, 1]$ and $|t - s| \leq 1/n$. Using this and the fact that g is periodic with period 1 we see that

$$U_j := \bigcup_{k=-\infty}^{\infty} [k + (j-1)/n, k + j/n], \quad j = 1, \dots, n$$

is an ε-partition for g. □

Lemma C.2.3. *If g and h are almost periodic, then so is their sum.*

Proof. If $\{U_1, \dots, U_n\}$ and $\{V_1, \dots, V_m\}$ are ε-partitions for g and h, respectively, then the sets

$$U_i \cap V_j, \quad i = 1, \dots, n; \; j = 1, \dots, m$$

form a 2ε-partition for $g + h$. □

Lemma C.2.4. *If g is the uniform limit of almost periodic functions g_n, $n \in \mathbb{N}$, then g is almost periodic.*

Proof. For each $\varepsilon > 0$ there exists $n \in \mathbb{N}$ such that

$$|g(t) - g_n(t)| < \varepsilon/3, \quad t \in \mathbb{R}.$$

Since

$$\begin{aligned}
&|g(s+x) - g(s+y)| \\
&\quad \leq |g(s+x) - g_n(s+x)| + |g_n(s+x) - g_n(s+y)| + |g_n(s+y) - g(s+y)| \\
&\quad \leq |g_n(s+x) - g_n(s+y)| + 2\varepsilon/3
\end{aligned}$$

we see that every $\varepsilon/3$-partition for g_n is an ε-partition for g. □

Applying the previous lemmata we obtain the following corollary.

Corollary C.2.5. *Every function g of the form*

$$g(t) = \sum_{n=1}^{\infty} c_n \mathrm{e}^{\mathrm{i}\alpha_n t}, \quad t \in \mathbb{R}$$

where $\alpha_n \in \mathbb{R}$, $c_n \in \mathbb{C}$ and $\sum_{n=1}^{\infty} |c_n| < \infty$, is almost periodic.

Theorem C.2.6. *Let g be a continuous almost periodic function. Then for all $\varepsilon > 0$ there exists a positive number $L = L(\varepsilon)$ such that every closed interval of the form*

$[a, a+L]$, $a \in \mathbb{R}$, *contains a number* τ *satisfying*[6]

$$|g(s+\tau) - g(s)| < \varepsilon, \quad s \in \mathbb{R}. \tag{1}$$

Proof. Let $\{U_1, \ldots, U_n\}$ be an ε-partition for g. Further, let $a \in \mathbb{R}$ and $a_j \in U_j$ be arbitrary and write $L := 2\max(|a_1|, \ldots, |a_n|)$. Choose $j \in \{1, \ldots, n\}$ such that $a + L/2 \in U_j$ and define τ by $\tau := a + L/2 - a_j$. Then $\tau \in [a, a+L]$. Since a_j and $a + L/2$ are in U_j we have

$$|g(s + a + L/2) - g(s + a_j)| < \epsilon, \quad s \in \mathbb{R}.$$

Replacing here s by $s - a_j$ we see that τ satisfies the inequality (1). □

Corollary C.2.7. *If* g *is a continuous almost periodic function, then the relation*

$$\liminf_{t \to \infty} |g(t) - g(s)| = 0$$

holds for all $s \in \mathbb{R}$.

C.3 Fourier series

Let f be a Lebesgue integrable function on the interval $[0, 2\pi)$. The series

$$\sum_{n \in \mathbb{Z}} \hat{f}(n) \mathrm{e}^{\mathrm{i}nx}, \quad x \in [0, 2\pi)$$

where

$$\hat{f}(n) = \frac{1}{2\pi} \int_0^{2\pi} f(x) \mathrm{e}^{-\mathrm{i}nx} \, \mathrm{d}x, \quad n \in \mathbb{Z}$$

is called the *Fourier series* of f. If N is a nonnegative integer, the *N-th symmetric partial sum* of the Fourier series of f is

$$s_N f(x) = \sum_{|n| \le N} \hat{f}(n) \mathrm{e}^{\mathrm{i}nx}.$$

If $f = \mathbf{1}_{[a,b)}$ where $[a, b) \subset [0, 2\pi)$, then

$$s_N f(x) = \frac{1}{2\pi} \sum_{|j| \le N} \frac{\mathrm{e}^{-\mathrm{i}a} - \mathrm{e}^{-\mathrm{i}b}}{\mathrm{i}j} \cdot \mathrm{e}^{\mathrm{i}jt}.$$

6 It can be shown that this property implies almost periodicity; see Maak [41], p. 94.

Recall that the *total variation* $V(f)$ of a complex-valued function f defined on an interval $[a, b]$ is the quantity

$$V(f) := \sup \sum_{i=1}^{m} |f(x_i) - f(x_{i-1})|$$

where the supremum is taken over all $m \in \mathbb{N}$ and all $x_i \in [a, b]$ such that $0 = x_0 < x_1 < \cdots < x_m = b$. If $V(f)$ is finite, then f is said to be of *bounded variation.*

A real-valued function f is of bounded variation if and only if it can be expressed as the difference of two monotone functions.

A proof of the next theorem can be found, e.g., in Section 10.1 of [13].

Theorem C.3.1. *Let f be a Lebesgue integrable function on the interval $[0, 2\pi)$ and suppose that the total variation of f is finite. Then*

$$\lim_{N\to\infty} s_N f(x) = \frac{1}{2} \cdot [f(x+0) + f(x-0)], \quad x \in [0, 2\pi)$$

and

$$|s_N f(x)| \leq K$$

with some constant $K \geq 0$.[7]

Applying Lebesgue's theorem on dominated convergence we obtain the following corollary.

Corollary C.3.2. *Let f be as in the previous theorem and let σ be a complex Borel measure on $[0, 2\pi)$. If $\sigma(\{x\}) = 0$ for all discontinuity points*[8] *of f, then $s_N f$ converges to f in $L^2(\sigma)$.*

C.4 The Gamma function and the formulae of Stirling and Binet

Definition C.4.1. The *Gamma function* Γ is defined by

$$\Gamma(x) = \int_0^\infty t^{x-1} e^{-t}\, dt, \quad x \in (0, \infty).$$

Note that if Z is an exponentially distributed random variable with expectation one, then

$$\mathbb{E}(Z^x) = \int_0^\infty t^x e^{-t}\, dt = \Gamma(x+1), \quad x > -1.$$

[7] $f(0-0)$ is defined by the 2π-periodic extension of f.
[8] Discontinuity in 0 is defined by the 2π-periodic extension of f.

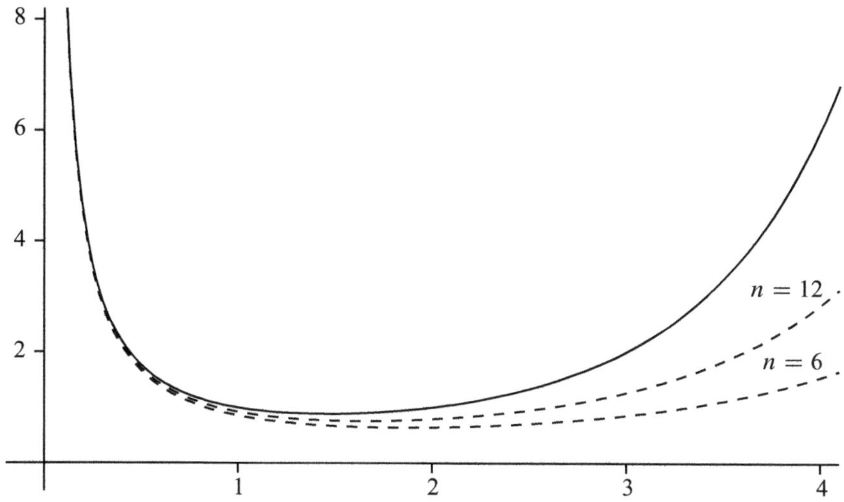

Figure C.1. The Gamma function Γ (continuous line) and the function $x \mapsto n!n^x/x(x+1)\cdots(x+n)$ from relation (C.4.2.v).

Lemma C.4.2. *We have*

(i) $\Gamma(x) < \infty$ *and* Γ *is continuous;*

(ii) $\Gamma(x+1) = x\Gamma(x)$*;*

(iii) $\Gamma(n+1) = n!$, $n \in \mathbb{N}_0$*;*

(iv) $\Gamma(1/2) = \sqrt{\pi}$*;*

(v) $\Gamma(x) = \lim_{n\to\infty} \frac{n!n^x}{x(x+1)\cdots(x+n)}$ *.*

The relation (C.4.2.v) is illustrated in Figure C.1.

Proof. (i) It is easy to check that $\Gamma(x) < \infty$ while the continuity can be proved by using Lebesgue's theorem on dominated convergence.

(ii) Integrating by parts we obtain

$$\Gamma(x+1) = \int_0^\infty t^x \mathrm{e}^{-t}\,\mathrm{d}t = -t^x\mathrm{e}^{-t}\big|_0^\infty + x\int_0^\infty t^{x-1}\mathrm{e}^{-t}\,\mathrm{d}t = x\Gamma(x).$$

(iii) Since $\Gamma(1) = 1$, (iii) follows from (ii) by induction on n.

(iv) We have

$$\begin{aligned}\Gamma(1/2) = \int_0^\infty t^{1/2-1}\mathrm{e}^{-t}\,\mathrm{d}t &= \int_0^\infty \frac{1}{\sqrt{t}}\,\mathrm{e}^{-t}\,\mathrm{d}t \\ &= 2\int_0^\infty \mathrm{e}^{-u^2}\,\mathrm{d}u = \sqrt{\pi}\end{aligned}$$

where we made the substitution $t = u^2$ and applied Lemma B.1.10.

(v) Repeated integration by parts gives

$$\begin{aligned} \int_0^n t^{x-1}\left(1-\frac{t}{n}\right)^n \mathrm{d}t &= \frac{1}{x}\int_0^n t^x \left(1-\frac{t}{n}\right)^{n-1} \mathrm{d}t \\ &\cdots \\ &= \frac{n!n^x}{x(x+1)\cdots(x+n)} \end{aligned}$$

from which (v) follows by dominated convergence.[9] □

Definition C.4.3. The *Beta function* B (see Figure C.2) is defined by the equation

$$B(a,b) = \int_0^1 (1-t)^{a-1} t^{b-1}\,\mathrm{d}t, \quad a,b > 0.$$

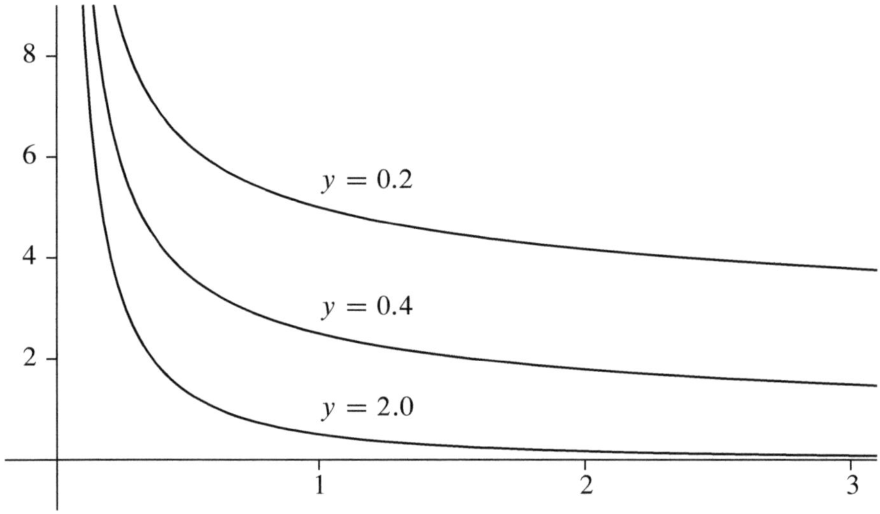

Figure C.2. The functions $B_y : x \mapsto B(x,y)$ where B is the Beta function from Definition C.4.3.

Proposition C.4.4. *We have*

$$B(a,b) = \frac{\Gamma(a)\Gamma(b)}{\Gamma(a+b)}, \quad a,b > 0.$$

[9] It is easy to check by differentiating that $\mathbf{1}_{[0,n]}(t)(1-t/n)^n$ is an increasing function of n.

Proof. It follows from the definition of the Gamma function that the function p_a defined by

$$p_a(x) := \mathbf{1}_{[0,\infty)}(x) \cdot \frac{x^{a-1}e^{-x}}{\Gamma(a)}, \qquad x \in \mathbb{R}$$

is a probability density. We have

$$\begin{aligned} p_a * p_b(x) &= \int_{-\infty}^{\infty} p_a(x-y)p_b(y)\,dy = \frac{e^{-x}}{\Gamma(a)\Gamma(b)} \cdot \int_0^x (x-y)^{a-1}y^{b-1}\,dy \\ &= \frac{x^{a+b-1}e^{-x}}{\Gamma(a)\Gamma(b)} \cdot \int_0^1 (1-t)^{a-1}t^{b-1}\,dt \end{aligned}$$

where we made the substitution $y = tx$. The relation above shows $p_a * p_b = cp_{a+b}$ with some constant c. Since $p_a * p_b$ and p_{a+b} are probability densities, we conclude that $c = 1$ from which the assertion follows. □

Proposition C.4.5 (Legendre's duplication formula). *We have*

$$\sqrt{\pi}\,\Gamma(2x) = 2^{2x-1}\Gamma(x)\Gamma\left(x+\frac{1}{2}\right), \qquad x > 0.$$

Proof. Let $f_1(x) = e^{-x}$, $f_2(x) = e^{-x}x^{1/2}$ and denote by g the multiplicative convolution of f_1 and f_2, i.e.,

$$g(x) = \int_0^{\infty} f_1(x/u)f_2(u)\frac{1}{u}\,du, \qquad x \in (0,\infty).$$

Then $Mf_1(s) = \Gamma(s)$, $Mf_2(s) = \Gamma(s+\frac{1}{2})$ and, by Lemma C.6.3,

$$Mg(s) = Mf_1(s) \cdot Mf_2(s)$$

where M denotes Mellin transformation. We prove that

$$Mg(s) = \frac{\sqrt{\pi}\,\Gamma(2s)}{2^{2s-1}}$$

from which the assertion follows. We have

$$g(x) = \int_0^{\infty} e^{-(x/u+u)}\frac{1}{\sqrt{u}}\,du = 2\int_0^{\infty} e^{-(x/t^2+t^2)}\,dt.$$

Changing t to $\frac{\sqrt{x}}{t}$ yields

$$g(x) = 2\int_0^{\infty} e^{-(x/t^2+t^2)}\frac{\sqrt{x}}{t^2}\,dt$$

and therefore

$$g(x) = \int_0^{\infty} e^{-(x/t^2+t^2)}\left(1+\frac{\sqrt{x}}{t^2}\right)dt.$$

Substituting $u = t - \frac{\sqrt{x}}{t}$ we obtain

$$g(x) = \int_{-\infty}^{\infty} \mathrm{e}^{-(u^2+2\sqrt{x})}\, du = \mathrm{e}^{-2\sqrt{x}} \int_{-\infty}^{\infty} \mathrm{e}^{-u^2}\, du = \sqrt{\pi}\, \mathrm{e}^{-2\sqrt{x}}.$$

Finally, set $y = 2\sqrt{x}$ and obtain

$$\begin{aligned} Mg(s) &= \sqrt{\pi} \int_0^{\infty} \mathrm{e}^{-2\sqrt{x}} x^{s-1}\, \mathrm{d}y \\ &= \sqrt{\pi} \int_0^{\infty} \mathrm{e}^{-y} \left(\frac{y}{2}\right)^{2s-1} \mathrm{d}y = \frac{\sqrt{\pi}\, \Gamma(2s)}{2^{2s-1}}. \end{aligned}$$

□

Theorem C.4.6 (Binet). *We have*

$$\Gamma(x+1) = \left(\frac{x}{\mathrm{e}}\right)^x \sqrt{2\pi x} \cdot \mathrm{e}^{\theta(x)}, \quad x > 0 \tag{1}$$

where[10]

$$\theta(x) = \int_0^{\infty} \left(\frac{1}{\mathrm{e}^t - 1} - \frac{1}{t} + \frac{1}{2}\right) \mathrm{e}^{-xt} \frac{1}{t}\, \mathrm{d}t.$$

C.4.7. For the proof of Binet's formula we define the function φ by the equation

$$\mathrm{e}^{\varphi(x)} = \frac{1}{\sqrt{2\pi}} \int_0^{\infty} \sqrt{x} \left(t\mathrm{e}^{1-t}\right)^x \mathrm{d}t, \quad x > 0$$

so that

$$\Gamma(x+1) = \left(\frac{x}{\mathrm{e}}\right)^x \sqrt{2\pi x} \cdot \mathrm{e}^{\varphi(x)}. \tag{1}$$

Binet's formula is equivalent with $\theta(x) = \varphi(x)$. For the proof of this we need that θ and φ satisfy a certain difference equation and that $\theta(\frac{1}{2}) = \varphi(\frac{1}{2})$. First we show these facts.

Lemma C.4.8. *For all $x > 0$ we have*

$$\varphi(x) - \varphi(x+1) = \theta(x) - \theta(x+1) = \left(x + \frac{1}{2}\right) \log\left(1 + \frac{1}{x}\right) - 1. \tag{1}$$

Proof. Denote by $g(x)$ the right-hand side of the relation above. The equation

$$\varphi(x) - \varphi(x+1) = g(x)$$

[10] That the integral below exists follows from the relation

$$\lim_{t \to 0} \left(\frac{1}{\mathrm{e}^t - 1} - \frac{1}{t} + \frac{1}{2}\right) \cdot \frac{1}{t} = \frac{1}{12}$$

which can be shown, e.g., by using the Bernoulli–l'Hospital rule.

follows immediately from (C.4.7.1) by using the relation

$$\Gamma(x+2) = (x+1)\Gamma(x+1).$$

To prove the statement on θ, first note that

$$\lim_{x\to\infty} \theta(x) - \theta(x+1) = \lim_{x\to\infty} g(x) = 0.$$

Moreover,[11]

$$\theta'(x) - \theta'(x+1) = \int_0^\infty \frac{e^{-xt} - e^{-(x+1)t}}{t} - \frac{e^{-xt} + e^{-(x+1)t}}{2}\, dt.$$

Applying Lemma B.1.12 we obtain

$$\theta'(x) - \theta'(x+1) = \ln\left(1 + \frac{1}{x}\right) - \frac{1}{2}\left(\frac{1}{x} + \frac{1}{x+1}\right) = g'(x)$$

completing the proof. □

Lemma C.4.9. $\varphi(\frac{1}{2}) = \theta(\frac{1}{2}) = \frac{1}{2} - \frac{1}{2}\log 2.$

Proof. Since $\Gamma(\frac{3}{2}) = \frac{1}{2}\sqrt{\pi}$, equation (C.4.7.1) shows that $\varphi(\frac{1}{2}) = \frac{1}{2} - \frac{1}{2}\log 2$.
By an obvious substitution,

$$\theta(1) = \int_0^\infty \left(\frac{1}{e^{\frac{1}{2}t} - 1} - \frac{2}{t} + \frac{1}{2}\right) e^{-\frac{1}{2}t} \frac{1}{t}\, dt.$$

Using this, we obtain

$$\begin{aligned}
\theta(1/2) &= (\theta(1/2) - \theta(1)) + \theta(1) \\
&= \int_0^\infty \left(\frac{e^{-\frac{1}{2}t}}{t} - \frac{1}{e^t - 1}\right)\frac{1}{t}\, dt + \int_0^\infty \left(\frac{1}{e^t - 1} - \frac{1}{t} + \frac{1}{2}\right) e^{-t}\frac{1}{t}\, dt \\
&= \int_0^\infty \left(\frac{e^{-\frac{1}{2}t} - e^{-t}}{t} - \frac{1}{2}e^{-t}\right)\frac{1}{t}\, dt \\
&= \int_0^\infty -\frac{d}{dt}\left(\frac{e^{-\frac{1}{2}t} - e^{-t}}{t}\right) - \frac{e^{-\frac{1}{2}t} - e^{-t}}{2t}\, dt.
\end{aligned}$$

Applying Lemma B.1.12 we obtain the desired result. □

Proof of Binet's formula. We have to show that $\varphi(x) = \theta(x)$. By Lemma C.4.8,

$$\theta(x) - \theta(x+1) = \varphi(x) - \varphi(x+1).$$

---

[11] The footnote to Lemma B.1.12 also holds for the differentiation below.

Applying this to $x, x+1, \dots, x+n-1$ and summing these equations we see that

$$\theta(x) - \theta(x+n) = \varphi(x) - \varphi(x+n).$$

Since $\lim_{n\to\infty} \theta(x+n) = 0$, we immediately obtain

$$\theta(x) = \varphi(x) - \lim_{n\to\infty} \varphi(x+n) =: \varphi(x) - h(x). \tag{1}$$

Next we show that the function h is decreasing. If $0 \le y \le x$ and $0 \le p \le 1$, then

$$\begin{aligned} \sqrt{x+n}\, p^{x+n} - \sqrt{y+n}\, p^{y+n} &\le \sqrt{x+n}\, p^{y+n} - \sqrt{y+n}\, p^{y+n} \\ &\le (\sqrt{x+n} - \sqrt{y+n})p \end{aligned}$$

for all $n \ge 1$. Noting that $0 \le t\mathrm{e}^{1-t} \le 1$, $t \ge 0$, and using the definition of φ, we conclude that

$$\mathrm{e}^{\varphi(x+n)} - \mathrm{e}^{\varphi(y+n)} \le (\sqrt{x+n} - \sqrt{y+n})\mathrm{e}^{\varphi(1)}.$$

Taking the limit as $n \to \infty$, we obtain $\mathrm{e}^{h(x)} - \mathrm{e}^{h(y)} \le 0$, i.e., $h(x) \le h(y)$. Since the function h is also periodic with period 1, it must be constant. Applying (1) and Lemma C.4.9, we now obtain that $h(x) = h(\frac{1}{2}) = 0$ for all $x > 0$. □

Since $\lim_{x\to\infty} \theta(x) = 0$, from Theorem C.4.6 we immediately obtain:

Corollary C.4.10 (Stirling's formula).

$$\lim_{n\to\infty} \frac{n!}{(n/\mathrm{e})^n \sqrt{2\pi n}} = 1.$$

To prove a more precise statement we will need the Bernoulli numbers.

C.4.11. The *Bernoulli numbers* B_n, $n \in \mathbb{N}_0$, are defined by[12]

$$\frac{z}{\mathrm{e}^z - 1} = \frac{1}{\frac{1}{1!} + \frac{z}{2!} + \frac{z^2}{3!} + \cdots} = \sum_{n=0}^{\infty} \frac{B_n}{n!} z^n, \quad z \in \mathbb{C},\ |z| < \pi. \tag{1}$$

It follows from this definition that

$$\left(\frac{1}{1!} + \frac{z}{2!} + \frac{z^2}{3!} + \cdots\right) \cdot \left(\frac{B_0}{0!} + \frac{B_1}{1!} z + \frac{B_2}{2!} z^2 + \cdots\right) = 1, \quad |z| < \pi.$$

Comparing the coefficients we see that $B_0 = 1$ and

$$\frac{1}{n!} \cdot \frac{B_0}{0!} + \frac{1}{(n-1)!} \cdot \frac{B_1}{1!} + \frac{1}{(n-2)!} \cdot \frac{B_2}{2!} + \cdots + \frac{1}{1!} \cdot \frac{B_{n-1}}{(n-1)!} = 0$$

[12] The expression on the left-hand side is defined by continuity at $z = 0$.

if $n > 1$. Multiplying both sides by $n!$ we obtain

$$\binom{n}{0}B_0 + \binom{n}{1}B_1 + \binom{n}{2}B_2 + \cdots + \binom{n}{n-1}B_{n-1} = 0. \tag{2}$$

In particular,

$$\begin{aligned} 0 &= 1 + 2B_1, \\ 0 &= 1 + 3B_1 + 3B_2, \\ 0 &= 1 + 4B_1 + 6B_2 + 4B_3, \\ 0 &= 1 + 5B_1 + 10B_2 + 10B_3 + 5B_4 \end{aligned}$$

and hence

$$B_1 = -\frac{1}{2}, \quad B_2 = \frac{1}{6}, \quad B_3 = 0, \quad B_4 = -\frac{1}{30}.$$

Using that $B_1 = -\frac{1}{2}$, relation (1) gives

$$1 + \frac{B_2}{2!}z^2 + \cdots = \frac{z}{e^z - 1} + \frac{z}{2} = \frac{z}{2}\frac{e^z + 1}{e^z - 1} = \frac{z}{2}\frac{e^{\frac{z}{2}} + e^{-\frac{z}{2}}}{e^{\frac{z}{2}} - e^{-\frac{z}{2}}}. \tag{3}$$

The function on the right-hand side is even, showing that $B_{2k+1} = 0$ for all $k \in \mathbb{N}$.

Theorem C.4.12. *For all $n, N \in \mathbb{N}$ we have*

$$\sum_{j=1}^{2N} \frac{B_{2j}}{2j(2j-1)n^{2j-1}} < \log\frac{n!}{(n/e)^n\sqrt{2\pi n}} < \sum_{j=1}^{2N+1} \frac{B_{2j}}{2j(2j-1)n^{2j-1}}.$$

Proof. From equation (C.4.11.3) we see that

$$\frac{1}{e^z - 1} - \frac{1}{z} + \frac{1}{2} = \sum_{j=1}^{\infty} \frac{B_{2j}}{(2j)!}z^{2j-1}.$$

By Problem 154 in Part I, Chapter 4 of [44], the series on the right-hand side has the so-called *enveloping property*, i.e.,

$$\sum_{j=1}^{2N} \frac{B_{2j}}{(2j)!}t^{2j-1} < \frac{1}{e^t - 1} - \frac{1}{t} + \frac{1}{2} < \sum_{j=1}^{2N+1} \frac{B_{2j}}{(2j)!}t^{2j-1}, \quad 0 < t < \pi.$$

Using the equation $j! = \int_0^\infty t^j e^{-t}\, dt$ and the definition of θ we immediately obtain the assertion. □

C.5 Bessel functions

Definition C.5.1. For $\nu \geq 0$ we define[13] the *Bessel function* J_ν by

$$J_\nu(x) = \frac{1}{\sqrt{\pi}\,\Gamma\left(\nu+\frac{1}{2}\right)}\left(\frac{x}{2}\right)^\nu \int_{-1}^{1} (1-t^2)^{\nu-1/2} \mathrm{e}^{\mathrm{i}tx}\,\mathrm{d}t, \quad x \geq 0. \tag{1}$$

For $x = \nu = 0$ the expression $\left(\frac{x}{2}\right)^\nu$ is equal to 1 by definition. Using that $\Gamma(1/2) = \sqrt{\pi}$ we see that

$$J_0(0) = \frac{1}{\pi}\int_{-1}^{1} \frac{1}{\sqrt{1-t^2}}\,\mathrm{d}t = \frac{1}{\pi} \arcsin t\,\Big|_{-1}^{1} = 1$$

while $J_\nu(0) = 0$ if $\nu > 0$.

If $\nu = 1/2$, then the integral above can be easily evaluated. We obtain:

$$J_{1/2}(x) = \sqrt{\frac{2}{\pi x}} \cdot \sin x.$$

Note that $(1-t^2)^{\nu-1/2}$ is even in t, hence one can replace $\mathrm{e}^{\mathrm{i}tx}$ by $\cos tx$ in Definition C.5.1.

Lemma C.5.2. *The function J_ν is continuous and*

$$|J_\nu(x)| \leq C_\nu \cdot x^\nu, \quad \nu, x \geq 0$$

where

$$C_\nu = \frac{1}{\sqrt{\pi}\,\Gamma\left(\nu+\frac{1}{2}\right) 2^\nu} \cdot \int_{-1}^{1} (1-t^2)^{\nu-1/2}\,\mathrm{d}t.$$

Proof. The continuity follows from the fact that the integral in Definition C.5.1 represents a characteristic function, up to a constant factor. The second statement is obvious. □

The substitution $t = \cos\varphi$ in Definition C.5.1 yields:

Lemma C.5.3. *For $\nu \geq 0$ we have*

$$J_\nu(x) = \frac{1}{\sqrt{\pi}\,\Gamma\left(\nu+\frac{1}{2}\right)}\left(\frac{x}{2}\right)^\nu \int_0^\pi \mathrm{e}^{\mathrm{i}x\cos\varphi} \sin^{2\nu}\varphi\,\mathrm{d}\varphi, \quad x \geq 0.$$

[13] We will extend the definition of J_ν for all $\nu \in \mathbb{R}$ in Remark C.5.6. Note that the extension is possible for all complex ν but we do not need it in this book. The functions J_0 and J_1 are shown in Figure C.3.

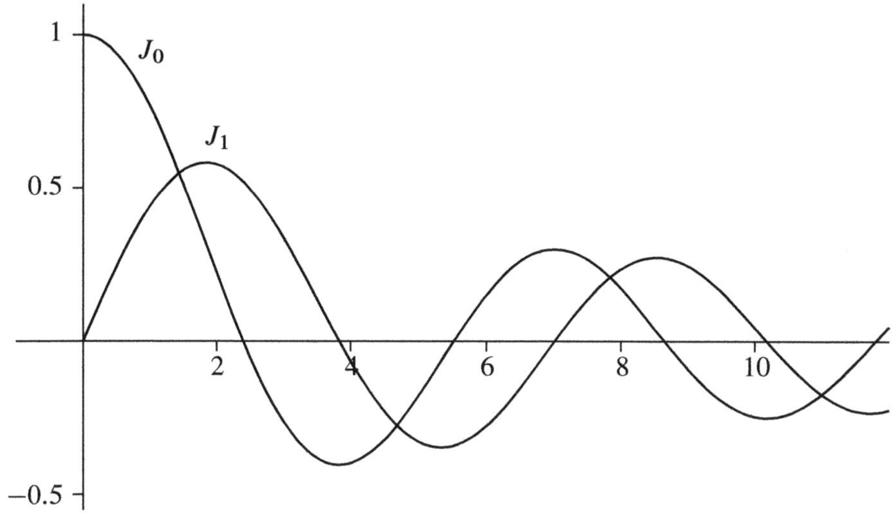

Figure C.3. The Bessel functions J_0 and J_1 from Definition C.5.1.

Theorem C.5.4. *The formula*

$$J_\nu(x) = \left(\frac{x}{2}\right)^\nu \sum_{k=0}^{\infty} \frac{(-1)^k}{k!\,\Gamma(k+\nu+1)} \cdot \left(\frac{x}{2}\right)^{2k}$$

holds for all $\nu \geq 0$ *and* $x \geq 0$.

Proof. Replacing e^{itx} by its power series expansion in Definition C.5.1 we see that $J_\nu(x)$ is equal to

$$\frac{1}{\sqrt{\pi}\,\Gamma\left(\nu+\frac{1}{2}\right)} \left(\frac{x}{2}\right)^\nu \sum_{k=0}^{\infty} \frac{1}{(2k)!} \int_{-1}^{1} (\mathrm{i}xt)^{2k} (1-t^2)^{\nu-1/2}\,\mathrm{d}t.$$

Proposition C.4.4 implies

$$\int_{-1}^{1} t^{2k}(1-t^2)^{\nu-1/2}\,\mathrm{d}t = \frac{\Gamma(k+1/2)\Gamma(\nu+1/2)}{\Gamma(k+\nu+1)}$$

so

$$J_\nu(x) = \frac{1}{\Gamma\left(\nu+\frac{1}{2}\right)} \left(\frac{x}{2}\right)^\nu \sum_{k=0}^{\infty} \frac{1}{(2k)!} (\mathrm{i}x)^{2k} \frac{\Gamma(k+1/2)\Gamma(\nu+1/2)}{\sqrt{\pi}\,\Gamma(k+\nu+1)}.$$

Using $(2k)! = \Gamma(2k+1)$, $\sqrt{\pi} = \Gamma(1/2)$, and the duplication formula (cf. Proposition C.4.5), which implies

$$\frac{\Gamma(k+1/2)}{\Gamma(1/2)\Gamma(2k+1)} = \frac{2^{-2k}}{\Gamma(k+1)}$$

we obtain the assertion. □

Proposition C.5.5. *We have*

$$\frac{\mathrm{d}}{\mathrm{d}x}(x^\nu J_\nu(x)) = x^\nu J_{\nu-1}(x), \quad \nu \geq 1,\ x \geq 0$$

and

$$\frac{\mathrm{d}}{\mathrm{d}x}(x^{-\nu} J_\nu(x)) = -x^{-\nu} J_{\nu+1}(x), \quad \nu \geq 0,\ x > 0$$

Proof. Applying Theorem C.5.3 we obtain

$$\begin{aligned}
\frac{\mathrm{d}}{\mathrm{d}x}(x^\nu J_\nu(x)) &= \frac{\mathrm{d}}{\mathrm{d}x}\sum_{k=0}^{\infty} \frac{(-1)^k}{k!\,\Gamma(k+\nu+1)}\frac{1}{2^{2k+\nu}}x^{2k+2\nu} \\
&= \sum_{k=0}^{\infty} \frac{(-1)^k(2k+2\nu)}{k!\,\Gamma(k+\nu+1)}\frac{1}{2^{2k+\nu}}x^{2k+2\nu-1} \\
&= \sum_{k=0}^{\infty} \frac{(-1)^k}{k!\,\Gamma(k+\nu)}\frac{1}{2^{2k+\nu-1}}x^{2k+2\nu-1} \\
&= x^\nu J_{\nu-1}(x).
\end{aligned}$$

The second equation can be proved in the same way:

$$\begin{aligned}
\frac{\mathrm{d}}{\mathrm{d}x}(x^{-\nu} J_\nu(x)) &= \frac{\mathrm{d}}{\mathrm{d}x}\sum_{k=0}^{\infty} \frac{(-1)^k}{k!\,\Gamma(k+\nu+1)}\frac{1}{2^{2k+\nu}}x^{2k} \\
&= \sum_{k=1}^{\infty} \frac{(-1)^k 2k}{k!\,\Gamma(k+\nu+1)}\frac{1}{2^{2k+\nu}}x^{2k-1} \\
&= -\sum_{k=0}^{\infty} \frac{(-1)^k(2k+2)}{(k+1)!\,\Gamma(k+\nu+2)}\frac{1}{2^{2k+2+\nu}}x^{2k+1} \\
&= -\sum_{k=0}^{\infty} \frac{(-1)^k}{k!\,\Gamma(k+\nu+2)}\frac{1}{2^{2k+1+\nu}}x^{2k+1} \\
&= -x^{-\nu} J_{\nu+1}(x).
\end{aligned}$$

□

Remark C.5.6. We use the first equation in Proposition C.5.5 to define $J_\nu(x)$ recursively for all $\nu \in \mathbb{R}$ and $x > 0$. Doing so, both equations in Proposition C.5.5 hold for all such ν and x. Note that the first equation can also be written in the form

$$\left(\frac{\mathrm{d}}{\mathrm{d}x} + \frac{\nu}{x}\right) J_\nu(x) = J_{\nu-1}(x), \quad x > 0. \tag{1}$$

From $J_{1/2}(x) = \sqrt{\frac{2}{\pi x}} \sin x$ we obtain

$$J_{-1/2}(x) = \sqrt{\frac{2}{\pi x}} \cos x, \quad x > 0.$$

The second equation in Proposition C.5.5 can also be written in the form

$$\left(\frac{\mathrm{d}}{\mathrm{d}x} - \frac{\nu}{x}\right) J_\nu(x) = -J_{\nu+1}(x), \quad x > 0. \tag{2}$$

Putting together (1) and (2) produces the equation

$$\left(\frac{\mathrm{d}}{\mathrm{d}x} - \frac{\nu - 1}{x}\right)\left(\frac{\mathrm{d}}{\mathrm{d}x} + \frac{\nu}{x}\right) J_\nu(x) = -J_\nu(x)$$

which is equivalent to

$$\left[\frac{\mathrm{d}^2}{\mathrm{d}x^2} + \frac{1}{x}\frac{\mathrm{d}}{\mathrm{d}x} + \left(1 - \frac{\nu^2}{x^2}\right)\right] J_\nu(x) = 0$$

known as *Bessel's equation*. Adding and subtracting (1) and (2) we obtain the identities

$$\begin{aligned} 2J_\nu'(x) &= J_{\nu-1}(z) - J_{\nu+1}(z) \\ \frac{2\nu}{z} J_\nu(x) &= J_{\nu-1}(z) + J_{\nu+1}(z)\,. \end{aligned}$$

Definition C.5.7. The *modified Bessel function of the second kind*[14] of index $\nu \in \mathbb{C}$ is defined by

$$K_\nu(z) = \int_0^\infty \mathrm{e}^{-z \cosh t} \cosh(\nu t)\, \mathrm{d}t, \quad z \in \mathbb{C},\ \operatorname{Re} z > 0.$$

It follows from the definition that K_ν is continuous, $K_\nu = K_{-\nu}$ if $\nu \in \mathbb{R}$, and $K_\nu(r) > 0$ if $\nu \in \mathbb{R}$ and $r > 0$. In this book we need only the following simple lemma.

Lemma C.5.8. *For all $b \in \mathbb{R}$ and $a > 0$ we have*

$$K_b(r) = \frac{1}{2a^b} \int_0^\infty \mathrm{e}^{-\frac{r}{2}\cdot\left(\frac{u}{a} + \frac{a}{u}\right)} u^{b-1}\, \mathrm{d}u, \quad r > 0.$$

---

[14] This function has also been called by the now-rare name *modified Bessel function of the third kind*. The function K_1 is shown in Figure C.4.

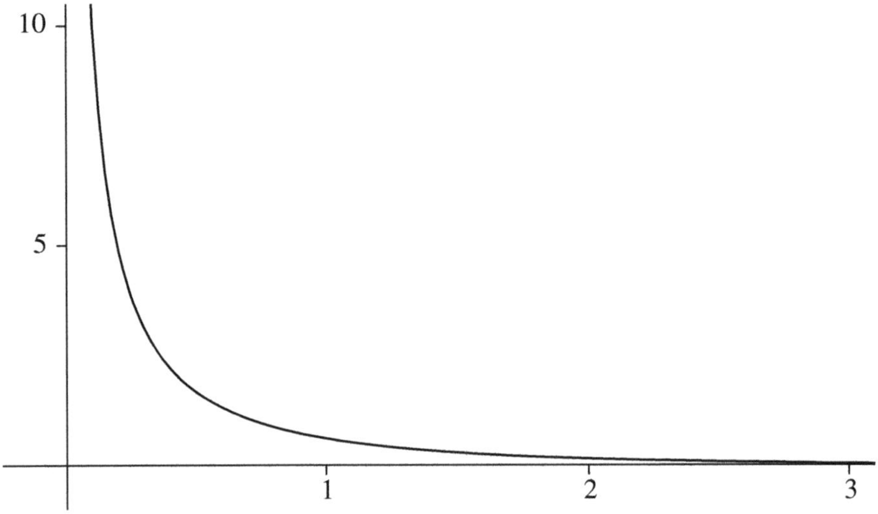

Figure C.4. The modified Bessel function K_1 from Definition C.5.7.

Proof. Using the substitution $u = ae^t$ we obtain

$$\begin{aligned}\frac{1}{2a^b}&\int_0^\infty e^{-\frac{r}{2}\cdot(\frac{u}{a}+\frac{a}{u})}u^{b-1}\,du\\&=\frac{1}{2}\int_{-\infty}^\infty e^{-\frac{r}{2}\cdot(e^t+e^{-t})}e^{bt}\,dt\\&=\frac{1}{2}\int_{-\infty}^\infty e^{-r\cosh t}e^{bt}\,dt = K_b(r).\end{aligned}$$

□

Note that setting $a = \frac{r}{2}$ in the previous lemma leads to the inequality

$$K_b(r) = 2^{b-1}r^{-b}\int_0^\infty e^{-u}e^{-r^2/(4u)}u^{b-1}\,du \le \Gamma(b)2^{b-1}r^{-b}, \quad a,b>0.$$

C.6 The Mellin transform

C.6.1. We consider $G = (0,\infty)$ as a group with multiplication as operation. The mapping $\varphi : r \mapsto \ln(r)$, $r \in G$, is then an algebraical and topological isomorphism between G and the additive group $\mathbb{R}$ of all real numbers.
Using this isomorphism we see that γ is a continuous character of G if and only if it has the form $\gamma(r) = r^{ix}$ with some $x \in \mathbb{R}$. Denoting by λ_* the image of the Lebesgue

measure by the mapping φ^{-1}, we have

$$\lambda_*((a,b)) = \ln(b) - \ln(a),\ 0 < a \le b < \infty.$$

More generally,

$$\int_0^\infty f(s)\,\mathrm{d}\lambda_*(s) = \int_0^\infty \frac{f(s)}{s}\,\mathrm{d}s, \quad f \in \mathcal{L}^1(\lambda_*).$$

Since λ is invariant under translations we have

$$\int_0^\infty f(rs)\,\mathrm{d}\lambda_*(s) = \int_0^\infty f(s)\,\mathrm{d}\lambda_*(s), \quad f \in \mathcal{L}^1(\lambda_*).$$

The *multiplicative convolution* on G corresponds to the convolution on $\mathbb{R}$. The multiplicative convolution on G is given by

$$h(r) = \int_0^\infty f(r/s)g(s)\,\mathrm{d}\lambda_*(s), \quad r \in G,\ f, g \in \mathcal{L}^2(\lambda_*).$$

Definition C.6.2. Let f be a Lebesgue measurable, complex-valued function on $(0,\infty)$. For all $z \in \mathbb{C}$ such that $\int_0^\infty |f(t)t^{z-1}|\,\mathrm{d}t < \infty$ we define the *Mellin transform* Mf of f by

$$Mf(z) := \int_0^\infty f(s)s^{z-1}\,\mathrm{d}s = \int_0^\infty f(s)s^z\,\mathrm{d}\lambda_*(s).$$

Lemma C.6.3. *If h is the multiplicative convolution of f and g, then*

$$Mh(z) = Mf(z) \cdot Mg(z)$$

for all $z \in \mathbb{C}$ for which the Mellin transforms $Mf(z)$ and $Mg(z)$ exist.

Proof. Indeed, using that λ_* is invariant under multiplications, we obtain

$$\begin{aligned} Mf(z)Mg(z) &= \int_0^\infty f(r)r^z\,\mathrm{d}\lambda_*(r) \cdot \int_0^\infty g(s)s^z\,\mathrm{d}\lambda_*(s) \\ &= \int_0^\infty \int_0^\infty f(r)g(s)(rs)^z\,\mathrm{d}\lambda_*(r)\,\mathrm{d}\lambda_*(s) \\ &= \int_0^\infty \int_0^\infty f(r/s)g(s)r^z\,\mathrm{d}\lambda_*(r)\,\mathrm{d}\lambda_*(s) = Mh(z). \end{aligned}$$ □

Remark C.6.4. Substituting $s = \mathrm{e}^{-t}$ and setting $g(t) := f(\mathrm{e}^{-t})$ in Definition C.6.2 we obtain

$$Mf(z) = \int_{-\infty}^\infty g(t)\mathrm{e}^{-tz}\,\mathrm{d}t. \tag{1}$$

Setting here $z = \mathrm{i}y,\ y \in \mathbb{R}$, we obtain

$$Mf(\mathrm{i}y) = \int_{-\infty}^{\infty} \mathrm{e}^{-\mathrm{i}yt} g(t)\,\mathrm{d}t = \hat{g}(y), \quad y \in \mathbb{R}.$$

In the same way as in Section 3.3 we see that there exists $a, b \in [-\infty, \infty]$, $a \le b$, such that the integral in (1) exists if $a < \operatorname{Re} z < b$ and Mf is holomorphic in this strip.

Example C.6.5. If $f(s) = \mathrm{e}^{-s}$, then

$$Mf(z) = \int_0^{\infty} \mathrm{e}^{-t} t^{z-1}\,\mathrm{d}t.$$

In this case, Mf is holomorphic in the region $\operatorname{Re}(z) > 0$ and Mf is equal to the Gamma function.

Let $f(s) = 1/(1+s)$. Using the substitution $s = x/(1-x)$ we see that

$$Mf(z) = \int_0^1 z^{s-1}(1-x)^{-z}\,\mathrm{d}x.$$

In this case, Mf is holomorphic in the region $\{z \in \mathbb{C} : 0 < \operatorname{Re}(z) < 1\}$ and $Mf(z) = B(z, 1-z)$.

C.7 The Laplace transform

Throughout this section μ denotes a nonnegative measure on $[0, \infty)$ such that the function $x \mapsto \mathrm{e}^{-tx}$ is μ-integrable for all $t > 0$.

Definition C.7.1. The function $L\mu$ defined by

$$L\mu(t) = \int_0^{\infty} \mathrm{e}^{-tx}\,\mathrm{d}\mu(x), \quad t > 0$$

is called the *Laplace transform* of μ.[15]

Example C.7.2. If $\mathrm{d}\mu(x) = x^a\,\mathrm{d}x,\ a > -1$, then

$$L\mu(t) = \int_0^{\infty} x^a \mathrm{e}^{-tx}\,\mathrm{d}x = \frac{1}{t^{a+1}} \int_0^{\infty} y^a \mathrm{e}^{-y}\,\mathrm{d}y = \frac{\Gamma(a+1)}{t^{a+1}}, \quad t > 0.$$

Lemma C.7.3. *The measure μ is finite if and only if its Laplace transform is bounded.*

[15] A classical reference for Laplace transforms is [60].

Proof. If μ is finite, then

$$L\mu(t) = \int_0^\infty e^{-tx}\, d\mu(x) \le \int_0^\infty 1\, d\mu(x) = \mu([0,\infty)) < \infty, \quad t > 0.$$

Assume that $L\mu \le K$ for some $K \in [0,\infty)$. By monotone convergence we have

$$K \ge \lim_{t\to 0} L\mu(t) = \lim_{t\to 0}\int_0^\infty e^{-tx}\, d\mu(x) = \int_0^\infty 1\, d\mu(x) = \mu([0,\infty))$$

i.e., μ is bounded. □

Lemma C.7.4. *The function $L\mu$ is infinitely differentiable and*

$$(L\mu)^{(k)}(t) = (-1)^k \int_0^\infty x^k e^{-tx}\, d\mu(x), \quad t > 0$$

for all $k \in \mathbb{N}_0$.

Proof. The lemma follows by successive differentiation which can be justified since

$$x^k e^{tx} = x^k e^{-tx/2} e^{-tx/2} \le K_{k,t} e^{-tx/2}, \quad x > 0$$

where $K_{k,t}$ is uniformly bounded for all t from a fixed compact subset of $(0,\infty)$. □

Theorem C.7.5. *The measure μ is uniquely determined by its Laplace transform.*

Proof. Writing $d\nu(x) := e^{-x}\, d\mu(x)$, the measure ν is finite. Since $L\nu(t) = L\mu(t+1)$, it suffices to show that ν is uniquely determined by its Laplace transform. By Lemma C.7.4 we have

$$(-1)^k (L\nu)^{(k)}(t) = \int_0^\infty x^k e^{-tx}\, d\nu(x), \quad t > 0,\ k \in \mathbb{N}_0$$

and hence

$$\sum_{0\le k\le ta} (-1)^k (L\nu)^{(k)}(t)\cdot \frac{t^k}{k!} = \int_0^\infty e^{-tx} \sum_{0\le k\le ta} \frac{(tx)^k}{k!}\, d\nu(x), \quad a, t > 0.$$

The integrand on the right-hand side is at most 1. Using this, dominated convergence and Lemma B.1.18 show that

$$\lim_{t\to\infty} \sum_{0\le k\le ta} (-1)^k (L\nu)^{(k)}(t)\cdot \frac{t^k}{k!} = \int_0^\infty \mathbf{1}_{[0,a]}(x)\, d\nu(x) = \nu([0,a]).$$

This relation shows that ν is uniquely determined by $L\nu$. □

C.8 Existence of continuous logarithms

Definition C.8.1. Let f be a complex-valued function defined on some topological space X. A continuous complex-valued function F on X is called a *continuous logarithm* of f if

$$f(x) = \mathrm{e}^{F(x)}, \quad x \in X.$$

By a *continuous argument* of f we mean any real-valued continuous function θ on X such that

$$f(x) = |f(x)| \cdot \mathrm{e}^{\mathrm{i}\theta(x)}, \quad x \in X.$$

Remark C.8.2. Note that if F is a continuous logarithm of f, then $\theta := \operatorname{Im} F$ is a continuous argument of f.

It is clear that functions which have a continuous logarithm cannot have zeros. If $X = \mathbb{T} \subset \mathbb{C}$ and $f(z) = z,\ z \in \mathbb{T}$, then f is continuous and has no zeros. However, f does not have a continuous logarithm. To see this, assume that

$$z = \mathrm{e}^{\mathrm{i}\theta(z)}, \quad z \in \mathbb{T}$$

with some continuous function $\theta : \mathbb{T} \to \mathbb{R}$. In particular, θ is injective and hence $\theta(z) - \theta(-z) \neq 0$. Consequently,

$$g(z) := \frac{\theta(z) - \theta(-z)}{|\theta(z) - \theta(-z)|}, \quad z \in \mathbb{T}$$

defines a continuous real-valued function on $\mathbb{T}$. Since $|g(z)| = 1$ we must have $g(\mathbb{T}) \subset \{-1, 1\}$. But $g(z) = -g(-z)$ and therefore $g(\mathbb{T}) = \{-1, 1\}$. This is a contradiction, since $\mathbb{T}$ is connected while $\{-1, 1\}$ is disconnected.

Lemma C.8.3. *Let X be a connected space and $f : X \to \mathbb{C} \setminus \{0\}$ be a continuous function. If θ_1 and θ_2 are continuous arguments of f, then there exists $k \in \mathbb{Z}$ such that*

$$\theta_2(x) = \theta_1(x) + 2k\pi\mathrm{i}, \quad x \in X.$$

Proof. From

$$f(x) = |f(x)| \cdot \mathrm{e}^{\mathrm{i}\theta_1(x)} = |f(x)| \cdot \mathrm{e}^{\mathrm{i}\theta_2(x)}$$

we see that $\theta_2(x) = \theta_1(x) + 2k(x)\pi\mathrm{i}$ with some function $k : X \to \mathbb{Z}$. Since θ_1 and θ_2 are continuous, so is k. The continuous image of connected sets is connected, therefore $k(X)$ consists of a single integer. □

Lemma C.8.4. *Every continuous function $f : X \to \mathbb{T}$ such that $f(X) \neq \mathbb{T}$ has a continuous argument.*

Proof. Choose $\alpha \in \mathbb{R}$ such that $\mathrm{e}^{\mathrm{i}\alpha} \in \mathbb{T} \setminus f(X)$. The mapping

$$t \mapsto \mathrm{e}^{\mathrm{i}t}, \quad t \in (\alpha, \alpha + 2\pi)$$

is continuous, injective and it maps $(\alpha, \alpha + 2\pi)$ onto $\mathbb{T} \setminus \{\mathrm{e}^{\mathrm{i}\alpha}\}$. Denoting by ψ the inverse mapping we have

$$z = \mathrm{e}^{\mathrm{i}\psi(z)}, \quad z \in \mathbb{T} \setminus \{\mathrm{e}^{\mathrm{i}\alpha}\}.$$

Thus, the function $\theta := \psi(f)$ is a continuous argument of f. □

Lemma C.8.5. *Let f_1 and f_2 be continuous functions on X with values in $\mathbb{T}$ such that $f_1(x) \neq -f_2(x)$ for all $x \in X$. If f_1 has a continuous argument, then so does f_2.*

Proof. Let ψ_1 be a continuous argument of f_1. The function $f := f_1/f_2$ is continuous, maps X into $\mathbb{T}$ and $f(x) \neq -1$ for all x. By Lemma C.8.4 it has a continuous argument ψ. If follows from the definition of f that $\psi_1 - \psi$ is a continuous argument of f_2. □

Proposition C.8.6. *Let X be a compact topological space and $f : X \times [0, 1] \to \mathbb{C} \setminus \{0\}$ be a continuous function. If the restriction of f to $X \times \{0\}$ has a continuous argument, then so does the restriction of f to $X \times \{1\}$.*

Proof. Replacing f by $f/|f|$ we may assume that $|f| = 1$. The function f is uniformly continuous on the compact set $X \times [0, 1]$. Therefore, there exists $n \in \mathbb{N}$ such that

$$|f(x,s) - f(x,t)| \leq 1, \quad x \in X \tag{1}$$

whenever $|s - t| \leq 1/n$. For $j = 0, 1, \ldots, n$ we define the function f_j by

$$f_j(x) := f(x, j/n), \quad x \in X.$$

Inequality (1) shows that $|f_j - f_{j+1}| \leq 1$. Since $|f_j| = |f_{j+1}| = 1$ we conclude that $f_j(x) \neq f_{j+1}(x)$ for all x. The statement of the proposition follows by applying Lemma C.8.5 n times. □

Theorem C.8.7. *Let $f : \mathbb{R}^d \to \mathbb{C} \setminus \{0\}$ be a continuous function such that $f(0) > 0$. Then f has a unique continuous argument θ such that $\theta(0) = 0$.*[16]

Proof. Replacing f by $f/|f|$ we may assume that $|f| = 1$ and $f(0) = 1$. Denote by f_n the restriction of f to the compact set $B_d^c(n)$, $n \in \mathbb{N}$. We claim that f_n has a continuous argument θ_n such that $\theta_n(0) = 0$. To show this define the function g_n by

$$g_n(x,t) := f_n(tx), \quad (x,t) \in B_d^c(n) \times [0, 1].$$

[16] See also Lemma C.1.11.

We have $g(x,0) = 1$ and $g(x,1) = f_n(x)$. Hence, the existence of a continuous argument for f_n follows from Proposition C.8.6.

Using Lemma C.8.3 and the fact that $0 = \theta_n(0) = \theta_{n+1}(0)$ we see that $\theta_n(x) = \theta_{n+1}(x)$, $x \in B_d^c(n)$. Let θ be the unique function on $\mathbb{R}^d$ such that $\theta(x) = \theta_n(x)$, $x \in B_d^c(n)$. It is clear that θ is a continuous argument of f while the uniqueness follows from Lemma C.8.3. □

C.9 Solutions of certain functional equations

Recall that $B^o(\delta) = B_d^o(\delta)$ denotes the open ball

$$\{t \in \mathbb{R}^d : \|t\| < \delta\}.$$

Lemma C.9.1 (Cauchy's functional equation). *Let f be a continuous real-valued function on $B^o(\delta)$ such that*

$$f(t+s) = f(t) + f(s), \quad t, s, t+s \in B^o(\delta). \tag{1}$$

Then there exists $c \in \mathbb{R}^d$ such that $f(t) = (c,t)$ for all $t \in B^o(\delta)$.

Proof. Assume first that $d = 1$. From $f(0) = f(0+0) = f(0) + f(0)$ we get $f(0) = 0$. Repeatedly applying equation (1) gives $f(t) = f(t/n + \cdots + t/n) = nf(t/n)$ for all $n \in \mathbb{N}$ and for all t. Now let $r = m/n$ be a rational number such that $rt \in B^o(\delta)$. Then

$$f(rt) = f(mt/n) = mf(t/n) = \frac{m}{n} f(t) = rf(t).$$

By continuity, the equation $f(rt) = rf(t)$ holds for all $r \in \mathbb{R}$ such that $rt, t \in B^o(\delta)$. We replace in this equation t by $t_0 \neq 0$ and r by t/t_0 and obtain

$$f(t) = \frac{f(t_0)}{t_0} \cdot t, \quad t \in B^o(\delta).$$

To prove the general case, let $e_1, \dots, e_d$ be the standard basis of $\mathbb{R}^d$. Then

$$f(t) = f(t_1 \cdot e_1 + \cdots + t_d \cdot e_d) = f_1(t_1) + \cdots + f_d(t_d)$$

where $f_j(t_j) = f(t_j \cdot e_j)$. We now apply the first part of the proof to the functions f_j. □

Theorem C.9.2. *A continuous complex-valued function $\varphi \neq 0$ on $B^o(\delta)$ satisfies the equation*

$$\varphi(t+s) = \varphi(t)\varphi(s), \quad t, s, t+s \in B^o(\delta) \tag{1}$$

if and only if there exists $c \in \mathbb{C}^d$ *such that*

$$\varphi(t) = \mathrm{e}^{(c,t)}, \quad t \in B^o(\delta).$$

Proof. It is clear that any function of the above form satisfies the equation (1). To prove the other direction we consider only the case $d = 1$. The general case can be reduced to this one in the same way as in the proof of Lemma C.9.1. From $\varphi(t+0) = \varphi(t)\varphi(0)$ we see that $\varphi(0) = 1$. Thus, $1 = \varphi(t - t) = \varphi(t)\varphi(-t)$ and therefore $\varphi(t) \neq 0$ for all t. The function $|\varphi|$ satisfies the same equation as φ. Applying Lemma C.9.1 to the logarithm of $|\varphi|$, we obtain $|\varphi(t)| = \mathrm{e}^{rt}$ with some $r \in \mathbb{R}$. The function $\gamma = \varphi/|\varphi|$ satisfies the same equation as φ. By continuity, there exists $\delta_0 > 0$ such that the real part of $\gamma(t)$ is positive whenever $t \in (-\delta_0, \delta_0)$. Setting $f(t) = \arctan(\operatorname{Im}\gamma(t)/\operatorname{Re}\gamma(t))$ we have $\gamma(t) = \mathrm{e}^{\mathrm{i}f(t)}$, $t \in (-\delta_0, \delta_0)$. The function f is continuous and satisfies equation (C.9.1.1). Applying Lemma C.9.1 and using $\varphi = |\varphi| \cdot \gamma$ we conclude that

$$\varphi(t) = \mathrm{e}^{ct}, \quad t \in (-\delta_0, \delta_0)$$

with some $c \in \mathbb{C}$. To show that this equation holds for arbitrary $t \in B^o(\delta)$, let $n \in \mathbb{N}$ be such that $t/n \in (-\delta_0, \delta_0)$. Repeated application of (1) shows that

$$\varphi(t) = \varphi(t/n + \cdots + t/n) = \varphi(t/n)^n = \mathrm{e}^{ct}. \qquad \square$$

Lemma C.9.3 (Pexider's functional equation). *Let f, g and h be real-valued functions on $B^o(\delta)$ such that f is continuous and*

$$f(t+s) = g(t) + h(s), \quad t, s, t+s \in B^o(\delta).$$

Then there exist $a, b \in \mathbb{R}$ and $c \in \mathbb{R}^d$ such that $f(t) = (c,t) + a + b$, $g(t) = (c,t) + a$, $h(t) = (c,t) + b$ for all $t \in B^o(\delta)$.

Proof. Let $a = g(0)$ and $b = h(0)$. Setting $s = 0$ in the above equation we get $g(t) = f(t) - b$. Similarly, $h(s) = f(s) - a$. Thus,

$$f(t+s) = f(t) + f(s) - a - b$$

and therefore the function $f - a - b$ satisfies Cauchy's functional equation. The assertion follows now from Lemma C.9.1. $\square$

Lemma C.9.4. *Let $k \in \mathbb{N}_0$ and let f be a continuous real-valued function on $B^o(\delta)$ such that for all $s \in B^o(\delta)$ the function*

$$t \mapsto f(t+s) - f(t), \quad t, t+s \in B^o(\delta)$$

is a polynomial of degree at most k. Then f is a polynomial of degree at most $k+1$.

Proof. First, let $d = 1$. By assumption,

$$f(t+s) - f(t) = a_k(s)t^k + \cdots + a_1(s)t + a_0(s), \quad t, s, t+s \in (-\delta, \delta) \qquad (1)$$

where $a_0, \dots, a_k$ are real-valued functions. First we show that the functions a_j are continuous. Let s be fixed and choose mutually different t_j such that $t_j, s + t_j \in (-\delta, \delta)$, $j = 0, \dots, k$. Replacing t by t_j in (1) we obtain $k + 1$ equations which can be written in the form $\Delta f = Va$, where V is the Vandermonde matrix corresponding to the t_j's and Δf and a are column vectors with coordinates $f(t_j + s) - f(t_j)$ and $a_j(s)$, respectively. Since V is invertible, we see that each a_j can be written as

$$a_j(s) = \sum_{i=0}^{k} r_i \cdot [f(t_i + s) - f(t_i)]$$

with some $r_i = r_i(t_0, \dots, t_k) \in \mathbb{R}$, from which the continuity of a_j follows.

We now prove the lemma by induction on k. The case $k = 0$ follows from Lemma C.9.3 with $g = f$. Assume that $k > 0$ and that the statement is true for $0, 1, \dots, k - 1$. Replacing s by $s + x$ in (1) gives

$$f(t + s + x) - f(t) = \sum_{j=0}^{k} a_j(s + x)t^j$$

while replacing t by $t + x$ gives

$$f(t + s + x) - f(t + x) = \sum_{j=0}^{k} a_j(s)(t + x)^j.$$

These equations hold whenever $s, t, x, t + x, s + x, s + t + x \in (-\delta, \delta)$. Subtracting the second equation from the first one yields

$$f(t + x) - f(t) = \sum_{j=0}^{k} \left[a_j(s + x)t^j - a_j(s)(t + x)^j\right].$$

On the other hand, (1) shows that

$$f(t + x) - f(t) = \sum_{j=0}^{k} a_j(x)t^j.$$

For fixed s and x these equations hold for all t in a neighborhood of zero. Comparing the coefficients of t^k we obtain

$$a_k(s + x) = a_k(s) + a_k(x), \quad s, x, s + x \in (-\delta, \delta).$$

Thus, in view of Lemma C.9.1, $a_k(s) = c \cdot s$ with some $c \in \mathbb{R}$. Define the function g by $g(t) = f(t) - \frac{c}{k+1} \cdot t^{k+1}$. Then

$$g(t + s) - g(t) = b_{k-1}(s)t^{k-1} + \cdots + b_0(s)$$

with some functions b_j. By the induction assumption g is a polynomial of degree at most k, completing the proof in the case $d = 1$.

Now let d be arbitrary. By assumption,

$$f(t+s) - f(t) = \sum_{|\alpha| \le k} a_\alpha(s) t^\alpha, \quad t, s, t+s \in B^o(\delta). \tag{2}$$

with some real-valued functions a_α. For fixed $t, s \in B^o(\delta)$ define the function h by

$$h(x) = f(t + x \cdot (s-t)), \quad x \in I := \{r \in \mathbb{R},\ t + r \cdot (s-t) \in B^o(\delta)\}.$$

Note that I is an open interval containing $[0, 1]$. If $x, y, x+y \in I$, then

$$\begin{aligned} h(x+y) - h(x) &= f(t + x \cdot (s-t) + y \cdot (s-t)) - f(t + x \cdot (s-t)) \\ &= \sum_{|\alpha| \le k} a_\alpha(y \cdot (s-t)) \cdot (t + x \cdot (s-t))^\alpha. \end{aligned}$$

The right-hand side is a polynomial in x of degree at most k. By the first part of the proof, h is a polynomial of degree at most $k+1$.[17] Application of Theorem B.6.6 completes the proof. □

Theorem C.9.5. *Let $f_1, \dots, f_m, g$ and h be continuous real-valued functions on $B^o(\delta)$. Further, let $A_i, B_i \in \mathbb{R}^{d \times d}$ be invertible matrices. If for some $\delta_0 \in (0, \delta]$ the equation*

$$\sum_{i=1}^m f_i(A_i t + B_i s) = g(t) + h(s), \quad t, s \in B^o(\delta_0) \tag{1}$$

holds, then in some neighborhood of zero the functions g and h are polynomials of degree at most m. If the matrices $B_i A_i^{-1} - B_j A_j^{-1}$ are invertible for all $i \neq j$, then all functions f_i are polynomials of degree at most m.

Proof. Without loss of generality we may assume that $A_i = I_d$ (the identity matrix) for all i. We prove the assertions by induction on m. The case $m = 1$ follows from Lemma C.9.3. Let $m > 1$ and assume that the assertions are true for $1, \dots, m-1$. For t, s and x near to the origin we have

$$\sum_{i=1}^m f_i(t + (B_m - B_i)x + B_i s) = \sum_{i=1}^m f_i(t + B_m x + B_i(s-x)) = g(t + B_m x) + h(s-x).$$

Subtracting from this equation (1) with $A_i = I_d$ we obtain

$$\sum_{i=1}^{m-1} [f_i(t + (B_m - B_i)x + B_i s) - f_i(t + B_i s)] = [g(t + B_m x) - g(t)] + [h(s-x) - h(s)].$$

[17] The first part remains valid if we replace $(-\delta, \delta)$ by the possibly nonsymmetric interval I.

We consider the functions

$$F_i(t) = f_i(t + (B_m - B_i)x) - f_i(t)$$

and

$$G(t) = g(t + B_m x) - g(t), \quad H(s) = h(s - x) - h(s).$$

These functions satisfy the equation

$$\sum_{i=1}^{m-1} F_i(t + B_i s) = G(t) + H(s). \tag{2}$$

The induction assumption (concerning g and h) and Lemma C.9.4 show that g and h are polynomials of degree at most m. If $B_m - B_i$ is nonsingular, then the induction assumption (concerning f_i) and Lemma C.9.4 show that f_i is a polynomial of degree at most m. □

Proposition C.9.6. *If $f : \mathbb{R}^d \to \mathbb{C}$ is a polynomial of degree $k > 0$, then for all $s \in \mathbb{R}^d$ the function*

$$t \mapsto f(t + s) - f(t)$$

is a polynomial of degree at most $k - 1$. More generally, if the measure

$$\mu = \sum_{i=1}^{n} c_i \delta_{x_i}, \quad c \in \mathbb{R}^n, \ x \in \mathbb{R}^n$$

is such that $\hat{\mu}(0) = \sum_{i=1}^{n} c_i = 0$, then $\mu * f$ is a polynomial of degree at most $k - 1$.

Proof. It is easy to check that the statement is true for polynomials of the form

$$f(t) = \prod_{j=1}^{d} t_j^{\alpha_j}, \quad t \in \mathbb{R}^d, \ \alpha \in \mathbb{N}_0^d$$

from which the general statement follows. □

C.10 Linear independence of exponential functions

C.10.1. Let P be a trigonometric polynomial of the form

$$P(t) = \sum_{j=1}^{n} c_j e^{\mathrm{i}(t, x_j)}, \quad t \in \mathbb{R}^d$$

where $n \in \mathbb{N}$, $c \in \mathbb{C}^n$ and the x_j's are pairwise distinct. Then P can be regarded as the inverse Fourier transform of the measure

$$\mu = \sum_{j=1}^{n} c_j \delta_{x_j}.$$

Since μ is uniquely determined by its inverse Fourier transform, we conclude that $P = 0$ implies $\mu = 0$. In other words, the functions $t \mapsto \mathrm{e}^{\mathrm{i}(t,x_j)}$, $t \in \mathbb{R}^d$, are linearly independent (over $\mathbb{C}$).

In this section we generalize this statement. We will use the notation

$$\gamma_x(t) := \mathrm{e}^{\mathrm{i}(x,t)}, \quad x, t \in \mathbb{R}^d.$$

Theorem C.10.2. *Suppose that $x_1, \dots, x_n \in \mathbb{R}^d$ are pairwise distinct and let $\emptyset \neq U \subset \mathbb{R}^d$ be open. Then the functions $\gamma_{x_j}|_U$, $1 \leq j \leq n$, are linearly independent.*

Proof. For $c \in \mathbb{C}^n$ write

$$P(t) := \sum_{j=1}^{n} c_j \gamma_{x_j}(t) = 0, \quad t \in \mathbb{R}^d$$

and assume that $P(t) = 0$ for all $t \in U$. For each $t_0 \in U$ the function φ defined by

$$\varphi(x) := P(x t_0), \quad x \in \mathbb{R}$$

is analytic on $\mathbb{R}$. Since φ vanishes on an open interval containing 1, we conclude that $\varphi = 0$. Consequently, $P(t) = 0$ for all $t \in \mathbb{R}^d$. The remarks in the preceding paragraph show that $c = 0$, from which the theorem follows. □

Theorem C.10.3. *Suppose that $x_1, \dots, x_n \in \mathbb{R}^d$, $d \geq 2$, are pairwise distinct and let $S \subset \mathbb{R}^d$ be a sphere of positive radius. Then the functions $\gamma_{x_j}|_S$, $1 \leq j \leq n$, are linearly independent.*[18]

Proof. We may suppose that $S = S^{d-1}$. We have to show that the relation

$$\sum_{j=1}^{n} c_j \gamma_{x_j}(t) = \sum_{j=1}^{n} c_j \mathrm{e}^{\mathrm{i}(t,x_j)} = 0, \quad t \in S^{d-1} \tag{1}$$

where $c_j \in \mathbb{C}$, implies that $c = 0$. Assume, on the contrary, that $c \neq 0$. Removing the zeros from the numbers c_j we suppose that $c_j \neq 0$ for all j. The case $n = 1$ being trivial, let $n \geq 2$. Write

$$R := \max_{1 \leq j \leq n} \|x_j\| > 0$$

[18] The proof of this result is taken from [24].

and let $e_1, \dots, e_d$ be the standard basis of $\mathbb{R}^d$. We may assume that $x_1 = Re_1$. Denoting by P the orthogonal projection onto $\operatorname{span}\{e_1, e_2\}$ we have

$$Px_j = r_j\,[\cos(\varphi_j)\,e_1 + \sin(\varphi_j)\,e_2]$$

where $r_j = \|Px_j\| \le \|x_j\| \le R$ and $0 \le \varphi_j < 2\pi$. Noting that $r_1 = R$ we may assume that for some $n' \in \mathbb{N}$ we have $r_j = R$ if $1 \le j \le n'$ and $r_j < R$ if $j > n'$. Then $\|Px_j\| = \|x_j\|$ and hence $Px_j = x_j$ for all $j \le n'$. Since the x_j's are pairwise distinct we may also assume that

$$0 = \varphi_1 < \varphi_2 < \cdots < \varphi_{n'} < 2\pi. \tag{2}$$

For all $\varphi \in \mathbb{R}$ we have $t := \cos(\varphi)\,e_1 + \sin(\varphi)\,e_2 \in S^{d-1}$. Putting this t into equation (1) and using that $(t, x_j) = (Pt, x_j) = (t, Px_j)$ we obtain

$$\sum_{j=1}^{n} c_j \mathrm{e}^{\mathrm{i}r_j \cos(\varphi-\varphi_j)} = 0, \qquad \varphi \in \mathbb{R}.$$

It follows from Theorem C.1.8 that this equation holds in the whole complex plane:

$$\sum_{j=1}^{n} c_j \mathrm{e}^{\mathrm{i}r_j \cos(z-\varphi_j)} = 0, \qquad z = x + \mathrm{i}y \in \mathbb{C}. \tag{3}$$

Since

$$\cos(z - \varphi_j) = \frac{1}{2}\big(\mathrm{e}^{\mathrm{i}(z-\varphi_j)} + \mathrm{e}^{-\mathrm{i}(z-\varphi_j)}\big)$$

we have

$$\begin{aligned}\big|\mathrm{e}^{\mathrm{i}r_j \cos(z-\varphi_j)}\big| &= \exp\Big(-\frac{r_j}{2}\operatorname{Im}\big(\mathrm{e}^{\mathrm{i}(z-\varphi_j)} + \mathrm{e}^{-\mathrm{i}(z-\varphi_j)}\big)\Big)\\ &= \mathrm{e}^{r_j \sinh(y)\sin(x-\varphi_j)}.\end{aligned}$$

Let $\psi_j(z) \in [0, 2\pi)$ be the argument of $\mathrm{e}^{\mathrm{i}r_j \cos(z-\varphi_j)}$. Then, by relation (3),

$$\sum_{j=1}^{n} c_j\, \mathrm{e}^{r_j \sinh(y)\sin(x-\varphi_j)}\, \mathrm{e}^{\mathrm{i}\psi_j(z)} = 0.$$

Multiplying this equation by $\mathrm{e}^{-R\sinh(y)}$, we obtain

$$\begin{aligned}&\sum_{j=1}^{n'} c_j\, \mathrm{e}^{R\sinh(y)[\sin(x-\varphi_j)-1]}\, \mathrm{e}^{\mathrm{i}\psi_j(z)}\\ &\qquad + \sum_{j=n'+1}^{n} c_j \mathrm{e}^{\sinh(y)[r_j \sin(x-\varphi_j)-R]}\, \mathrm{e}^{\mathrm{i}\psi_j(z)} = 0.\end{aligned} \tag{4}$$

Set $x := \varphi_1 + \frac{\pi}{2}$. Then $\sin(x - \varphi_j) - 1$ is equal to 0 if $j = 1$ and it is negative if $1 < j \leq n'$, in view of (2). Since $r_j < R$ if $j > n'$ we conclude that $r_j \sin(x - \varphi_j) - R$ is negative for all such j. Taking the limit $y \to \infty$ in (4), we obtain $c_1 = 0$. This contradiction shows that the relation (1) implies that $c = 0$. The theorem is proved. □

Appendix D

Functional analysis

D.1 Inner product spaces

Let V and W be linear spaces over $\mathbb{K}$ where $\mathbb{K} = \mathbb{R}$ or $\mathbb{K} = \mathbb{C}$. The *dimension* of V is denoted by $\dim(V)$.

A *linear operator* from V into W is a mapping $A : V \to W$ satisfying

$$A(av + bw) = aA(v) + bA(w), \quad a, b \in \mathbb{K}, \ v, w \in V.$$

If $W = V$ then we say that A is a linear operator in V. If $W = \mathbb{K}$ then A is called a *linear functional* on V. A mapping $(\cdot,\cdot)$ from $V \times V$ into $\mathbb{K}$ satisfying

(i) $(c_1 v_1 + c_2 v_2, w) = c_1 (v_1, w) + c_2 (v_2, w)$

(ii) $(w, c_1 v_1 + c_2 v_2) = \overline{c_1}(w, v_1) + \overline{c_2}(w, v_2)$

for all $v, v_1, v_2, w \in V$ and $c_1, c_2 \in \mathbb{K}$, is called a *sesquilinear form* on V.

A *positive semidefinite inner product* is a sesquilinear form such that $(v, w) = \overline{(w, v)}$ and $(v, v) \geq 0$. If $(v, v) > 0$ whenever $v \neq 0$, then the inner product $(\cdot,\cdot)$ is called *positive definite*. The linear space V together with the inner product $(\cdot,\cdot)$ is called a (positive definite or positive semidefinite, respectively) *inner product space.*

Two elements v, w of an inner product space V are called *orthogonal* if $(v, w) = 0$. Orthogonality of two sets is defined correspondingly. Orthogonality of elements and sets is denoted by the symbol $\perp$. The *orthogonal complement* $M^\perp$ of a set $M \subset V$ is the set of all elements of V which are orthogonal to M : $M^\perp = \{v \in V : v \perp M\}$. The orthogonal complement is a linear space. If $L, M \subset V$ are linear spaces, then the notation

$$V = L \oplus M$$

means that $V = L + M$ and $V \perp M$.

If V is finite dimensional then any linear functional l on V can be written as

$$l(v) = (v, w), \quad v \in V$$

with some $w \in V$.[1] In case of a positive definite inner product, w is uniquely determined. Any positive semidefinite inner product $\langle \cdot,\cdot \rangle$ on the linear space $\mathbb{K}^d$ can be

[1] See also Theorem D.3.4.

written as $\langle v, w\rangle = (Av, w)$ where $A \in \mathbb{K}^{d\times d}$ is positive semidefinite and

$$(v, w) = \sum_{j=1}^{d} v_j \overline{w_j}$$

is the standard inner product of $\mathbb{K}^d$.

Writing

$$\|v\| = \sqrt{(v, v)}, \quad v \in V$$

for all $v, w \in V$ and $a \in \mathbb{K}$ we have

(i) $\|v\| \geq 0$

(ii) $\|av\| = |a| \cdot \|v\|$

(iii) $|(v, w)| \leq \|v\| \cdot \|w\|$ *(Cauchy–Schwarz inequality)*

(iv) $\|v + w\| \leq \|v\| + \|w\|$ *(triangle inequality)*

(v) $\|v + w\|^2 + \|v - w\|^2 = 2\|v\|^2 + 2\|w\|^2$ *(parallelogram law).*

Any mapping $\|\cdot\| \mapsto [0, \infty)$ satisfying (ii) and (iv) is called a *seminorm*. If $\|v\| > 0$ for all $v \in V\backslash\{0\}$ then $\|\cdot\|$ is called a *norm*. A linear space equipped with a norm is called a *normed linear space*. Defining the distance $\rho(v, w)$ of v and w by $\rho(v, w) := \|v - w\|$, the linear space V becomes a metric space. If this metric space is complete, i.e., if every Cauchy sequence is convergent, then V is called a *Banach space*.

The inner product $(\cdot, \cdot)$ of a semidefinite inner product space can be expressed in terms of the corresponding seminorm:

$$(v, w) = \frac{1}{4}\left(\|v + w\|^2 - \|v - w\|^2\right), \quad \text{if } \mathbb{K} = \mathbb{R} \tag{1}$$

$$(v, w) = \frac{1}{4}\sum_{k=0}^{3} \mathrm{i}^k \|v + \mathrm{i}^k w\|^2, \quad \text{if } \mathbb{K} = \mathbb{C}. \tag{2}$$

The two equations above are called *polarization identity*. A result, attributed to M. R. Fréchet, J. von Neumann and P. Jordan, states that if a seminorm (norm) $\|\cdot\|$ satisfies the parallelogram law, then the equations above define a semidefinite (positive definite, respectively) inner product $(\cdot, \cdot)$.

D.2 Matrices and kernels

Basic facts and notation D.2.1. The *identity matrix* is $I = I_n = (\delta_{ij}) \in \mathbb{R}^{n\times n}$, where $\delta_{ii} = 1$ and $\delta_{ij} = 0$ if $i \neq j$. A *permutation matrix* is obtained by reordering the columns of the identity matrix. The symbol diag $(c_1, \ldots, c_n)$ where $c_j \in \mathbb{C}$ denotes the *diagonal matrix* (d_{ij}) with $d_{ii} = c_i$ and $d_{ij} = 0,\ i \neq j$. More generally, if

$D_1, \ldots, D_n$ are matrices, then we write

$$\operatorname{diag}(D_1, \ldots, D_n) = \begin{bmatrix} D_1 & & \\ & \ddots & \\ & & D_n \end{bmatrix}$$

where we use the convention that blank entries in a matrix denote zeros.

For a matrix $A = (a_{ij}) \in \mathbb{C}^{m \times n}$, A^T denotes the *transpose* $(a_{ji}) \in \mathbb{C}^{n \times m}$ and A^* denotes the *conjugate transpose* $(\overline{a}_{ji})$. The *rank* of A is defined by $\operatorname{rank}(A) = \dim(\operatorname{range}(A))$, where[2]

$$\operatorname{range}(A) = A\mathbb{C}^n = \{Ax : x \in \mathbb{C}^n\} \subset \mathbb{C}^m.$$

The rank of A is equal to the maximum number of linearly independent column vectors of A, which is the same as the maximum number of linearly independent row vectors. If the product AB is defined then

$$\operatorname{rank}(AB) \leq \min(\operatorname{rank}(A), \operatorname{rank}(B)).$$

Let $A \in \mathbb{C}^{n \times n}$ be a quadratic matrix for the rest of this paragraph. A is *diagonalizable* if there exists an invertible matrix $X \in \mathbb{C}^{n \times n}$ such that XAX^{-1} is diagonal. A is *symmetric* if $A = A^T$ and A is *Hermitian* if $A = A^*$. A is called *unitary* if $A^* = A^{-1}$. If A is real and $A^T = A^{-1}$, then A is called *orthogonal*. We denote by $\mathcal{O}(n)$ the set of all $n \times n$ orthogonal matrices. A is said to be *normal* if $A^*A = AA^*$. A matrix is normal if and only if it is diagonalizable by a unitary matrix.

A is *positive semidefinite* if $x^*Ax \geq 0$ for all $x \in \mathbb{C}^n$ and *positive definite* if $x^*Ax > 0$ for all nonzero $x \in \mathbb{C}^n$. We show below (cf. Theorem D.2.3) that positive semidefinite matrices are Hermitian.

The *determinant* of A is denoted by $\det(A)$. Key properties are

$$\det(AB) = \det(A) \cdot \det(B)$$
$$\det(cA) = c^n \det(A), \quad c \in \mathbb{C}.$$

The complex number λ is an *eigenvalue* of A with corresponding *eigenvector* $x \in \mathbb{C}^n \setminus \{0\}$ if $Ax = \lambda x$. The eigenvalues are the zeros of the *characteristic polynomial*

$$q(t) = \det(tI - A), \quad t \in \mathbb{C}$$

which has degree n. The *eigenspace* of A corresponding to an eigenvalue λ is the linear space $\{x \in \mathbb{C}^n : Ax = \lambda x\}$. The *algebraic multiplicity* of λ is its multiplicity as a zero of the characteristic polynomial. The *geometric multiplicity* of λ is the dimension of the eigenspace of A corresponding to λ. If A is Hermitian, then the eigenvalues are real; if A is positive semidefinite, then they are nonnegative.

[2] Recall our convention from A.2 that in expressions involving matrix operations, e.g., as in Ax, we consider $x \in \mathbb{C}^n$ as a column vector.

The *trace* of A is defined by

$$\operatorname{trace}(A) = \sum_{i=1}^{n} a_{ii}.$$

It is easy to check that $\operatorname{trace}(AB) = \operatorname{trace}(BA)$ and hence $\operatorname{trace}(X^{-1}AX) = \operatorname{trace}(A)$ for an arbitrary invertible matrix $X \in \mathbb{C}^{n \times n}$. The trace of A is the sum of all eigenvalues of A while the determinant is the product of all eigenvalues of A.

A *Vandermonde matrix* V has the form

$$V = V(c_1, \dots, c_n) = \begin{bmatrix} 1 & c_1 & c_1^2 & \dots & c_1^{n-1} \\ 1 & c_2 & c_2^2 & \dots & c_2^{n-1} \\ \vdots & \vdots & \vdots & \dots & \vdots \\ 1 & c_n & c_n^2 & \dots & c_n^{n-1} \end{bmatrix}$$

its determinant is equal to

$$\prod_{1 \le i < j \le n} (c_j - c_i)$$

and hence V is invertible if and only if the c_j's are mutually distinct.

Let f be a complex-valued function defined on some set $S \subset \mathbb{C}$ such that S contains all eigenvalues of A. Assume that A is diagonalizable, then A can be expressed in the form $A = X^{-1}DX$ with some invertible matrix X and a diagonal matrix D. The diagonal entries $\lambda_1, \dots, \lambda_n$ of D are the eigenvalues of A. The matrix $f(A)$ is defined by

$$f(A) = f(X^{-1}DX) = X^{-1}\operatorname{diag}(f(\lambda_1), \dots, f(\lambda_n))X.$$

The diagonal matrix D is unique up to the ordering of the eigenvalues, but the matrix X is not unique. However, $f(A)$ can be shown to be independent of the particular representation of A.[3]

Kernels D.2.2. Let X be a nonempty set. A *kernel* on $X \times X$ is a complex-valued function on $X \times X$. A kernel K is called *Hermitian* if $K(y, x) = \overline{K(x, y)}$ for all $x, y \in X$ and it is called *positive definite* if the matrix $(K(x_j, x_k))_{j,k=1}^{n}$ is positive definite for every choice of $n \in \mathbb{N}$, $x_1, \dots, x_n \in X$ such that $x_j \neq x_k$ for $j \neq k$. The kernel K is *positive semidefinite* if the same matrices are always positive semidefinite. In this case, they are positive semidefinite even without the assumption that $x_j \neq x_k$ for $j \neq k$. Indeed, if $x_1, \dots, x_n \in X$ are arbitrary and $x_{l_1}, \dots, x_{l_p}$ are

[3] There are several equivalent ways to define $f(A)$ for arbitrary quadratic matrices; a detailed treatment is given in [29].

the mutually different elements among the x_j's, then

$$\sum_{j,k=1}^{n} c_j \overline{c_k} K(x_j, x_k) = \sum_{j,k=1}^{p} d_j \overline{d_k} K(x_{l_j}, x_{l_k})$$

where

$$d_k = \sum_{i:\, x_i = x_{l_k}} c_i, \quad k = 1, \ldots, p.$$

Next we give proofs of some results on matrices which are used in the book.

Theorem D.2.3. *Every positive semidefinite matrix $A = (a_{jk})$ is Hermitian and $a_{jj} \geq 0$ for all j.*

Proof. For $u, v \in \mathbb{C}$ we have

$$0 \leq |u|^2 a_{jj} + |v|^2 a_{kk} + u\bar{v} a_{jk} + v\bar{u} a_{kj}.$$

Taking (u, v) equal to $(1, 0)$ or $(0, 1)$, we see that a_{jj} and a_{kk} are nonnegative and, in particular, real. Thus,

$$u\bar{v} a_{jk} + v\bar{u} a_{kj} \in \mathbb{R}, \quad u, v \in \mathbb{C}.$$

Taking $(u, v) = (1, 1)$ we get $a_{jk} + a_{kj} \in \mathbb{R}$. Taking $(u, v) = (1, \mathrm{i})$ we get that $\mathrm{i}(a_{jk} - a_{kj}) \in \mathbb{R}$. Writing $a = a_{jk} + a_{kj} \in \mathbb{R}$ and $b = -\mathrm{i}(a_{jk} - a_{kj}) \in \mathbb{R}$, we infer $a_{jk} = \frac{a + \mathrm{i}b}{2}$ and $a_{kj} = \frac{a - \mathrm{i}b}{2}$, from which the desired equality is immediate. □

We remark that the matrix $(a_{jk}) = \begin{bmatrix} 1 & 1 \\ -1 & 1 \end{bmatrix}$ is not Hermitian but we have

$$\sum_{j,k=1}^{2} a_{jk} r_j r_k = r_1^2 + r_2^2 > 0$$

for arbitrary real numbers r_1 and r_2, not both zero.

Theorem D.2.4. *A real matrix $A = (a_{jk})_{j,k=1}^{n}$ is positive semidefinite if and only if it is symmetric and*

$$\sum_{j,k=1}^{n} a_{jk} r_j r_k \geq 0$$

for all real numbers $r_1, \ldots, r_n$.

Proof. The "only if" part follows immediately from Theorem D.2.3. If A is real and symmetric, then

$$\sum_{j,k=1}^{n} a_{jk} c_j \overline{c_k} = \sum_{j,k=1}^{n} a_{jk} \mathrm{Re}(c_j \overline{c_k}) \in \mathbb{R}$$

for all $c_j \in \mathbb{C}$. Using this and writing $c_j = x_j + \mathrm{i} y_j,\ x_j, y_j \in \mathbb{R}$ we have

$$\begin{aligned}\sum_{j,k=1}^{n} a_{jk} c_j \overline{c_k} &= \sum_{j,k=1}^{n} a_{jk}(x_j x_k + y_j y_k) + \mathrm{i} \sum_{j,k=1}^{n} a_{jk}(y_j x_k - x_j y_k) \\ &= \sum_{j,k=1}^{n} a_{jk} x_j x_k + \sum_{j,k=1}^{n} a_{jk} y_j y_k\end{aligned}$$

from which the "if part" follows. □

Definition D.2.5. Suppose $A = (a_{jk})_{j,k=1}^{n}$ is a Hermitian matrix. Then A has n real eigenvalues, counted with multiplicity, and there exists an orthonormal basis of $\mathbb{C}^n$ consisting of eigenvectors of A. The *number of negative squares* of A is the number of negative eigenvalues of A, counted with multiplicity. A *negative* (respectively *nonnegative,* respectively *positive*) *subspace* for A is a linear subspace E of $\mathbb{C}^n$ such that

$$\sum_{j,k=1}^{n} a_{jk} x_j \overline{x_k} = (Ax, x) < 0$$

(respectively ≥ 0, respectively > 0) for all $x \in E \setminus \{0\}$.

Note that, by orthogonality, the linear space spanned by eigenvectors corresponding to negative (nonnegative, positive, nonpositive) eigenvalues is negative (nonnegative, positive, nonpositive, respectively).

Theorem D.2.6. *If A is a Hermitian matrix, then the number of negative squares of A is equal to the maximal dimension of a negative subspace for A.*

Proof. Let the order of A be n and let k be the number of negative squares of A. In view of the remark after Definition D.2.5, the matrix A has a negative subspace of dimension k and A has a nonnegative subspace P of dimension $n - k$. Now let N be an arbitrary negative subspace for A. Since N cannot intersect P except in 0 we must have $\dim(N) \leq k$. □

Theorem D.2.7. *If $A \in \mathbb{C}^{n \times n}$ is an invertible Hermitian, matrix then the number of negative squares of A is equal to the number of changes of sign in the sequence*

$$(1, \det A_1, \ldots, \det A_n)$$

where $A_k = (a_{ij})_{i,j=1}^{k}$.

Proof. For $1 \leq k \leq n$ we identify $\mathbb{C}^k$ with the linear subspace

$$\{ (x_1, \ldots, x_k, 0, \ldots, 0) : (x_1, \ldots, x_k) \in \mathbb{C}^k \}$$

of $\mathbb{C}^n$. Further, let m_k be the number of negative squares of A_k. It follows from Theorem D.2.6 that $m_{k-1} \leq m_k$ for $k = 2, \ldots, n$. If $k \geq 2$ and if E is a negative subspace for A_k of dimension m_k, then $E \cap \mathbb{C}^{k-1}$ is a negative subspace for A_{k-1} of co-dimension at most 1 in E, hence of dimension at least $m_k - 1$. Thus, $m_{k-1} \leq m_k \leq m_{k-1} + 1$ for $k = 2, \ldots, n$. The determinant of a Hermitian matrix is equal to the product of all eigenvalues (considered with multiplicities). Therefore, the sign of $\det A_k$ is $(-1)^{m_k}$, from which the statement follows. □

Noting that positive definite matrices are nonsingular, we obtain the following corollary.

Corollary D.2.8. *An invertible Hermitian matrix $A \in \mathbb{C}^{n \times n}$ is positive definite if and only if* $\det A_k > 0$ *for* $k = 1, \ldots, n$.

As the example

$$A = \begin{bmatrix} 0 & 0 \\ 0 & -1 \end{bmatrix}$$

shows, there exist matrices with $\det A_k \geq 0$ for all k that are not positive semidefinite. However, the following result holds.

Theorem D.2.9. *A Hermitian matrix $A \in \mathbb{C}^{n \times n}$ is positive semidefinite if and only if all subdeterminants of A are nonnegative.*

Proof. If $A = (a_{jk})$ is positive semidefinite, then the eigenvalues and hence the determinant of A are nonnegative.

On the other hand, let us assume that all subdeterminants are nonnegative. For $\epsilon > 0$ we define the matrix $A_\epsilon = A + \epsilon \cdot I_n$. Then

$$\det(A_\epsilon) = \sum_{l=0}^{n} d_l \epsilon^l$$

where $d_n = 1$ and

$$d_l = \sum_{S_l} \det((a_{jk})_{j,k \in S_l}) \geq 0, \quad l = 0, 1, \ldots, n-1.$$

Here the sum is over all sets $S_l \subset \{1, \ldots, n\}$ having $n - l$ elements. From this we conclude that $\det(A_\epsilon) \geq \epsilon^n > 0$. Corollary D.2.8 shows that A_ϵ is positive definite. Consequently, the pointwise limit $A = \lim_{\epsilon \to 0} A_\epsilon$ is positive semidefinite. □

As an application of Theorem D.2.9 we present the following result:

Lemma D.2.10. *The kernels*

$$K_1(x, y) = \min(x, y), \quad x, y \in [0, \infty)$$

and

$$K_2(x,y) = \frac{1}{2}\cdot(|x|+|y|-|x-y|), \quad x,y\in\mathbb{R}$$

are positive semidefinite.[4]

Proof. We consider first the kernel K_1 and show that the matrix

$$A = \big(\min(x_i,x_j)\big)_{i,j=1}^n$$

is positive semidefinite for all $x_1,\dots,x_n \in [0,\infty)$. We may assume that $x_1 \le x_2 \le \dots \le x_n$. We then have

$$D_n(x_1,\dots,x_n) := \det(A) = \begin{vmatrix} x_1 & x_1 & x_1 & \dots & x_1 \\ x_1 & x_2 & x_2 & \dots & x_2 \\ x_1 & x_2 & x_3 & \dots & x_3 \\ \vdots & \vdots & \vdots & & \vdots \\ x_1 & x_2 & x_3 & \dots & x_n \end{vmatrix}.$$

Subtracting the first row from all other rows and expanding the determinant along the first column we see that

$$D_n(x_1,x_2,\dots,x_n) = x_1\cdot D_{n-1}(x_2-x_1,\dots,x_n-x_1).$$

Induction on n leads to

$$D_n(x_1,x_2,\dots,x_n) = x_1(x_2-x_1)(x_3-x_2)\cdots(x_n-x_{n-1})$$

from which $\det A \ge 0$ follows. Since all subdeterminants of A have the same structure, they are also nonnegative. Theorem D.2.9 shows that A is positive semidefinite.

We now consider the kernel K_2. If $xy \le 0$ then $K_2(x,y) = 0$, otherwise $K_2(x,y) = K_1(|x|,|y|)$. Thus, for $x_1 \le \dots \le x_n$,

$$\big(K_2(x_i - x_j)\big)_{i,j=1}^n = \operatorname{diag}(A_1, A_2)$$

where A_1 and A_2 are, by the first part of the lemma, positive semidefinite matrices. This completes the proof. □

Alternative proof. We present an alternative proof for the positive semidefiniteness of the matrix $A = (\min(x_i,x_j))_{i,j=1}^n$, where $0 \le x_1 \le x_2 \le \dots \le x_n$. By continuity, it suffices to consider positive rational x_j's. Multiplying A by a positive integer we may even assume that the x_j's are positive integers. Then for some m the matrix A is a submatrix of $(\min(i,j))_{i,j=1}^m$ which is positive definite since it is the square of the symmetric matrix $B = (b_{ij})$ where $b_{ij} = 0$ if $i+j \le m$ and $b_{ij} = 1$ if $i+j > m$. For example,

$$\begin{bmatrix} 1 & 1 & 1 & 1 \\ 1 & 2 & 2 & 2 \\ 1 & 2 & 3 & 3 \\ 1 & 2 & 3 & 4 \end{bmatrix} = \begin{bmatrix} 0 & 0 & 0 & 1 \\ 0 & 0 & 1 & 1 \\ 0 & 1 & 1 & 1 \\ 1 & 1 & 1 & 1 \end{bmatrix}^2.$$

□

[4] Note that $K_1(x,y) = K_2(x,y)$ if $x,y \ge 0$.

Theorem D.2.11. *A matrix $A \in \mathbb{C}^{n\times n}$ is positive semidefinite if and only if there exists a Hermitian matrix $B \in \mathbb{C}^{n\times n}$ such that $A = B^2$. If A is real then B can be chosen real as well.*

Proof. If $A = B^2$ for some $n \times n$ Hermitian matrix B, then

$$(Ax, x) = (B^2x, x) = (Bx, Bx) \geq 0, \quad x \in \mathbb{C}^n.$$

Thus, A is positive semidefinite.

For the converse, let $\{e_1, \dots, e_n\}$ be an orthonormal basis of $\mathbb{C}^n$ (or of $\mathbb{R}^n$ if A is real), where e_j is an eigenvector of A with corresponding eigenvalue $r_j \geq 0$. Consider e_j as a column vector and let $Q = [e_1, \dots, e_n]$. Using orthonormality we see that Q is unitary, i.e., Q^*Q is the identity matrix. Moreover,

$$Q \operatorname{diag}(r_1, \dots, r_n)\, Q^* e_j = r_j e_j = A e_j$$

and hence $A = Q \operatorname{diag}(r_1, \dots, r_n)\, Q^*$. It is now easy to check that the square of the Hermitian matrix

$$B = Q \operatorname{diag}(\sqrt{r_1}, \dots, \sqrt{r_n})\, Q^*$$

is equal to A. □

Theorem D.2.12 (Schur). *Let $A = (a_{ij})$ and $B = (b_{ij})$ be $n\times n$ positive semidefinite matrices. Then the matrix $C = (a_{ij}b_{ij})_{i,j=1}^n$ is positive semidefinite. If $A \neq 0$ and B is positive definite, then C is positive definite, as well.*

Proof. Suppose first that A and B are positive semidefinite. We assume that $A \neq 0$. By Theorem D.2.11 there exists an $n \times n$ Hermitian matrix D such that $A = D^2$. Writing $D = (d_{ij})$, we have for $x = (x_i) \in \mathbb{C}^n$

$$\begin{aligned}
(Cx, x) &= \sum_{i,j=1}^{n} a_{ij} b_{ij} x_j \overline{x_i} \\
&= \sum_{i,j=1}^{n} \sum_{l=1}^{n} \overline{d_{li}} d_{lj} b_{ij} x_j \overline{x_i} \\
&= \sum_{l=1}^{n} \sum_{i,j=1}^{n} b_{ij} y_{lj} \overline{y_{li}} \\
&= \sum_{l=1}^{n} (By_l, y_l) \geq 0
\end{aligned}$$

by the positive semidefiniteness of B, where $y_l = (y_{l1}, \dots, y_{ln}) \in \mathbb{C}^n$ with components $y_{li} = d_{li} x_i$, $i, l = 1, \dots, n$.

The second statement can be proved in the same way. □

Corollary D.2.13. *Let $A = (a_{ij})$ be an $n \times n$ positive semidefinite matrix and let p_k, $k \in \mathbb{N}_0$, be nonnegative numbers such that the series*

$$\varphi(z) := \sum_{k=0}^{\infty} p_k z^k$$

is convergent for all $z \in \mathbb{C}$. Then the matrix $(\varphi(a_{ij}))$ is positive semidefinite.

Lemma D.2.14. *Let $C = (c_{jk})_{j,k=1}^{n}$ be a positive semidefinite matrix and $A = (a_{jk}) = \operatorname{Re} C$, $B = (b_{jk}) = \operatorname{Im} C$. Then the $2n \times 2n$ matrix*

$$D = (d_{jk}) = \begin{bmatrix} A & -B \\ B & A \end{bmatrix}$$

is positive semidefinite as well.

Proof. Since $c_{jk} = \overline{c_{kj}}$ we conclude that $A^T = A$, $B^T = -B$ and hence $D = D^T$. In view of Theorem D.2.4 it suffices to show that

$$\sum_{j,k=1}^{2n} d_{jk} r_j r_k \geq 0 \tag{1}$$

for all $r_1, \ldots, r_{2n} \in \mathbb{R}$. By assumption,

$$\begin{aligned} 0 &\leq \sum_{j,k=1}^{n} c_{jk}(r_j - \mathrm{i} r_{n+j})(r_k + \mathrm{i} r_{n+k}) \\ &= \sum_{j,k=1}^{n} c_{jk}(r_j r_k + r_{n+j} r_{n+k}) - \mathrm{i} \sum_{j,k=1}^{n} c_{jk}(r_{n+j} r_k - r_j r_{n+k}) =: S_1 - \mathrm{i} S_2. \end{aligned}$$

Using the relation $c_{jk} = \overline{c_{kj}}$ we see that S_1 is real, while S_2 is purely imaginary. Consequently,

$$\begin{aligned} S_1 - \mathrm{i} S_2 &= \sum_{j,k=1}^{n} a_{jk}(r_j r_k + r_{n+j} r_{n+k}) + \sum_{j,k=1}^{n} b_{jk}(r_{n+j} r_k - r_j r_{n+k}) \\ &= \sum_{j,k=1}^{2n} d_{jk} r_j r_k \end{aligned}$$

showing the inequality (1). □

Definition D.2.15. Let V be a positive definite inner product space over $\mathbb{K}$ where $\mathbb{K} = \mathbb{C}$ or $\mathbb{K} = \mathbb{R}$ and let $v_1, \dots, v_n \in V$. The *Gram matrix* G associated with $v_1, \dots, v_n$ is defined by

$$G = G(v_1, \dots, v_n) = \Big((v_i, v_j)\Big)_{i,j=1}^{n}$$

where $(\cdot, \cdot)$ denotes the inner product of V.

Choosing an orthonormal basis in the linear span of the vectors v_j and building a matrix A by writing the coordinates of v_j as the j-th row of A we have $G = AA^*$.

Theorem D.2.16. *(With the notation of the previous definition.)*

(i) *The Gram matrix $G(v_1, \dots, v_n)$ is positive semidefinite.*

(ii) *If $w_1, \dots, w_m \in V$ are any vectors for which*

$$v_i = \sum_{j=1}^{m} t_{ij} w_j, \quad t_{ij} \in \mathbb{K},\ i = 1, \dots, n$$

then

$$G(v_1, \dots, v_n) = TG(w_1, \dots, w_m)T^* \tag{1}$$

where $T = (t_{ij})$.

(iii) *The dimension of the linear space L spanned by the vectors $v_1, \dots, v_n$ is equal to the rank of $G(v_1, \dots, v_n)$.*

Proof. The first statement follows from[5]

$$\sum_{i,j=1}^{n} (v_i, v_j) c_i \overline{c_j} = \left(\sum_{i=1}^{n} v_i c_i, \sum_{i=1}^{n} v_i c_i \right) \geq 0, \quad c_i \in \mathbb{K}$$

while the second one is a consequence of

$$(v_i, v_k) = \sum_{j,l=1}^{m} t_{ij} (w_j, w_l) \overline{t_{kl}}.$$

To prove the last statement choose arbitrary vectors $w_1, \dots, w_n$ which span the same linear space L. Since the rank of the product of matrices is not greater than that of any factor, equation (1) shows that

$$\operatorname{rank} G(v_1, \dots, v_n) \leq \operatorname{rank} G(w_1, \dots, w_n).$$

Exchanging the role of the v's and w's we conclude that we have equality in the above display. The result now follows by choosing the w's such that $w_1, \dots, w_k$ form an orthonormal basis of L and $w_j = 0$ for $j > k = \dim L$. □

[5] The positive semidefiniteness also follows from the representation $G = AA^*$ after Definition D.2.15.

Lemma D.2.17. *Let $C_n \in \mathbb{C}^{d\times d}$, $n \in \mathbb{N}$, be positive semidefinite matrices such that the limit*

$$q(t) = \lim_{n\to\infty} (C_n t, t)$$

exists for all $t \in \mathbb{R}^d$. Then there exists a positive semidefinite matrix $C \in \mathbb{C}^{d\times d}$ such that $q(t) = (Ct, t)$, $t \in \mathbb{R}^d$. Moreover, the sequence $\{C_n\}_1^\infty$ converges to C entrywise.

Proof. It is clear that the above limit also exists for all $t \in \mathbb{C}^d$, we denote it in this case also with $q(t)$. Since C_n is positive semidefinite, $\|t\|_n = \sqrt{(C_n t, t)}$ is a seminorm on $\mathbb{C}^d$ satisfying the parallelogram law (cf. (D.1.v)). The same is true for the limit $\sqrt{q}$. By the theorem of Fréchet, von Neumann and Jordan, q is generated by a positive semidefinite inner product $\langle\cdot,\cdot\rangle$ on $\mathbb{C}^d$:

$$q(t) = \langle t, t\rangle, \quad t \in \mathbb{C}^d.$$

The first statement of the lemma follows from the fact that $\langle t, t\rangle = (Ct, t)$ with some positive semidefinite matrix $C \in \mathbb{C}^{d\times d}$. Let $e_1, \dots, e_d$ be the standard basis of $\mathbb{C}^d$. Using the polarization identity (D.1.2) with $v = e_j$ and $w = e_k$, we obtain the second statement. □

Lemma D.2.18. *If A is a positive semidefinite matrix. then $A^\epsilon := A + \epsilon I$ is positive definite for all $\epsilon > 0$.*

Proof. The matrix A^ϵ is obviously positive semidefinite. If it was not positive definite, then it would have an eigenvector with eigenvalue 0. This vector would be an eigenvector of A with eigenvalue $-\epsilon$, contradicting the positive semidefiniteness of A. □

Theorem D.2.19. *Let $n \in \mathbb{N}$ and let S be a subset of the square*

$$S_n := \{(i, j) \in \mathbb{Z}^2 : 1 \le i, j \le n\}$$

such that

(i) *$(i, i) \in S$ whenever $1 \le i \le n$;*

(ii) *if $(l, k) \in S$ then $\{(i, j) : \min(l, k) \le i, j \le \max(l, k)\} \subset S$.*

Suppose that for each $(i, j) \in S$ a complex number c_{ij} is given such that

(iii) *for all $(l, k) \in S$ with $l \le k$ the matrix $(c_{ij})_{i,j=l}^k$ is positive semidefinite.*

Then there exists a positive semidefinite matrix $A = (a_{ij})_{i,j}^n$ such that $a_{ij} = c_{ij}$ for $(i, j) \in S$.

Proof. The case $n = 1$ being trivial suppose that $n > 1$.

We consider first the special case where $S = S_n \setminus \{(1,n),(n,1)\}$. For $z \in \mathbb{C}$ we define the matrix $A(z)$ by

$$A(z) := \begin{bmatrix} c_{11} & c_{12} & \dots & c_{1,n-1} & z \\ c_{21} & c_{22} & \dots & c_{2,n-1} & c_{2n} \\ & & \ddots & & \\ c_{n-1,1} & c_{n-1,2} & \dots & c_{n-1,n-1} & c_{n-1,n} \\ \overline{z} & c_{n2} & \dots & c_{n,n-1} & c_{nn} \end{bmatrix}.$$

Set

$$A_1 := (c_{ij})_{i,j=1}^{n-1}, \quad A_3 := (c_{ij})_{i,j=2}^{n}, \quad A_5 := (c_{ij})_{i,j=2}^{n-1}$$

and let $A_2(z)$ ($A_4(z)$) denote the matrix obtained by canceling the first (last) column and the last (first) row of $A(z)$. Assume that all matrices in (iii) are positive definite. By Corollary D.2.8 it suffices to prove the existence of a complex number z for which the determinant $\det(A(z))$ is positive. We have

$$\det(A(z)) = \frac{\det(A_1)\det(A_3) - \det(A_2(z))\det(A_4(z))}{\det(A_5)}.$$

Since $A(z)$ is Hermitian $\det(A_4(z)) = \overline{\det(A_2(z))}$ and hence the inequality $\det(A(z)) > 0$ holds if and only if

$$|\det(A_2(z))|^2 < \det(A_1)\det(A_3). \tag{1}$$

By assumption, the right-hand side of this inequality is positive. Expanding the determinant $\det(A_2(z))$ along the first row we see that

$$\det(A_2(z)) = \det(A_5)z + c$$

where $\det(A_5) > 0$ and c is some complex number not depending on z. We conclude that there exist infinitely many $z \in \mathbb{C}$ satisfying (1).

If the matrices in (iii) are not all positive definite, then we replace $A(z)$ by the matrices $A^{\epsilon}(z) := A(z) + \epsilon I$ where I denotes the $n \times n$ identity matrix and $\epsilon > 0$. By Lemma D.2.18 the corresponding matrices are positive definite. For $m \in \mathbb{N}$ we choose $z_m \in \mathbb{C}$ such that $A^{1/m}(z_m)$ is positive semidefinite. Then

$$\det \begin{bmatrix} c_{11} + \frac{1}{m} & z_m \\ \overline{z}_m & c_{nn} + \frac{1}{m} \end{bmatrix} \geq 0$$

and hence the sequence $\{z_m\}$ is bounded. Consequently, there exists a subsequence converging to some $z \in \mathbb{C}$. The matrix $A(z)$ is then positive semidefinite.

We now turn to the general case. We may suppose that $S \neq S_n$. Let k be the greatest integer with the property that $(1,1),\dots,(1,k-1) \in S$ and let l be the greatest integer with $(1,k),\dots,(l,k) \notin S$. Using (i) and (ii) we see that $1 \leq l < k \leq n$ and

$$S \cap \{(i,j) : l \leq i,j \leq k\} = \{(i,j) : l \leq i,j \leq k\} \setminus \{(l,k),(k,l)\}.$$

The first part of the proof shows the existence of a complex number z such that the matrix $(c_{ij})_{i,j=l}^{k}$ is positive semidefinite, where $c_{kl} := z$, $c_{lk} := \overline{z}$. Replacing S by $S \cup \{(l,k),(k,l)\}$, this new index set and the corresponding complex numbers c_{ij} also satisfy the conditions of the theorem. Thus, repeating the arguments above we obtain the desired matrix A. □

D.3 Hilbert spaces and linear operators

Hilbert spaces D.3.1. Let V be a normed linear space. We say that a sequence $\{v_n\}_1^\infty$ of elements of V *converges strongly to* or simply *converges to* $v \in V$ if

$$\lim_{n\to\infty} \|v - v_n\| = 0.$$

In case of strong convergence we write $v = \lim_{n\to\infty} v_n$ or $v_n \to v$. When speaking about convergence, continuity, closed or open sets, etc., we always refer to the corresponding notion for the topology generated by the norm.[6]

A *Hilbert space* $\mathcal{H}$ over $\mathbb{K}$ where $\mathbb{K} = \mathbb{R}$ or $\mathbb{K} = \mathbb{C}$ is a positive definite inner product space over $\mathbb{K}$ which is complete with respect to the norm

$$\|h\| = \sqrt{(h,h)},\ h \in \mathcal{H}$$

where $(\cdot,\cdot)$ denotes the inner product of $\mathcal{H}$. Examples of Hilbert spaces are $\mathbb{R}^d$, $\mathbb{C}^d$ and $L^2(\Omega, \mathcal{A}, P)$. Throughout the rest of this section the symbol $\mathcal{H}$ denotes a Hilbert space.

We call a nonempty set $L \subset \mathcal{H}$ a *linear manifold* if for any two vectors $g, h \in L$ and any $a, b \in K$ we have $af + bg \in L$. A linear manifold is called a *subspace* if it is closed. The orthogonal complement $M^\perp$ of a nonempty subset $M \subset \mathcal{H}$ is a subspace of $\mathcal{H}$. If L is a subspace of $\mathcal{H}$, then

$$\mathcal{H} = L \oplus L^\perp.$$

Thus, each $h \in \mathcal{H}$ can be represented in the form

$$h = h_L + h_{L^\perp}, \quad h_L \in L,\ h_{L^\perp} \in L^\perp.$$

The element h_L is called the *orthogonal projection* of h onto L. The orthogonal projection h_L is the (unique) element of L having minimal distance to h:

$$\|h - h_L\| = \inf_{v\in L} \|h - v\|.$$

[6] See also D.5.1 for the definition of weak convergence.

Lemma D.3.2 (Cauchy criterion). *For a sequence $\{h_n\}_1^\infty$ of elements of $\mathcal{H}$ the following conditions are equivalent:*

(i) $\{h_n\}_1^\infty$ *is convergent;*

(ii) $\lim_{n,m\to\infty} \|h_n - h_m\| = 0$;

(iii) *The limit* $\lim_{n,m\to\infty}(h_n, h_m)$ *exists.*

If the limit in (iii) exists, then it is equal to $\| \lim_{n\to\infty} h_n \|^2$.

Linear operators D.3.3. Let V and W be normed linear spaces. A linear operator A from V into W is called *bounded* if there exists a $C \geq 0$ such that $\|Av\| \leq C\|v\|$ for all $v \in V$. We denote by $\mathcal{B}(V, W)$ the set of all bounded linear operators from V to W. A linear operator is bounded if and only if it is continuous. The *norm* $\|A\|$ of a bounded operator is defined by

$$\|A\| := \inf\{C \geq 0 : \ \|Av\| \leq C\|v\| \text{ for all } v \in V\}. \tag{1}$$

If $\|A\| \leq 1$, then A is called a (linear) *contraction*. We have

$$\begin{aligned} \|A\| &= \sup\{\|Ah\| : \ h \in \mathcal{H}, \ \|h\| = 1\} \\ &= \sup\{\|Ah\| : \ h \in \mathcal{H}, \ \|h\| \leq 1\} \end{aligned}$$

and

$$\|Av\| \leq \|A\| \cdot \|v\| \quad \text{and} \quad \|AB\| \leq \|A\| \cdot \|B\|$$

where B is a bounded linear operator with values in V.

If $W = \mathcal{H}$ is a Hilbert space, then

$$\|A\| = \sup\{|(Ag, h)| : \ g \in V, \ h \in \mathcal{H}, \ \|g\| \leq 1, \ \|h\| \leq 1\}.$$

A linear operator A in $\mathcal{H}$ is called *isometric* if $\|Ah\| = \|h\|$ for all $h \in \mathcal{H}$. An isometric operator A is obviously bounded and its norm is equal to 1. If A is an isometric operator in $\mathcal{H}$ and $A(\mathcal{H}) = \mathcal{H}$, then A is called *unitary*.

For an arbitrary bounded linear operator $A : \ \mathcal{H} \to \mathcal{H}$ there exists a uniquely determined bounded linear operator $A^* : \mathcal{H} \to \mathcal{H}$, the *adjoint of* A, such that

$$(Ah, g) = (h, A^*g), \quad h, g \in \mathcal{H}.$$

Key properties are $\|A\| = \|A^*\|$ and $(AB)^* = B^*A^*$. The operator A is called *self-adjoint* if $A^* = A$. It is called *normal* if $AA^* = A^*A$. If U is unitary, then U is invertible and $U^* = U^{-1}$. Unitary and self-adjoint operators are normal. Two eigenvectors of a normal operator are orthogonal whenever they correspond to different eigenvalues. All eigenvalues of a self-adjoint operator are real, while the eigenvalues of a unitary operator have modulus 1.

For a self-adjoint operator A, the inner product (Ah, h) is real for all $h \in \mathcal{H}$. If, in addition, $(Ah, h) \geq 0$ for all $h \in \mathcal{H}$, then A is said to be *nonnegative*. The eigenvalues of nonnegative operators are nonnegative.

A linear manifold $L \subset \mathcal{H}$ is said to be *invariant* under an operator A, or A-*invariant*, if $A(L) \subset L$. If L is A-invariant, then $L^\perp$ is invariant under A^*.

Theorem D.3.4 (Riesz–Fréchet). *For each bounded linear functional l on $\mathcal{H}$ there exists a unique $v \in \mathcal{H}$ such that*

$$l(h) = (h, v), \quad h \in \mathcal{H}$$

and $\|l\| = \|v\|$.

Theorem D.3.5. *Let $\langle \cdot, \cdot \rangle_L$ be a sesquilinear form on a dense linear manifold $L \subset \mathcal{H}$ such that*

$$|\langle g, h \rangle_L| \le C\|g\| \cdot \|h\|, \ h \in L$$

with some $C \in [0, \infty)$. This form can be uniquely extended to a sesquilinear form $\langle \cdot, \cdot \rangle$ on $\mathcal{H}$ such that

$$|\langle g, h \rangle| \le C\|g\| \cdot \|h\|, \ h \in H.$$

Moreover, there exists a linear operator A on $\mathcal{H}$ such that $\|A\| \le C$ and

$$\langle g, h \rangle = (Ag, h), \quad g, h \in \mathcal{H}.$$

Theorem D.3.6. *Let $\mathcal{H}_1$ and $\mathcal{H}_2$ be two Hilbert spaces and let $I : L \to \mathcal{H}_2$ be an isometric linear operator defined on a linear manifold $L \subset \mathcal{H}_1$. Then I can be uniquely extended to an isometric linear operator from the closure of L onto the closure of $I(L)$.*

Theorem D.3.7. *For every bounded nonnegative operator P on $\mathcal{H}$ there exists a unique bounded nonnegative operator $\sqrt{P}$ such that $P = \sqrt{P}\sqrt{P}$. If a bounded linear operator commutes with P, then it commutes with $\sqrt{P}$, as well.*

Theorem D.3.8 (Banach–Steinhaus, principle of uniform boundedness). *Let B be a Banach space and V be a normed linear space. If $\mathcal{A} \subset \mathcal{B}(B, V)$ is such that for each $v \in V$*

$$\sup\{\|Av\| : A \in \mathcal{A}\} < \infty$$

then

$$\sup\{\|A\| : A \in \mathcal{A}\} < \infty.$$

Definition D.3.9. A bounded linear operator K in $\mathcal{H}$ is called *compact* if $\{K(x_n)\}_1^\infty$ contains a convergent subsequence for all bounded sequences $\{x_n\}_1^\infty$ in $\mathcal{H}$.

The next lemma states basic properties of compact operators which follow immediately from the definition of compactness.

Lemma D.3.10.

(i) *If K_1 and K_2 are compact operators, then so is $a_1K_1+a_2K_2$ for all $a_1, a_2 \in \mathbb{K}$.*

(ii) *If K is a compact and A is a bounded linear operator, then AK and KA are compact.*

Theorem D.3.11. *If K is a compact self-adjoint operator in $\mathcal{H}$, then K has an eigenvalue $\lambda \in \mathbb{R}$ such that $|\lambda| = \|K\|$.*

Proof. The case $K = 0$ being trivial, we may assume that $\|K\| \neq 0$. We choose a sequence $\{h_n\}$ in $\mathcal{H}$ such that $\|h_n\| = 1$ and $\|Kh_n\|$ converges to $\|K\|$. Since K is self-adjoint we have $(K^2h_n, h_n) = (Kh_n, Kh_n) = \|Kh_n\|^2$. Using this we obtain

$$\begin{aligned} 0 &\le \|K^2h_n - \|Kh_n\|^2 \cdot h_n\|^2 \\ &= \|K^2h_n\|^2 - 2\|Kh_n\|^2 \cdot (K^2h_n, h_n) + \|Kh_n\|^4 \\ &= \|K^2h_n\|^2 - \|Kh_n\|^4 \\ &\le \|K\|^2 \cdot \|Kh_n\|^2 - \|Kh_n\|^4. \end{aligned}$$

Since the sequence $\{\|Kh_n\|\}$ converges to $\|K\|$ we see that

$$\lim_{n\to\infty} \|K^2h_n - \|Kh_n\|^2 \cdot h_n\| = 0. \tag{1}$$

By Lemma D.3.10, the operator K^2 is compact. Thus, we can choose a subsequence $\{K^2h_{n_k}\}$ converging to an element, which we write in the form $\|K\|^2 \cdot g$. Relation (1) shows that $\{h_{n_k}\}$ converges to g and

$$K^2g = \|K\|^2g.$$

To complete the proof we observe that $\|g\| = 1$ and

$$(K - \|K\|)(K + \|K\|)g = 0.$$

If $h := (K + \|K\|)g = 0$, then $-\|K\|$ is an eigenvalue of K. Otherwise we have $(K - \|K\|)h = 0$, i.e., $\|K\|$ is an eigenvalue of K. □

Theorem D.3.12. *Let $K \neq 0$ be a self-adjoint compact operator in $\mathcal{H}$. For all $h \in \mathcal{H}$ we have*

$$h = \sum_{n\in S} (h, \varphi_n)\varphi_n \tag{1}$$

and

$$Kh = \sum_{n\in S} (h, \varphi_n)\lambda_n\varphi_n \tag{2}$$

where

(i) $S = \{1, \dots, N\}$ *with some* $N \in \mathbb{N}$ *or* $S = \mathbb{N}$;

(ii) $\{\lambda_n : n \in S\} \subset \mathbb{R}$ *is the set of all nonzero eigenvalues of* K;

(iii) $\{\varphi_n : n \in S\}$ *is an orthonormal system in* $\mathcal{H}$ *and* φ_n *is an eigenvector of* K *with eigenvalue* λ_n.

(iv) *If* $S = \mathbb{N}$, *then* $\lim_{n\to\infty} \lambda_n = 0$.

Proof. By Theorem D.3.11 there exists an eigenvalue $\lambda_1 \in \mathbb{R}$ of K with $|\lambda_1| = \|K\|$. Let φ_1 be an eigenvector with eigenvalue λ_1 such that $(\varphi_1, \varphi_1) = 1$ and denote by $\mathcal{H}_1$ the subspace generated by φ_1. We have

$$\mathcal{H} = \mathcal{H}_1 \oplus \mathcal{H}_1^\perp$$

where $\mathcal{H}_1$ is K-invariant. Since K is self-adjoint, $\mathcal{H}_1^\perp$ is K-invariant as well. The restriction of K to $\mathcal{H}_1^\perp$ is a compact self-adjoint operator with norm at most $\|K\|$. If this restriction is not the zero operator, then we apply Theorem D.3.11 to it. Proceeding in this way we obtain a finite or infinite sequence $\lambda_1, \lambda_2, \dots$ of eigenvalues such that $|\lambda_n| \geq |\lambda_{n+1}| > 0$ and a sequence of corresponding subspaces $\mathcal{H}_n$ spanned by eigenvectors of K with eigenvalue λ_n. If this sequence is finite, then

$$\mathcal{H} = \mathcal{H}_1 \oplus \cdots \oplus \mathcal{H}_N \oplus \mathcal{N}$$

where $N \in \mathbb{N}$ and $K(\mathcal{N}) = \{0\}$, from which (1) and (2) follow.

Assume that this sequence is infinite and suppose that the sequence $\{|\lambda_n|\}$ has a positive lower bound. Then the sequence $\{\varphi_n/\lambda_n\}$ is bounded but the sequence $\{K\varphi_n/\lambda_n\} = \{\varphi_n\}$ does not contain a convergent subsequence. For $\|\varphi_n - \varphi_m\|^2 = 2$ if $m \neq n$ so that no subsequence can be a Cauchy sequence. Thus, $\lim_n \lambda_n = 0$. We have

$$\mathcal{H} = \left(\bigoplus_{n=1}^{\infty} \mathcal{H}_n\right) \oplus \mathcal{N}$$

where $\mathcal{N}$ is a K-invariant subspace. By our procedure, $\|K|_{\mathcal{N}}\| \leq |\lambda_n|$ for all n, where $K|_{\mathcal{N}}$ denotes the restriction of K to $\mathcal{N}$. Thus $K(\mathcal{N}) = \{0\}$. The decompositions (1) and (2) follow as in the finite case. □

Remark D.3.13. If the Hilbert space $\mathcal{H}$ in the previous theorem is separable and infinite-dimensional, then there exists an orthonormal basis $\{\varphi_n\}_1^\infty$ of $\mathcal{H}$ consisting of eigenvectors φ_n of K with (possibly zero) eigenvalues λ_n. This follows from the previous proof by selecting an orthonormal basis of the subspace $\mathcal{N}$.

D.4 Convex sets and the theorem of Kreĭn and Milman

A short introduction to the theory of locally convex spaces can be found in Chapter 1 of [3].

D.4.1. Throughout this section the symbol L will denote a linear space over a field $\mathbb{K}$ which is either $\mathbb{R}$ or $\mathbb{C}$. We assume that L is equipped with a Hausdorff topology $\mathcal{O}$ such that the following hold:

(i) The mappings $(x, y) \mapsto x + y$ of $L \times L$ into L and $(r, x) \mapsto rx$ of $\mathbb{K} \times L$ into L are continuous.[7]

(ii) Each neighborhood of $0 \in L$ contains a convex neighborhood of 0.

The linear space L together with such a topology $\mathcal{O}$ is called *locally convex*. Let A and B be subsets of L such that $A \subset B$. Then A is called an *extreme subset* of B if for all $x, y \in B$ and $p \in (0, 1)$ the relation

$$px + (1 - p)y \in A$$

implies that $x, y \in A$. In the special case $A = \{a\}$ we say that a is an *extreme point* of B. The set of all extreme points of B will be denoted by $\operatorname{ex}(B)$.

The next separation result is often useful when dealing with convex sets. A proof can be found, e.g., in [3] (cf. Theorem 2.3 in Chapter 1).

Theorem D.4.2. *Let E be a locally convex space over $\mathbb{K}$. Further, let F and C be disjoint nonempty convex subsets of E such that F is closed and C is compact. Then there exists a continuous linear functional $l : E \to \mathbb{K}$ such that*

$$\sup_{x \in C} \operatorname{Re} l(x) < \inf_{x \in F} \operatorname{Re} l(x).$$

Since a locally convex space over $\mathbb{C}$ can also be considered as a locally convex space over $\mathbb{R}$, we obtain:

Corollary D.4.3. *Under the assumptions of the preceding theorem there exists a continuous $\mathbb{R}$-linear functional $l : E \to \mathbb{R}$ such that*

$$\sup_{x \in C} l(x) < \inf_{x \in F} l(x).$$

Theorem D.4.4 (Hahn–Banach). *Let E be a linear space over $\mathbb{K}$, let p be a seminorm on E and let $L \subset E$ be a linear manifold. For every linear functional l on L such that $|l(x)| \le p(x)$ for all $x \in L$ there exists a linear functional e on E such that $e(x) = l(x)$ for all $x \in L$ and $|e(x)| \le p(x)$ for all $x \in E$.*

---

[7] Here $\mathbb{K}$ is equipped with its usual topology and the product sets are equipped with the product topology.

In the proof of the next theorem we will repeatedly use the following simple fact: If A is an extreme subset of B and B is an extreme subset of C then A is an extreme subset of C.

Theorem D.4.5 (Kreĭn–Milman). *Every compact convex subset K of a locally convex space L is the closed convex hull of its extreme points.*

Proof. First we prove that an arbitrary non-void compact set $C \subset L$ has at least one extreme point. Applying Zorn's lemma we see that C contains a closed extreme subset M which is minimal with respect to inclusion. We claim that M has only one element. This element is then an extreme point of C. Assume, on the contrary, that M has more than one element. Then, by Corollary D.4.3, there exists a continuous real linear functional l on L such that l is not constant on M. It is easy to check that the set $N = \{x \in M : l(x) = \sup l(M)\}$ is a proper, closed and extreme subset of M. This implies that N is a closed extreme subset of C contradicting the minimality of M.

Denote by K_0 the closed convex hull of the extreme points of K. Assume that there exists a $y \in K \setminus K_0$. Corollary D.4.3 with $C = K_0$ and $F = \{y\}$ shows the existence of a continuous real linear functional l on L such that

$$\sup l(K) \geq l(y) > \sup l(K_0). \tag{1}$$

By the first part of the proof, the compact convex set

$$N = \{x \in K : l(x) = \sup l(K)\}$$

which is obviously not empty, contains at least one extreme point x_0. As N is an extreme subset of K, x_0 is an extreme point of K as well and hence $x_0 \in K_0$. Since this contradicts (1), we must have $K = K_0$. □

Corollary D.4.6. *Let K be a compact convex subset of L. For every $x \in K$ there exists a Radon probability measure μ_x on $\overline{\mathrm{ex}\,(K)}$ such that*

$$l(x) = \int l(y)\, \mathrm{d}\mu_x(y)$$

holds for every continuous linear functional l on L.

Proof. By the previous theorem there exists a net $\{x_\alpha\}$ of points of K of the form

$$x_\alpha = \sum_{j=1}^{n_\alpha} p_\alpha^j \cdot x_\alpha^j, \quad n_\alpha < \infty$$

where $x_\alpha^j \in \mathrm{ex}(K)$, $p_\alpha^j \geq 0$, $\sum_j p_\alpha^j = 1$, such that $\lim_\alpha x_\alpha = x$. We define the probability measures μ_α on $\overline{\mathrm{ex}(K)}$ by

$$\mu_\alpha = \sum_{j=1}^{n_\alpha} p_\alpha^j \cdot \delta_{x_\alpha^j}.$$

Since $\overline{\mathrm{ex}(K)}$ is compact, Theorem E.1.13 shows the existence of a subnet $\{\mu_\beta\}$ converging weakly to a probability measure μ_x on $\overline{\mathrm{ex}(K)}$. For an arbitrary continuous linear functional l we then have

$$\int l(y)\,\mathrm{d}\mu_x(y) = \lim_\beta \int l(y)\,\mathrm{d}\mu_\beta(y) = \lim_\beta \sum_j p_\beta^j \cdot l(x_\beta^j) = l(x)$$

completing the proof. □

D.5 Weak topologies

D.5.1. Let V be a normed linear space and denote by V^* the linear space of all bounded linear functionals on V. Equipped with the norm (D.3.3.1), V^* is a Banach space.

The *weak topology* on V is the coarsest topology (the topology with the fewest open sets) such that each element of V^* is a continuous function. The *weak-$*$ topology* on V^* is the coarsest topology such that for all $v \in V$ the maps $l \mapsto l(v)$, $l \in V^*$, remain continuous. Thus, a net $\{v_\alpha\}$ in V *converges weakly* to $v \in V$ if and only if

$$\lim_\alpha l(v_\alpha) = l(v), \quad l \in V^*$$

while a net $\{l_\alpha\}$ in V^* converges in the weak-$*$ topology to $l \in V^*$ if and only if

$$\lim_\alpha l_\alpha(v) = l(v), \quad v \in V.$$

In case of weak and weak-$*$ convergence we write $v_\alpha \xrightarrow{w} v$ and $l_\alpha \xrightarrow{w*} l$, respectively. The limits of weakly and weak-$*$ convergent sequences are unique. Moreover, strong convergence implies weak and weak-$*$ convergence.

If $V = \mathcal{H}$ is a Hilbert space then, by Theorem D.3.4, V^* can be identified with $\mathcal{H}$. Moreover, the weak and weak-$*$ topologies on $\mathcal{H}$ coincide and a net $\{h_\alpha\}$ converges weakly to h if and only if

$$\lim_\alpha\,(h_\alpha, g) = (h, g), \quad g \in \mathcal{H}.$$

In $\mathcal{H}$ any orthonormal sequence $\{e_n\}_{n=1}^\infty$ converges weakly to 0. This follows from *Bessel's inequality*

$$\sum_{n=1}^\infty |(e_n, h)|^2 \le \|h\|^2, \quad h \in \mathcal{H}.$$

Since $\|e_n\| = 1$, the sequence $\{e_n\}$ does not converge strongly to 0.

Let $\mathcal{H}_1$ and $\mathcal{H}_2$ be Hilbert spaces. The *weak operator topology* on the set $\mathcal{B}(\mathcal{H}_1, \mathcal{H}_2)$ is the coarsest topology rendering all of the maps[8]

$$A \mapsto (Ag, h), \quad g \in \mathcal{H}_1,\ h \in \mathcal{H}_2$$

[8] Recall that $\mathcal{B}(\mathcal{H}_1, \mathcal{H}_2)$ denotes the set of all bounded linear operators from $\mathcal{H}_1$ to $\mathcal{H}_2$ (cf. D.3.3).

continuous, where $(\cdot,\cdot)$ denotes the inner product of $\mathcal{H}_2$. A net $\{A_\alpha\}$ of operators $A_\alpha \in \mathcal{B}(\mathcal{H}_1,\mathcal{H}_2)$ converges in the weak operator topology to an operator $A \in \mathcal{B}(\mathcal{H}_1,\mathcal{H}_2)$ if and only if

$$\lim_\alpha (A_\alpha g, h) = (Ag, h), \quad g \in \mathcal{H}_1,\ h \in \mathcal{H}_2.$$

The relation above is equivalent to

$$A_\alpha g \stackrel{w}{\to} Ag, \quad g \in \mathcal{H}_1.$$

In case of convergence in the weak operator topology we write $A_\alpha \stackrel{wo}{\to} A$.

Lemma D.5.2. *Multiplication of bounded operators in a Hilbert space $\mathcal{H}$ is separately continuous in the weak operator topology, i.e., if $A_\alpha \stackrel{wo}{\to} A$ then $A_\alpha B \stackrel{wo}{\to} AB$ and $BA_\alpha \stackrel{wo}{\to} BA$.*

Proof. Assume that $A_\alpha \stackrel{wo}{\to} A$ so that

$$\lim_\alpha (A_\alpha g, h) = (Ag, h), \quad g, h \in \mathcal{H}.$$

Replacing g by Bg we see that $A_\alpha B \stackrel{wo}{\to} AB$. The second statement follows from

$$\lim_\alpha (BA_\alpha g, h) = \lim_\alpha (A_\alpha g, B^* h) = (Ag, B^* h) = (BAg, h). \qquad \square$$

Remark D.5.3. Multiplication of operators is not jointly continuous in the weak operator topology. To see an example, let $\mathcal{H}$ be a Hilbert space with orthonormal basis $\{e_n\}_{n=-\infty}^{\infty}$ and write $Ue_n := e_{n+1}$. It is easy to check that U can be extended to a unitary operator in $\mathcal{H}$. Using that any orthonormal sequence in $\mathcal{H}$ converges weakly to 0 (cf. D.5.1) we see that $U^n \stackrel{wo}{\to} 0$ and $U^{-n} \stackrel{wo}{\to} 0,\ n \to \infty$. On the other hand, $U^n U^{-n}$ is the identity operator for all n.

Theorem D.5.4. *Convergence of a net in the weak, weak-$*$ or weak operator topology implies that the net is bounded.*

Proof. Let V be a normed linear space and let $\{l_\alpha\}$ be a net in V^* with $l_\alpha \stackrel{w*}{\to} l$. Then for each $v \in V$ the net $\{l_\alpha(v)\}$ converges to $l(v)$ and hence it is bounded. Theorem D.3.8 shows that the net $\{l_\alpha\}$ is bounded.

Assume that $\{v_\alpha\}$ is a weakly convergent net in V and define the linear functional L_α on V^* by $L_\alpha(l) := l(v_\alpha)$. Then $\|L_\alpha\| = \|v_\alpha\|$ and the net $\{L_\alpha\}$ converges in the weak-$*$ topology. By the first part of the proof $\{L_\alpha\}$, and hence also $\{v_\alpha\}$, is bounded.

Let $\{A_\alpha\}$ be a net in $\mathcal{B}(\mathcal{H}_1,\mathcal{H}_2)$ where $\mathcal{H}_1$ and $\mathcal{H}_2$ are Hilbert spaces. If $A_\alpha \stackrel{wo}{\to} A$, then $A_\alpha g \stackrel{w}{\to} Ag$ for all $g \in \mathcal{H}_1$. It follows that $\{A_\alpha g\}$ is bounded for all $g \in \mathcal{H}_1$. Applying again Theorem D.3.8, we conclude that $\{A_\alpha\}$ is bounded. $\square$

Theorem D.5.5. *Let V be a normed linear space, $\mathcal{H}_1, \mathcal{H}_2$ be Hilbert spaces, $\{v_\alpha\}$, $\{l_\alpha\}$ and $\{A_\alpha\}$ be nets in V, V^* and $\mathcal{B}(\mathcal{H}_1, \mathcal{H}_2)$, respectively.*[9]

(i) *If $v_\alpha \overset{w}{\to} v$, then $\|v\| \le \liminf_\alpha \|v_\alpha\|$.*

(ii) *If $l_\alpha \overset{w*}{\to} l$, then $\|l\| \le \liminf_\alpha \|l_\alpha\|$.*

(iii) *If $A_\alpha \overset{wo}{\to} A$, then $\|A\| \le \liminf_\alpha \|A_\alpha\|$.*

Proof. (i) Using the linear functionals L_α from Theorem D.5.4 we see that (i) follows from (ii).

(ii) Let $\{l_\beta\}$ be a subnet of $\{l_\alpha\}$ such that $\lim_\beta \|l_\beta\| = \liminf_\alpha \|l_\alpha\|$. For all $v \in V$ we have

$$|l(v)| = \lim_\beta |l_\beta(v)| \le \lim_\beta \|l_\beta\| \cdot \|v\| = \liminf_\alpha \|l_\alpha\| \cdot \|v\|$$

from which (ii) follows.

(iii) The proof is similar to that of (ii). We choose a subnet $\{A_\beta\}$ of $\{A_\alpha\}$ such that $\lim_\beta \|A_\beta\| = \liminf_\alpha \|A_\alpha\|$. For all $g \in \mathcal{H}_1$ and $h \in \mathcal{H}_2$ we have

$$|(Ag, h)| = \lim_\beta |(A_\beta g, h)| \le \lim_\beta \|A_\beta\| \cdot \|g\| \cdot \|h\| = \liminf_\alpha \|A_\alpha\| \cdot \|g\| \cdot \|h\|$$

from which the inequality in (iii) follows. □

Theorem D.5.6 (Banach–Alaoglu). *Let V be a normed linear space and $K \subset V^*$ be a bounded set which is closed in the weak-$*$ topology. Then K is compact in the weak-$*$ topology.*

Proof. It suffices to consider the special case $K = \{l \in V^* : \|l\| \le r\}$, $r > 0$. This set is weak-$*$ closed in view of (D.5.5.ii) and $|l(v)| \le r\|v\|$ for all $l \in K$ and all $v \in V$. The weak-$*$ topology is the topology of pointwise convergence. Hence, by Tychonoff's theorem, each net in K has a subnet converging pointwise to some function l_0 on V. This function is obviously linear and $|l_0(v)| \le r\|v\|$, i.e., $l_0 \in K$. Thus, K is compact in the weak-$*$ topology. □

Corollary D.5.7. *Let $\mathcal{H}_1$ and $\mathcal{H}_2$ be Hilbert spaces and let $S \subset \mathcal{B}(\mathcal{H}_1, \mathcal{H}_2)$ be closed in the weak operator topology. Then S is compact in the weak operator topology if and only if Sg is bounded for all $g \in \mathcal{H}_1$.*

Proof. The "if part" follows by the same argument as in the proof of the previous theorem. Here we use the fact that the weak operator topology is the topology of pointwise convergence, considering the weak-$*$ topology on $\mathcal{H}_2$.

To show the "only if part", assume that Sg is unbounded for some $g \in \mathcal{H}_1$. Then for all $n \in \mathbb{N}$ there exists $s_n \in S$ such that $\|s_n g\| \ge n$. By compactness we can choose a subsequence $\{s_{n_k}\}$ which is convergent in the weak operator topology. In view of

[9] For a net $\{r_\alpha\}$ of real numbers $\liminf_\alpha r_\alpha$ is defined by $\liminf_\alpha r_\alpha = \lim_\alpha \inf\{r_\beta : \beta \ge \alpha\}$.

Theorem D.5.4 this subsequence is bounded and hence $\{s_{n_k} g\}$ is bounded, as well. This contradiction completes the proof. □

Theorem D.5.8. *Let V be a normed linear space and let $K \subset V$ be convex. Then the weak and strong closures of K coincide.*

Proof. It suffices to prove that every strongly closed convex set $E \subset V$ is weakly closed. By Corollary D.4.3, for all $x \notin E$ there exists a continuous real linear functional l_x on V such that

$$m_x := \inf\{l_x(h) : y \in E\} > l_x(x).$$

Then

$$E = \bigcap_{x \notin E} \{y \in V : l_x(y) \geq m_x\}.$$

Since each set $\{y \in V : l_x(y) \geq m_x\}$ is weakly closed, the set E is weakly closed as well. □

Theorem D.5.9. *If $S \subset \mathcal{B}(\mathcal{H}_1, \mathcal{H}_2)$ is compact in the weak operator topology, then so is the weak operator closure of its convex hull.*

Proof. Corollary D.5.7 and Theorem D.3.8 show that S is bounded and hence so is its convex hull. By (D.5.5.iii) the weak operator closure Q of the convex hull is bounded as well. Therefore, the compactness of Q follows from Corollary D.5.7. □

Lemma D.5.10. *Let B be a Banach space such that B (B^*) is strictly convex.*[10] *If K is a convex subset of B (B^*) and K is closed in the weak (weak-$*$) topology, then there exists a unique element of K with minimal norm.*

Proof. We consider only B, the statement about B^* can be proved in the same way.
Choose a sequence $\{v_n\}$ in K such that

$$\lim_{n \to \infty} \|v_n\| = \inf\{\|v\| : v \in K\}.$$

By Theorem D.5.6 this sequence has a subsequence, converging weakly to some element $v \in V$. Since K is weakly closed we have $v \in K$. Theorem D.5.5 shows that the norm of v is minimal. The uniqueness of v follows from the fact that B is strictly convex. □

Theorem D.5.11. *Let S be a convex semigroup of linear contractions in $\mathcal{H}$. If S is compact in the weak operator topology, then there exists a unique orthogonal projection $s_0 \in S$ such that $s_0 s = s s_0 = s_0$ holds for all $s \in S$.*[11]

---

[10] A normed space V is called *strictly convex* if the relations $\|v\| = \|w\|$ and $v \neq w$ imply the inequality $\|(v + w)/2\| < \|v\|$. Hilbert spaces are strictly convex.

[11] A more general result can be found in [48].

Proof. Denote by $\mathcal{H}_0$ the set of common fixed points of the operators in S. Then $0 \in \mathcal{H}_0$ and $\mathcal{H}_0$ is an S-invariant subspace of $\mathcal{H}$. Write

$$S^* := \{s^* : s \in S\}.$$

It is easy to check that S^* is a compact convex semigroup of contractions. Moreover, the subspace $\mathcal{H}_0^\perp$ is S^*-invariant. Indeed, if $h_0 \in \mathcal{H}_0$ and $h \in \mathcal{H}_0^\perp$ are arbitrary then

$$(s^*h, h_0) = (h, sh_0) = (h, h_0) = 0, \quad s \in S.$$

For all $h \in \mathcal{H}$ the set S^*h is convex. Using the fact that S is compact we see that S^*h is weakly compact. By Lemma D.5.10 there exists a unique $h_0 \in S^*h$ having minimal norm. Since S^* is a semigroup of contractions, s^*h_0 is in S^*h and $\|s^*h_0\| \leq h_0$ for all $s \in S$. By the uniqueness of h_0 we must have $s^*h_0 = h_0$.

We prove that $h_0 = 0$ if $h \in \mathcal{H}_0^\perp$. Let $g \in \mathcal{H}$ be arbitrary and consider the weakly compact convex set Sg. The same arguments as above show the existence of $g_0 \in Sg$ such that $sg_0 = g_0$ for all $s \in S$. Thus, $g_0 \in \mathcal{H}_0$ and hence $(g_0, h_0) = 0$. Choose $s_0 \in S$ such that $g_0 = s_0 g$. Then

$$(g, h_0) = (g, s_0^* h_0) = (s_0 g, h_0) = (g_0, h_0) = 0, \quad g \in \mathcal{H}$$

and hence $h_0 = 0$. We have thus shown that for all $h_0 \in \mathcal{H}_0^\perp$ the set S^*h_0 contains 0.

Now let $h_1, \ldots, h_n \in \mathcal{H}_0^\perp$ be arbitrary and choose $s_1 \in S$ such that $s_1^* h_1 = 0$. Next we choose $s_2 \in S$ such that $s_2^*(s_1^* h_2) = 0$. Continuing this process we obtain an operator $s \in S$ with $s^* h_j = 0$ for all $j = 1, \ldots, n$. A simple compactness argument shows the existence of an operator $s_0 \in S$ such that $s_0^* h = 0$ for all $h \in \mathcal{H}_0^\perp$. It follows that $s_0 \mathcal{H} \subset \mathcal{H}_0$ and hence $ss_0 = s_0$.

It remains to prove that $s_0 s = s_0$. Note first that the equations $s_0 s s_0 = s_0$ and $s_0 s s_0 s = s_0 s$ hold because of $ss_0 = s_0$. We have

$$\|s^* s_0^* h\| = \|s^* s_0^* s^* s_0^* h\| \leq \|s_0^* s^* s_0^* h\| \leq \|s^* s_0^* h\|, \quad h \in \mathcal{H}.$$

Using this and the relation $s_0^* s^* s_0^* = s_0^*$ we see that $\|s^* s_0^* h\| = \|s_0^* h\|$. If $s_0^* h \neq s^* s_0^* h$ for some $h \in \mathcal{H}$, then

$$\|s_0^* h\| = \|s^* s_0^* h\| = \frac{1}{2}\|s_0^*(s^* s_0^* h + s_0^* h)\| \leq \frac{1}{2}\|s^* s_0^* h + s_0^* h\| < \|s_0^* h\|.$$

This contradiction shows that $s_0^* = s^* s_0^*$ and therefore $s_0 s = s_0$. The uniqueness of s_0 follows at once from $ss_0 = s_0 s = s_0$. □

Appendix E
Measure theory

E.1 Borel measures, weak and vague convergence

Mainly we need only measures on $\mathbb{R}^d$ or on $\mathbb{Z}^d$. However, at a few places we use Borel measures on topological spaces in general. In this section we collect some basic facts that will be sufficient for our purposes. We refer to Chapter 2 of [3] or to §§30–31 of [2] for the proofs.[1]

Throughout this section X will denote a Hausdorff topological space. The measures considered will be defined on the σ-algebra $\mathcal{B}(X)$ of all *Borel subsets* of X, i.e., the σ-algebra generated by the open subsets of X. A measure (nonnegative, real or complex) defined on $\mathcal{B}(X)$ will be called a *Borel measure*. In our terminology a measure need not be nonnegative. If μ and ν are finite Borel measures and $c \in \mathbb{C}$, then the measures $c\cdot\mu$ and $\mu+\nu$ are defined by $(c\cdot\mu)(B) = c\cdot\mu(B)$ and $(\mu+\nu)(B) = \mu(B)+\nu(B)$, $B \in \mathcal{B}(X)$.

Definition E.1.1. A *nonnegative Radon measure* μ on X is a nonnegative Borel measure such that

(i) $\mu(K) < \infty$ for each compact set $K \subset X$;

(ii) $\mu(B) = \sup\{\mu(K) : K \subset B,\ K \text{ compact}\}$ for each Borel set $B \subset X$.

E.1.2. The set of all nonnegative Radon measures is denoted by $\mathcal{M}_+(X)$ while $\mathcal{M}^b_+(X)$ denotes the set of all finite measures in $\mathcal{M}_+(X)$. A *complex Radon measure* ν on X is a complex measure of the form

$$\nu = (\mu_1 - \mu_2) + \mathrm{i}\cdot(\mu_3 - \mu_4)$$

where the μ_j's are in $\mathcal{M}^b_+(X)$. If a Borel measurable function f is μ_j-integrable for all j, then we say that f is *ν-integrable*. The integral $\int f\,\mathrm{d}\nu$ is then defined by

$$\int f\,\mathrm{d}\nu = \int f\,\mathrm{d}\mu_1 - \int f\,\mathrm{d}\mu_2 + \mathrm{i}\cdot\int f\,\mathrm{d}\mu_3 - \mathrm{i}\cdot\int f\,\mathrm{d}\mu_4.$$

We write $\mathcal{M}^b(X)$ for the set of all complex Radon measures on X while $\mathcal{M}^f(X)$ stands for the set of all complex measures with finite support.

---

[1] We remark that there are some small differences in the terminology of these books, e.g., in the definitions of Borel and Radon measures.

We define the measure $\overline{\mu}$ by $\overline{\mu}(B) = \overline{\mu(B)},\ B \in \mathcal{B}(X)$. If $\mu \in \mathcal{M}^b(\mathbb{R}^d)$, then the measures $\mu^{\triangle}$ and $\tilde{\mu}$ are defined by

$$\mu^{\triangle}(B) = \mu(-B) \quad \text{and} \quad \tilde{\mu}(B) = \overline{\mu(-B)}, \qquad B \in \mathcal{B}(\mathbb{R}^d)$$

respectively. Note that on $\mathbb{R}^d$ each nonnegative Borel measure that is finite on compact sets is a Radon measure.

If μ and ν are complex Radon measures on a locally compact space X and

$$\int f \,\mathrm{d}\mu = \int f \,\mathrm{d}\nu, \quad f \in \mathrm{C}_{00}(X)$$

then $\mu = \nu$.

Let μ be a complex Radon measure on a locally compact space X. Then there exist a finite nonnegative Borel measure $|\mu|$ and a Borel measurable function g on X such that[2] $|g(x)| = 1$ for all $x \in X$ and

$$\int f \,\mathrm{d}\mu = \int fg \,\mathrm{d}|\mu| \tag{1}$$

for every μ-integrable function f. This equation implies the inequality

$$\left|\int f \,\mathrm{d}\mu\right| \leq \int |f| \,\mathrm{d}|\mu|.$$

Theorem E.1.3 (Riesz representation theorem). *Let X be locally compact. Then there is a bijection between all nonnegative linear functionals*[3] *L on $\mathrm{C}_{00}(X)$ and all Radon measures μ on X given by*

$$L(f) = \int f \,\mathrm{d}\mu, \quad f \in \mathrm{C}_{00}(X).$$

Theorem E.1.4 (Riesz–Markov theorem). *Let X be locally compact. For any continuous*[4] *linear functional L on $\mathrm{C}_0(X)$, there is a unique complex Radon measure μ on X such that*

$$L(f) = \int f \,\mathrm{d}\mu, \quad f \in \mathrm{C}_0(X).$$

The norm of L as a linear functional is $|\mu|(X)$.

Lemma E.1.5. *Let $\mathcal{F}$ be a linear space of real-valued bounded functions on some set X such that $1 := \mathbf{1}_X \in \mathcal{F}$. Further, let L be a real linear functional on $\mathcal{F}$ such that $L(1) = 1$ and*

$$|L(f)| \leq \sup_{x \in X} |f(x)|, \quad f \in \mathcal{F}.$$

Then L is nonnegative, i.e., $L(p) \geq 0$ whenever $p \geq 0$.

---

[2] See Theorem 14.14 and Corollary 11.41 in [27].

[3] Nonnegativity means that $L(f) \geq 0$ whenever $f \geq 0$.

[4] We consider the supremum norm on $\mathrm{C}_0(X)$

Proof. If L was not nonnegative, then we could find a function $p \in \mathcal{F}$ such that $0 \le p \le 1$ and $L(p) < 0$. From $1 = L(1) = L(p) + L(1-p)$ it follows that $L(1-p) > 1$. This contradicts the assumptions since $0 \le 1 - p \le 1$. □

Definition E.1.6. The *weak topology* on $\mathcal{M}_+^b(X)$ is the coarsest topology (the topology with the fewest open sets) such that the functions $\mu \mapsto \int f \, d\mu$ become lower semicontinuous for every bounded lower semicontinuous real-valued function f on X.[5]

The space $\mathcal{M}_+^b(X)$ is a Hausdorff space in the weak topology. Convergence in the weak topology is characterized by the next theorem.

Theorem E.1.7. *For $\mu \in \mathcal{M}_+^b(X)$ and a net $\{\mu_\alpha\}$ in $\mathcal{M}_+^b(X)$ the following properties are equivalent:*

(i) $\mu_\alpha \to \mu$ *in the weak topology;*

(ii) $\liminf \mu_\alpha(G) \ge \mu(G)$ *for all open* $G \subset X$ *and* $\lim \mu_\alpha(X) = \mu(X)$*;*

(iii) $\limsup \mu_\alpha(F) \le \mu(F)$ *for all closed* $F \subset X$ *and* $\lim \mu_\alpha(X) = \mu(X)$*;*

(iv) $\liminf \int f \, d\mu_\alpha \ge \int f \, d\mu$ *for all bounded lower semicontinuous* $f : X \to \mathbb{R}$.

(v) $\limsup \int f \, d\mu_\alpha \le \int f \, d\mu$ *for all bounded upper semicontinuous* $f : X \to \mathbb{R}$.

If (i)–(v) *are fulfilled, then*

(vi) $\lim \int f \, d\mu_\alpha = \int f \, d\mu$ *for all bounded continuous* $f : X \mapsto \mathbb{R}$.

Finally, if X is a completely regular space,[6] *then property* (vi) *implies* (i)–(v).

We say that a net $\{\mu_\alpha\}$ of complex Radon measures on a completely regular space X *converges weakly* to some complex measure μ if the relation (E.1.7.vi) holds.

The set of finitely supported nonnegative measures is a dense subset of $\mathcal{M}_+^b(X)$ with respect to the weak topology. This set is dense even in a stronger sense:

Theorem E.1.8. *For every $\mu \in M_+^b(X)$ there exists a net $\{\mu_\alpha\}$ of nonnegative finitely supported measures such that*

$$\lim_\alpha \mu_\alpha(B) = \mu(B), \qquad B \in \mathcal{B}(X).$$

[5] Note that a real-valued function f on X is called lower (upper) semicontinuous if the set $\{x \in X : f(x) > r\}$ (the set $\{x \in X : f(x) < r\}$, respectively) is open for every $r \in \mathbb{R}$.

[6] A topological space X is *completely regular* if, given any closed set F and any point $x \in X \setminus F$, there is a continuous real-valued function f on X such that $f(x) = 0$ and $f(y) = 1$ for every $y \in F$. Locally compact Hausdorff spaces and metric spaces are completely regular.

The next simple result shows that finitely supported measures on $\mathbb{R}^d$ can be approximated by absolutely continuous measures with compact support.

Lemma E.1.9. *For every finitely supported nonnegative measure μ on $\mathbb{R}^d$ there exists a sequence $\{h_n\}$ of nonnegative functions $h_n \in C_{00}(\mathbb{R}^d)$ such that the sequence $\{\nu_n\}$ where $\mathrm{d}\nu_n(t) = h_n(t)\,\mathrm{d}\lambda(t)$ converges weakly to μ.*

Proof. Consider first the special case $\mu = \delta_x$, $x \in \mathbb{R}^d$, and denote by $B_n(x)$ the open ball with center x and radius $1/n$, $n \in \mathbb{N}$. For each n we choose a nonnegative function h_n such that $\int h_n\,\mathrm{d}\lambda = 1$ and $h_n(t) = 0$, $t \notin B_n(x)$, and define ν_n as in the statement of the lemma. Now let f be an arbitrary continuous real-valued function on $\mathbb{R}^d$. For an arbitrary $\epsilon > 0$ there exists $N(\epsilon) \in \mathbb{N}$ such that

$$f(x) - \epsilon < f(t) < f(x) + \epsilon, \quad t \in B_n(x),\ n \geq N(\epsilon).$$

Integrating both sides with respect to ν_n gives

$$f(x) - \epsilon < \int f(t)\,\mathrm{d}\nu_n(t) < f(x) + \epsilon, \quad n \geq N(\epsilon).$$

Noting that $\int f\,\mathrm{d}\mu = f(x)$ and applying Theorem E.1.7 we see that $\{\nu_n\}$ converges weakly to μ.

The general case follows by linearity. □

Definition E.1.10. The *vague topology* on $\mathcal{M}_+(X)$ is the coarsest topology such that the functions $\mu \mapsto \int f\,\mathrm{d}\mu$ are continuous for every function $f \in C_{00}(X)$.

Convergence in the vague topology is characterized by the next theorem.

Theorem E.1.11. *Let X be locally compact. For $\mu \in \mathcal{M}_+(X)$ and a net $\{\mu_\alpha\}$ in $\mathcal{M}_+(X)$ the following properties are equivalent:*

(i) $\mu_\alpha \to \mu$ *in the vague topology;*

(ii) $\limsup \mu_\alpha(K) \leq \mu(K)$ *for all compact $K \subset X$ and* $\liminf \mu_\alpha(G) \geq \mu(G)$ *for all relatively compact*[7] *open sets $G \subset X$;*

(iii) $\lim \mu_\alpha(B) = \mu(B)$ *for all relatively compact sets $B \in \mathcal{B}(X)$ such that* $\mu(\partial B) = 0$.[8]

The following relationship holds between the weak and the vague topologies:

---

[7] A set is called *relatively compact* if its closure is compact.

[8] The symbol ∂B denotes the *boundary* of B. It consists of all points of the closure of B which are not inner points of B.

Theorem E.1.12. *Suppose that X is locally compact and let $\{\mu_\alpha\}$ be a net in $\mathcal{M}_+^b(X)$. Then $\{\mu_\alpha\}$ converges weakly to a measure $\mu \in \mathcal{M}_+^b(X)$ if and only if it converges vaguely to μ and* $\lim \mu_\alpha(X) = \mu(X)$.

The next theorem is very useful. A special case of it is known as Helly's selection theorem.

Theorem E.1.13. *Suppose that X is locally compact and let $p > 0$. The set*

$$\{\mu \in \mathcal{M}_+^b(X) : \ \mu(X) \le p\}$$

is vaguely compact. If X is compact, then the set of all Radon probability measures on X is compact in the weak topology.

If μ and ν are Radon measures on spaces X and Y, then there is a unique Radon measure $\mu \times \nu$ on $X \times Y$, called *product measure*, satisfying

$$(\mu \times \nu)(A \times B) = \mu(A) \cdot \nu(B) \quad \text{for all compact } A \subset X, \ B \subset Y.$$

The next theorem states that the product measure depends continuously on both of its arguments.

Theorem E.1.14. *Let X and Y be Hausdorff spaces. Then $(\mu, \nu) \mapsto \mu \times \nu$ is weakly continuous as a mapping from $\mathcal{M}_+^b(X) \times \mathcal{M}_+^b(Y)$ to $\mathcal{M}_+^b(X \times Y)$. If X and Y are locally compact, then $(\mu, \nu) \mapsto \mu \times \nu$ is vaguely continuous as a mapping from $\mathcal{M}_+(X) \times \mathcal{M}_+(Y)$ to $\mathcal{M}_+(X \times Y)$.*

Theorem E.1.15. *Let X and Y be Hausdorff spaces and let $f : X \to Y$ be a continuous mapping. Then for any $\mu \in \mathcal{M}_+^b(X)$ the image measure μ^f belongs to $\mathcal{M}_+^b(Y)$. Moreover, the transformation $\mu \mapsto \mu^f$ from $\mathcal{M}_+^b(X)$ to $\mathcal{M}_+^b(Y)$ is continuous in the weak topology.*

The previous theorem has a useful corollary.

Corollary E.1.16. *Let $f : \mathbb{R}^d \to \mathbb{R}^k$ be continuous. If a sequence $\{X_n\}$ of d-dimensional random vectors converges in distribution*[9] *to a random vector X, then $\{f(X_n)\}$ converges in distribution to $f(X)$.*

Theorem E.1.17. *Let $(O_\alpha)_{\alpha \in I}$ be an open covering of X and on each O_α let a Radon measure μ_α be given such that $\mu_\alpha(B) = \mu_\beta(B)$ for each pair of indices $\alpha, \beta \in I$ and for each Borel set $B \subset O_\alpha \cap O_\beta$. Then there is a uniquely determined Radon measure μ on X such that $\mu(B) = \mu_\alpha(B)$ for every Borel set $B \subset O_\alpha$.*

[9] Recall that, by definition, a sequence of random vectors converges in distribution to a random vector X if the sequence of the corresponding distributions converges weakly to the distribution of X.

Lemma E.1.18. *Let f be a Borel measurable function on a locally compact Hausdorff space X and μ be a complex Radon measure on X. Suppose that there exists $K \geq 0$ such that the inequality*

$$\left| \int_S f \, \mathrm{d}\mu \right| \leq K$$

holds for all compact sets $S \subset X$. Then f is μ-integrable and the same inequality holds with S replaced by X.

Proof. Let g be as in (E.1.2.1) and write $h = fg$. Then

$$\left| \int_S f \, \mathrm{d}\mu \right| = \left| \int_S h \, \mathrm{d}|\mu| \right| \leq K$$

holds for all compact sets $S \subset X$. We show first that h is $|\mu|$-integrable. Since $|\mathrm{Re}\, z| \leq |z|$ and $|\mathrm{Im}\, z| \leq |z|$, $z \in \mathbb{C}$, we may suppose that h is real-valued. Setting $h^+ = \max(0, h)$, $h^- = -\min(0, h)$ and $S_h^+ = \{x \in X : h(x) \geq 0\}$, we have $h = h^+ - h^-$ and

$$0 \leq \int_S h^+ \, \mathrm{d}|\mu| = \int_{S \cap S_h^+} h \, \mathrm{d}|\mu| \leq K.$$

The validity of the last inequality follows from the fact that μ is a Radon measure and hence the integral over $S \cap S_h^+$ can be approximated by integrals over compact sets. From this we conclude that h^+ is $|\mu|$-integrable. The integrability of h^- follows in the same way. Thus h is integrable. Since $|f| = |h|$, the function f is integrable as well. As μ is a Radon measure, there exist compact sets $K_n \subset X$, $n \in \mathbb{N}$, such that

$$\lim_{n \to \infty} \int_{K_n} f \, \mathrm{d}\mu = \int_S f \, \mathrm{d}\mu$$

from which the required inequality follows. □

Lemma E.1.19 (Scheffe). *Let μ be a nonnegative measure on a measurable space $(\Omega, \mathcal{A})$ and let f, f_n be nonnegative, μ-integrable functions such that*

$$\lim_{n \to \infty} f_n = f$$

μ-almost everywhere and

$$\lim_{n \to \infty} \int_\Omega f_n \, \mathrm{d}\mu = \int_\Omega f \, \mathrm{d}\mu.$$

Then

$$\lim_{n \to \infty} \int_\Omega |f_n - f| \, \mathrm{d}\mu = 0.$$

Proof. The function $f_n + f - |f_n - f|$ is obviously nonnegative. Applying Fatou's lemma we obtain

$$\begin{aligned} 2\int_\Omega f \, d\mu &= \int_\Omega \liminf_{n\to\infty} (f_n + f - |f_n - f|) \, d\mu \\ &\le \liminf_{n\to\infty} \int_\Omega (f_n + f - |f_n - f|) \, d\mu \\ &= 2\int_\Omega f \, d\mu - \limsup_{n\to\infty} \int_\Omega |f_n - f| \, d\mu \end{aligned}$$

from which the statement follows. □

E.2 Convolution of measures and functions

This section contains basic facts about the convolution of functions and measures. Proofs and more details can be found in the books [2, 3, 27, 28]. All measures in this section will be complex Borel measures on $\mathbb{R}^d$ (cf. Section E.1).

Definition E.2.1. The *additive convolution* or simply *convolution* $\mu * \nu$ of two finite Borel measures μ and ν is defined as the image of the product measure $\mu \times \nu$ under the mapping $A : \mathbb{R}^d \times \mathbb{R}^d \to \mathbb{R}^d$ where $A(x, y) = x + y$.

If $d = 1$, then the *multiplicative convolution* $\mu \star \nu$ is the image of the product measure $\mu \times \nu$ under the mapping $M : \mathbb{R} \times \mathbb{R} \to \mathbb{R}$ where $M(x, y) = xy$.

E.2.2. The convolution $\mu * \nu$ and the multiplicative convolution $\mu \star \nu$ are again finite Borel measures satisfying

$$\int f \, d\mu * \nu = \iint f(x + y) \, d\mu(x) \, d\nu(y)$$

and

$$\int g \, d\mu \star \nu = \iint g(xy) \, d\mu(x) \, d\nu(y)$$

for arbitrary bounded continuous functions f on $\mathbb{R}^d$ and g on $\mathbb{R}$, respectively. In the sequel we consider only the additive convolution, analogous formulae hold for the multiplicative convolution as well.

We have

(i) $\mu * \nu = \nu * \mu$;

(ii) $(\mu * \nu) * \kappa = \mu * (\nu * \kappa)$;

(iii) $(c \cdot \mu) * \nu = c \cdot (\nu * \mu), \quad c \in \mathbb{C}$;

(iv) $\mu * (\nu + \kappa) = \mu * \nu + \mu * \kappa$;

(v) $\delta_x * \delta_y = \delta_{x+y}$.

The last three equations imply that the convolution of the finitely supported measures $\mu = \sum_{j=1}^{n} c_j \delta_{x_j}$ and $\nu = \sum_{k=1}^{m} d_k \delta_{y_k}$, where $x_j, y_k \in \mathbb{R}^d$, $c_j, d_k \in \mathbb{C}$, is given by

$$\mu * \nu = \sum_{j=1}^{n} \sum_{k=1}^{m} c_j d_k \delta_{x_j + y_k}.$$

E.2.3. Let $\nu \in \mathcal{M}^b(\mathbb{R}^d)$ and $f, g \in \mathcal{L}^1(\mathbb{R}^d)$. Recall that $\lambda = \lambda_d$ denotes the Lebesgue measure on $\mathbb{R}^d$. We define the functions $f * \nu$ and $f * g$ for $x \in \mathbb{R}^d$ by

$$f * \nu(x) = \int f(x - y) \, \mathrm{d}\nu(y)$$

and

$$f * g(x) = \int f(x - y) g(y) \, \mathrm{d}\lambda(y)$$

if the integrals exist, otherwise we set $f * \nu(x) = 0$ or $f * g(x) = 0$, respectively. It can be shown that the integrals above exist λ-almost everywhere (see also Theorem E.2.4 below). Moreover, the analogues of (E.2.2.i–iv) hold for convolutions of functions and measures and for convolutions of functions as well (λ-almost everywhere).

If μ is absolutely continuous with respect to the Lebesgue measure, i.e., $\mathrm{d}\mu(x) = f(x) \, \mathrm{d}\lambda(x)$ with some $f \in \mathcal{L}^1(\mathbb{R}^d)$, then $\mu * \nu$ is absolutely continuous, as well, and

$$\mathrm{d}\mu * \nu(x) = f * \nu(x) \, \mathrm{d}\lambda(x).$$

If ν is absolutely continuous, too, i.e., $\mathrm{d}\nu(x) = g(x) \, \mathrm{d}\lambda(x)$ with some $g \in \mathcal{L}^1(\mathbb{R}^d)$, then

$$\mathrm{d}\mu * \nu(x) = f * g(x) \, \mathrm{d}\lambda(x).$$

Theorem E.2.4. *If $f, g \in \mathcal{L}^2(\mathbb{R}^d)$ then the integral*

$$f * g(x) := \int f(x - y) g(y) \, \mathrm{d}\lambda(y)$$

exists for all $x \in \mathbb{R}^d$ and $f * g \in \mathrm{C}_0(\mathbb{R}^d)$.

If $f \in \mathcal{L}^1(\mathbb{R}^d)$ and $g \in \mathcal{L}^p(\mathbb{R}^d)$, $1 \le p \le \infty$, then the integral above exists λ-almost everywhere, $f * g \in \mathcal{L}^1(\mathbb{R}^d)$ and $\|f * g\|_p \le \|f\|_1 \cdot \|g\|_p$.

The next theorem states that the convolution depends continuously on both of its arguments.

Theorem E.2.5. *If the nets $\{\mu_\alpha\}$, $\{\nu_\alpha\}$ converge weakly to μ and ν, respectively, then $\{\mu_\alpha * \nu_\alpha\}$ converges weakly to $\mu * \nu$.*

Appendix F
Probability

In this chapter we collect some basic definitions, facts, and notation of probability theory. For more details see, e.g., [10, 16, 17, 37, 45].

F.1 Basic notions

A *probability space* is a triple $(\Omega, \mathcal{A}, P)$, where Ω is a nonempty set, $\mathcal{A}$ is a σ-algebra of subsets of Ω, and P is a *probability measure*[1] on $\mathcal{A}$. A d-dimensional *random vector* on $(\Omega, \mathcal{A}, P)$ is an $(\mathcal{A}, \mathcal{B}_d)$-measurable function $X \colon \Omega \to \mathbb{R}^d$, where $\mathcal{B}_d$ denotes the σ-algebra of Borel subsets of $\mathbb{R}^d$. If $d = 1$, then we call X a *random variable.* If we mention two or more random vectors in a theorem, definition, etc., we implicitly assume that all random vectors are defined on the same probability space.

Two random vectors X and Y are called *independent* if the equality

$$P(X \in B, Y \in D) = P(X \in B) \cdot P(Y \in D)$$

holds for all $B, D \in \mathcal{B}_d$. Let g and h be $(\mathcal{B}_d, \mathcal{B}_n)$-measurable functions from $\mathbb{R}^d$ into $\mathbb{R}^n$. If X and Y are independent, then the random vectors $g(X)$ and $h(Y)$ are independent as well. *Complex-valued random variables, complex random vectors* and their independence are defined in the same way by considering the σ-algebra of Borel subsets of $\mathbb{C}^d$. If not stated otherwise, random vectors are supposed to be real.

The probability measure μ_X defined by

$$\mu_X(B) := P(X \in B), \quad B \in \mathcal{B}_d$$

is called the *distribution* of the random vector X, while the *distribution function* F_X of X is defined by

$$F_X(t) = \mu_X((-\infty, t_1) \times \cdots \times (-\infty, t_d)), \quad t \in \mathbb{R}^d.$$

A random vector X is said to be *continuous* if μ_X is continuous, i.e., if the equation $\mu_X(\{t\}) = P(X = t) = 0$ holds for all $t \in \mathbb{R}^d$. If there exists a nonnegative Borel measurable function p on $\mathbb{R}^d$ such that

$$\mu_X(B) = \int_B p(x)\, \mathrm{d}\lambda(x), \quad B \in \mathcal{B}_d$$

---

[1] That is, a nonnegative measure such that $P(\Omega) = 1$.

i.e., if μ_X is absolutely continuous with respect to the Lebesgue measure, then p is called the *density* of X (or of μ_X). In this case we say that X is an *absolutely continuous* random vector. If there exists a Borel subset S of $\mathbb{R}^d$ such that $\lambda(S) = 0$ and $\mu_X(S) = 1$, then μ_X and X are said to be *singular* (with respect to the Lebesgue measure). If there exists a denumerable subset S of $\mathbb{R}^d$ such that $\mu_X(S) = 1$, then μ_X and X are said to be *discrete*. Note that discrete distributions are singular.

For arbitrary X the distribution μ_X can be written as

$$\mu_X = p_1\mu_d + p_2\mu_{ac} + p_3\mu_{sc}$$

where p_1, p_2, p_3 are nonnegative numbers summing to 1 and $\mu_d, \mu_{ac}, \mu_{sc}$ are probability distributions such that μ_d is discrete, μ_{ac} is absolutely continuous, and μ_{sc} is continuous and singular.

If X and Y are independent, then

$$\mu_{X+Y} = \mu_X * \mu_Y$$

where $*$ denotes *convolution*[2] of measures, respectively. We have

$$\int_{\mathbb{R}^d} h(t)\,\mathrm{d}(\mu_X * \mu_Y)(t) = \int_{\mathbb{R}^d}\int_{\mathbb{R}^d} h(x+y)\,\mathrm{d}\mu_X(x)\,\mathrm{d}\mu_Y(y) \tag{1}$$

for every bounded, continuous complex-valued function h on $\mathbb{R}^d$. If Y is absolutely continuous, then $X + Y$ is absolutely continuous as well, and its density p_{X+Y} is given by

$$p_{X+Y}(y) = \int_{\mathbb{R}^d} p_Y(y - x)\,\mathrm{d}\mu_X(x) \tag{2}$$

where p_Y denotes the density of Y.

If $d = 1$, then

$$\mu_{XY} = \mu_X \star \mu_Y$$

where $\star$ denotes *multiplicative convolution*. We have

$$\int_{\mathbb{R}^d} h(t)\,\mathrm{d}\mu_X \star \mu_Y(t) = \int_{\mathbb{R}^d}\int_{\mathbb{R}^d} h(xy)\,\mathrm{d}\mu_X(x)\,\mathrm{d}\mu_Y(y) \tag{3}$$

for every bounded, continuous complex-valued function h on $\mathbb{R}$.

Let X be a P-integrable real- or complex-valued random variable. The number

$$\mathbb{E}(X) := \int_\Omega X(\omega)\,\mathrm{d}P(\omega)$$

is called the *expectation* of X while the *variance* $\mathrm{Var}(X)$ of X is defined by

$$\mathrm{Var}(X) := \mathbb{E}(|X - \mathbb{E}(X)|^2).$$

Note that this definition allows the variance to be ∞.

---

[2] For properties of convolution of functions and measures we refer to Section E.2.

If X and Y are independent and integrable, then XY is integrable and we have

$$\mathbb{E}(XY) = \mathbb{E}(X) \cdot \mathbb{E}(Y).$$

More generally, let $X = (X_1, \ldots, X_d)$ be a d-dimensional real or complex random vector such that X_j is P-integrable for all j. Then the *expectation* of X is defined by

$$\mathbb{E}\,(X) := (\mathbb{E}X_1, \ldots, \mathbb{E}X_d).$$

Suppose that X_j is square integrable for all j and put $m_j = \mathbb{E}X_j$. The matrix $\operatorname{cov}(X) = (c_{jk})_{j,k=1}^d$ where

$$c_{jk} = \mathbb{E}[(X_j - m_j)(\overline{X_k - m_k})] = \mathbb{E}[X_j \overline{X_k}] - m_j \overline{m_k}$$

is called the *covariance matrix* of X. The covariance matrix is positive semidefinite (cf. Section D.2), since

$$\sum_{j,k=1}^{d} c_{jk} z_j \overline{z_k} = \mathbb{E}\left|\sum_{j=1}^{d} z_j \cdot (X_j - m_j)\right|^2 \geq 0, \quad z_j \in \mathbb{C}.$$

If the entries of a matrix $Y = (Y_{jk})$ are integrable random variables, then we write

$$\mathbb{E}(Y) := (\mathbb{E}Y_{jk}).$$

Using this notation, the covariance matrix can also be written in the form[3]

$$\operatorname{cov}(X) = \mathbb{E}\,[(X - \mathbb{E}(X)) \cdot (X - \mathbb{E}(X))^*].$$

Let $A \in \mathbb{C}^{d \times d}$ be a square matrix. Then, as it is easy to verify,

$$\operatorname{cov}(AX) = A \operatorname{cov}(X)\, A^*.$$

Let $\alpha = (\alpha_1, \ldots, \alpha_d) \in \mathbb{N}_0^d$ be a d-tuple of nonnegative integers and let μ be the distribution of a d-dimensional random vector X. If the function $x \mapsto x^\alpha$ is μ-integrable, then we call the number

$$M_\alpha = M_\alpha(\mu) = M_\alpha(X) = \int_{\mathbb{R}^d} x^\alpha \, \mathrm{d}\mu(x)$$

the *moment of order* α of μ (of X). In this case we also say that the moment of order α of μ exists. Note that M_α exists if and only if

$$A_\alpha = A_\alpha(\mu) = A_\alpha(X) = \int_{\mathbb{R}^d} |x^\alpha| \, \mathrm{d}\mu(x) < \infty.$$

[3] Recall our convention from Section A.2 saying that in expressions involving matrix operations the elements of $\mathbb{R}^d$ and $\mathbb{C}^d$ are considered as column vectors.

The number A_α is called the *absolute moment of order* α. The existence of M_α implies the existence of M_β for all $\beta \le \alpha$. In the case $d = 1$ the inequality

$$|x|^{2k-1} \le \frac{x^{2k} + x^{2k-2}}{2}, \qquad x \in \mathbb{R},\ k \ge 1$$

shows that

$$A_{2k-1} \le \frac{M_{2k} + M_{2k-2}}{2}, \qquad k \in \mathbb{N}. \tag{4}$$

Let $n > k \ge 1$ and suppose that A_n exists. By Hölder's inequality,

$$A_k = \int_{\mathbb{R}} |x|^k \cdot 1 \,\mathrm{d}\mu(x) \le \left(\int_{\mathbb{R}} |x|^{pk} \,\mathrm{d}\mu(x) \right)^{1/p}$$

for every $p > 1$. Setting $p = \frac{n}{k}$ we find

$$A_k^{1/k} \le A_n^{1/n}, \qquad 1 \le k \le n. \tag{5}$$

We will also use the simple but useful inequality

$$P(|X - \mathbb{E}\,(X)| \ge \epsilon\sigma) \le \frac{1}{\epsilon^2}, \qquad \epsilon > 0 \tag{6}$$

where $\sigma = \sqrt{\mathrm{Var}\,(X)}$. This inequality is called *Chebyshev's inequality*, it holds for every random variable X with finite variance.

F.2 Convergence of random vectors

A sequence $\{X_n\}$ of random vectors is said to *converge in distribution*, or *converge weakly*, or *converge in law* to a random vector X if the sequence $\{\mu_n\}$ of their distributions converges weakly (see Section E.1) to the distribution μ of X. We say that the sequence $\{X_n\}$ *converges in probability* to X if for all $\epsilon > 0$

$$\lim_{n\to\infty} P(\|X - X_n\| \ge \epsilon) = 0.$$

If $g : \mathbb{R}^d \to \mathbb{R}^n$ is continuous and $\{X_n\}$ is a sequence of d-dimensional random vectors which converges in distribution (in probability) to X, then $\{g(X_n)\}$ converges in distribution[4] (in probability, respectively) to $g(X)$. Finally, $\{X_n\}$ *converges to* X *almost surely* if

$$P(\{\omega \in \Omega : \lim_{n\to\infty} X_n(\omega) = X(\omega)\}) = 1.$$

Note that the set of all ω for which $\lim X_n(\omega)$ exists belongs to $\mathcal{A}$.

Almost sure convergence implies convergence in probability which implies convergence in distribution. The converse is not true, but if a sequence $\{X_n\}$ converges in

[4] See Corollary E.1.16 for the convergence in distribution.

distribution to a constant random vector, then the sequence is convergent in probability, as well.

Theorem F.2.1 (zero–one law of Borel–Cantelli). *Let $(\Omega, \mathcal{A}, P)$ be a probability space and $\{A_n\}_{n=1}^{\infty}$ be a sequence of events in $\mathcal{A}$.*

(i) *If $\sum P(A_n) < \infty$, then the probability that infinitely many of the A_n's occur is* 0.

(ii) *If the events are pairwise independent and $\sum P(A_n) = \infty$, then the probability that infinitely many of them occur is* 1.

Lemma F.2.2. *Let $\{X_n\}$ and $\{Y_n\}$ be sequences of random vectors such that*

$$\sum P(X_n \neq Y_n) < \infty.$$

Then $\sum X_n$ is almost surely convergent if and only if $\sum Y_n$ is almost surely convergent (and they converge to the same limit).

Proof. The lemma follows immediately from the fact that, by the zero–one law of Borel–Cantelli (see Theorem F.2.1), the probability that the event $[X_n \neq Y_n]$ happens for infinitely many n is zero. □

Theorem F.2.3. *Let $X_1, X_2, \ldots$ be random variables and $c_1, c_2, \ldots$ be complex numbers such that*

$$\sup_{n \in \mathbb{N}} \mathbb{E}\,|X_n| < \infty \quad \text{and} \quad \sum_{n=1}^{\infty} |c_n| < \infty.$$

Then the series

$$\sum_{n=1}^{\infty} c_n X_n \tag{1}$$

converges absolutely with probability one. If in addition $\sup_{n \in \mathbb{N}} \mathbb{E}\,|X_n|^2 < \infty$, then the series converges in mean square to the same limit.[5]

Proof. The monotone convergence theorem gives

$$\begin{aligned} \mathbb{E}\left(\sum_{n=1}^{\infty} |c_n X_n|\right) &= \lim_{N \to \infty} \mathbb{E}\left(\sum_{n=1}^{N} |c_n X_n|\right) \\ &\leq \lim_{N \to \infty} \left(\sum_{n=1}^{N} |c_n|\right) \cdot \sup_n \mathbb{E}\,|X_n| < \infty \end{aligned}$$

[5] Note that $\mathbb{E}\,|X| \leq (\mathbb{E}\,|X|^2)^{\frac{1}{2}}$ in view of (F.1.5) and therefore the condition $\sup_n \mathbb{E}\,|X_n|^2 < \infty$ implies that $\sup_n \mathbb{E}\,|X_n| < \infty$.

from which it follows that $\sum_n |c_n X_n(\omega)|$ is finite with probability one. Thus, the sequence of partial sums of (1) converges almost surely to a limit X.

If $K := \sup_n \mathbb{E}\,|X_n|^2 < \infty$ and $0 < m < n$, then

$$\mathbb{E}\left|\sum_{m<j\le n} c_j X_j\right|^2 = \sum_{m<j\le n}\sum_{m<k\le n} c_j \overline{c_k}\,\mathbb{E}\,(X_j \overline{X}_k) \le K \cdot \Bigg(\sum_{m<j\le n} |c_j|\Bigg)^2.$$

The right-hand side tends to zero as $m, n \to \infty$ and so by the Cauchy criterion (cf. D.3.2) the sequence of partial sums of (1) converges in mean square to a limit Y. Since mean square convergence and almost sure convergence imply convergence in probability, we conclude that $X = Y$ with probability one. □

F.2.4. For a d-dimensional random vector X we denote by $\sigma(X)$ the smallest σ-algebra containing all events of the form $[X \in B]$, where $B \in \mathcal{B}_d$. Similarly, if $\{X_n\}_1^\infty$ is a sequence of d-dimensional random vectors on the same probability space $(\Omega, \mathcal{A}, P)$, then $\sigma(X_1, X_2, \ldots)$ denotes the smallest σ-algebra containing $\sigma(X_n)$ for all n. An event $A \in \mathcal{A}$ is called a *tail event* (with respect to $\{X_n\}_1^\infty$) if

$$A \in \bigcap_{n=1}^{\infty} \sigma(X_n, X_{n+1}, \ldots).$$

Theorem F.2.5 (zero–one law of Kolmogorov). *If A is a tail event with respect to a sequence of independent random vectors, then the probability of A is* 0 *or* 1.

F.3 Products of probability spaces

Suppose that we are given a nonempty index set I and a family $(\Omega_i, \mathcal{A}_i, P_i)$, $i \in I$, of probability spaces. We consider the Cartesian product

$$\Omega := \prod_{i \in I} \Omega_i.$$

If $I = \mathbb{N}$, then we also use the notation

$$\Omega = \Omega_1 \times \Omega_2 \times \cdots \quad \text{and} \quad \omega = (\omega_1, \omega_2, \ldots) \in \Omega.$$

Let $\mathcal{A}$ be the smallest σ-algebra in Ω containing all sets A of the form

$$A = \prod_{i \in I} A_i$$

where $A_i \in \mathcal{A}_i$ and $A_i = \Omega_i$ for all but a finite number of $i \in I$.

Theorem F.3.1. *There exists a unique probability measure P on $\mathcal{A}$ such that*

$$P(A) = \prod_{i \in I} P_i(A_i)$$

for all $A \in \mathcal{A}$ of the form $A = \prod_{i \in I} A_i$ where $A_i \in \mathcal{A}_i$ and $A_i = \Omega_i$ for all but a finite number of $i \in I$.

The probability space $(\Omega, \mathcal{A}, P)$ is called the *product* of the probability spaces $(\Omega_i, \mathcal{A}_i, P_i)$, $i \in I$. The measure P is usually written as $P = \prod_{i \in I} P_i$, and, in case of $I = \mathbb{N}$, also as $P = P_1 \times P_2 \times \cdots$.

The next result is an immediate corollary of the previous theorem.

Corollary F.3.2. *For an arbitrary family $\{\mu_\alpha\}$ of d-dimensional distributions there exists a family $\{X_\alpha\}$ of independent d-dimensional random vectors such that μ_α is the distribution of X_α.*

A more general existence theorem than F.3.1 was given by Kolmogorov.

Theorem F.3.3 (Kolmogorov's existence theorem). *Let $T \neq \emptyset$ be an arbitrary set and $d \in \mathbb{N}$. For each finite subset $\{t_1, \ldots, t_k\} \subset T$ let $\mu_{t_1,\ldots,t_k}$ be a probability measure on $(\mathbb{R}^d)^k$. Suppose that these measures satisfy the so-called* consistency conditions*:*

(i) *For all permutations π of $\{1, \ldots, k\}$ and all Borel sets $B_1, \ldots, B_k \subset \mathbb{R}^d$*

$$\mu_{t_1,\ldots,t_k}(B_1 \times \cdots \times B_k) = \mu_{t_{\pi(1)},\ldots,t_{\pi(k)}}(B_{\pi(1)} \times \cdots \times B_{\pi(k)}) .$$

(ii) *For all $n \in \mathbb{N}$ and Borel sets $B_1, \ldots, B_k \subset \mathbb{R}^d$*

$$\mu_{t_1,\ldots,t_k}(B_1 \times \cdots \times B_k) = \mu_{t_1,\ldots,t_k,t_{k+1},\ldots,t_{k+n}}(B_1 \times \cdots \times B_k \times (\mathbb{R}^d)^n) .$$

Then there exist a probability space $(\Omega, \mathcal{A}, P)$ and d-dimensional random vectors X_t on Ω such that $\mu_{t_1,\ldots,t_k}$ is the distribution of the random vector $(X_{t_1}, \ldots, X_{t_k})$.

Remark F.3.4. Let $\mu_{t_1,\ldots,t_k}$ be as in Theorem F.3.3 and denote by $f_{t_1,\ldots,t_k}$ the corresponding characteristic function. Using Theorem 1.1.8 and Corollary 1.1.9 we see that the consistency conditions can be formulated in terms of the characteristic functions as follows:

(i) For all permutations π of $\{1, \ldots, k\}$ and all $x_1, \ldots, x_k \in \mathbb{R}^d$

$$f_{t_1,\ldots,t_k}(x_1, \ldots, x_k) = f_{t_{\pi(1)},\ldots,t_{\pi(k)}}(x_{\pi(1)}, \ldots, x_{\pi(k)}) .$$

(ii) For all $n \in \mathbb{N}$ and $x_1, \ldots, x_k \in \mathbb{R}^d$

$$f_{t_1,\ldots,t_k}(x_1, \ldots, x_k) = f_{t_1,\ldots,t_k,t_{k+1},\ldots,t_{k+n}}(x_1, \ldots, x_k, 0, \ldots, 0).$$

F.4 Conditional expectation

Let $(\Omega, \mathcal{A}, P)$ be a probability space and $\mathcal{B} \subset \mathcal{A}$ be a σ-algebra. For any P-integrable random variable X there exists a $\mathcal{B}$-measurable P-integrable random variable Z such that

$$\int_B Z \, \mathrm{d}P = \int_B X \, \mathrm{d}P, \quad B \in \mathcal{B}.$$

This random variable Z, which is P-almost everywhere uniquely determined, is called the *conditional expectation* of X with respect to $\mathcal{B}$ and is denoted by $\mathbb{E}(X|\mathcal{B})$ or $\mathbb{E}_{\mathcal{B}}(X)$. If $\mathcal{B} = \sigma(Y)$ where $Y = (Y_1, \ldots, Y_d)$ is a random vector we also write $\mathbb{E}(X|Y_1, \ldots, Y_d)$ or $\mathbb{E}(X|Y)$ instead of $\mathbb{E}(X|\mathcal{B})$. For an arbitrary $\sigma(Y_1, \ldots, Y_d)$-measurable random variable Z there exists a real-valued Borel measurable function g on $\mathbb{R}^d$ such that $Z = g(Y_1, \ldots, Y_d)$ holds P-almost everywhere. In particular, if $Z = \mathbb{E}(X|Y_1, \ldots, Y_d)$, then

$$\mathbb{E}(X|Y_1, \ldots, Y_d) = g(Y_1, \ldots, Y_d). \tag{1}$$

For this g the notation

$$\mathbb{E}(X|Y = y) := g(y), \quad y \in \mathbb{R}^d$$

is used and $\mathbb{E}(X|Y = y)$ is called the *conditional expectation of X given* $[Y = y]$. $\mathbb{E}(X|Y = y)$ is determined uniquely up to sets of μ_Y-measure 0 where μ_Y denotes the distribution of Y. The function g is μ_Y-integrable and for all Borel sets $B \subset \mathbb{R}^d$ we have

$$\int_B g(y) \, \mathrm{d}\mu_Y(y) = \int_{Y^{-1}(B)} X(\omega) \, \mathrm{d}P(\omega). \tag{2}$$

Conversely, if a Borel measurable function g satisfies this equation then (1) holds. From (2) it is easy to deduce the more general equation

$$\int_{\mathbb{R}^d} h(y) g(y) \, \mathrm{d}\mu_Y(y) = \int_\Omega h(Y(\omega)) X(\omega) \, \mathrm{d}P(\omega) \tag{3}$$

which holds for any complex-valued Borel measurable function h on $\mathbb{R}^d$ such that hg is μ_Y-integrable.

The *conditional variance* of a square integrable random variable X with respect to a σ-algebra $\mathcal{B} \subset \mathcal{A}$ is defined by

$$\operatorname{Var}(X|\mathcal{B}) := \mathbb{E}\left[(X - \mathbb{E}(X|\mathcal{B}))^2 | \mathcal{B}\right] = \mathbb{E}(X^2|\mathcal{B}) - (\mathbb{E}(X|\mathcal{B}))^2.$$

In the next two theorems we collect some basic properties of the conditional expectation. Equations and inequalities for conditional expectations are meant to hold almost surely.

Theorem F.4.1. *Let X and Y be integrable random variables. The following relations hold:*

(i) $\mathbb{E}_{\mathcal{B}}(aX + bY) = a\mathbb{E}_{\mathcal{B}}(X) + b\mathbb{E}_{\mathcal{B}}(Y), \quad a, b \in \mathbb{R}$ *or* $\mathbb{C}$;

(ii) *if* $X \geq Y$, *then* $\mathbb{E}_{\mathcal{B}}(X) \geq \mathbb{E}_{\mathcal{B}}(Y)$;

(iii) *if* X *is* $\mathcal{B}$*-measurable, then* $\mathbb{E}_{\mathcal{B}}(X) = X$;

(iv) $\mathbb{E}(\mathbb{E}_{\mathcal{B}}(X)) = \mathbb{E}(X)$;

(v) $|\mathbb{E}_{\mathcal{B}}(X)| \leq \mathbb{E}_{\mathcal{B}}(|X|)$ *and* $\|\mathbb{E}_{\mathcal{B}}(X)\|_1 \leq \|X\|_1$;

(vi) *if* Z *is a* $\mathcal{B}$*-measurable random variable such that* $Z \cdot X$ *is integrable, then* $\mathbb{E}_{\mathcal{B}}(Z \cdot X) = Z \cdot \mathbb{E}_{\mathcal{B}}(X)$;

(vii) *if* $\mathcal{B}_1$ *and* $\mathcal{B}_2$ *are sub-*σ*-algebras of* $\mathcal{A}$ *and* $\mathcal{B}_1 \subset \mathcal{B}_2$, *then* $\mathbb{E}_{\mathcal{B}_2}(\mathbb{E}_{\mathcal{B}_1}(X)) = \mathbb{E}_{\mathcal{B}_1}(\mathbb{E}_{\mathcal{B}_2}(X)) = \mathbb{E}_{\mathcal{B}_1}(X)$;

(viii) *if* $\mathcal{B}$ *and* $\sigma(X)$ *are independent,*[6] *then* $\mathbb{E}_{\mathcal{B}}(X) = \mathbb{E}(X)$.

By the previous theorem, $\mathbb{E}_{\mathcal{B}}$ is a linear operator from $\mathcal{L}^1(P)$ into $\mathcal{L}^1(P)$. This operator is nonnegative, contractive and idempotent, the $\mathcal{B}$-measurable random variables are fixed points of $\mathbb{E}_{\mathcal{B}}$.

Lemma F.4.2. *Let X_0 be an integrable random variable, $X = (X_1, \dots, X_d)$ be a random vector and g be a real-valued Borel measurable function g on $\mathbb{R}^d$. The equation*

$$\mathbb{E}(X_0|X) = g(X) \tag{1}$$

holds if and only if

$$\mathbb{E}\big(e^{i(t,X)} X_0\big) - \mathbb{E}\big(e^{i(t,X)} \cdot g(X)\big), \quad t \in \mathbb{R}^d. \tag{2}$$

Proof. Multiplying both sides of (1) by $e^{i(t,X)}$ and taking expectations yields

$$\mathbb{E}\big(e^{i(t,X)} \cdot \mathbb{E}(X_0|X)\big) = \mathbb{E}\big(e^{i(t,X)} \cdot g(X)\big).$$

Equation (2) now follows from (F.4.1.vi) and (F.4.1.iv).

For the proof of the other direction, let μ denote the distribution of X. By (F.4.3) equation (2) can be rewritten as

$$\int_{\mathbb{R}^d} e^{i(t,x)} \cdot \mathbb{E}(X_0|X = x)\, d\mu(x) = \int_{\mathbb{R}^d} e^{i(t,x)} \cdot g(x)\, d\mu(x), \quad t \in \mathbb{R}^d.$$

[6] Two subsets $\mathcal{B}_1$ and $\mathcal{B}_2$ of $\mathcal{A}$ are called *independent* if the events B_1 and B_2 are independent for arbitrary $B_1 \in \mathcal{B}_1$, $B_2 \in \mathcal{B}_2$.

By the uniqueness theorem for the Fourier–Stieltjes transform (cf. Theorem 1.8.3) we have $\mathbb{E}(X_0|X = x)\,\mathrm{d}\mu(x) = g(x)\,\mathrm{d}\mu(x)$. Thus, $\mathbb{E}(X_0|X = x) = g(x)$ for μ-almost all x from which (1) follows. □

Bibliography

[1] N. I. Akhiezer, *The Classical Moment Problem*, Oliver & Boyd, Edinburgh, 1965.

[2] H. Bauer, *Measure and Integration Theory*, De Gruyter, Berlin-New York, 2001.

[3] C. Berg, J. P. R. C. Christensen, and P. Ressel, *Harmonic Analysis on Semigroups*, Springer, Berlin, 1984.

[4] G. Berschneider, Spectral representation of intrinsically stationary fields, *Stoch. Proc. Appl.* **122** (2012), 3837–3851.

[5] G. Berschneider and Z. Sasvári, On a theorem of Karhunen and related moment problems and quadrature formulae, in: *Spectral Theory, Mathematical System Theory, Evolution Equations, Differential and Difference Equations.* 21st International Workshop on Operator Theory and Applications. Berlin, July 2010, Operator Theory: Advances and Applications 221, pp. 171–185, Springer, Basel, 2012.

[6] T. M. Bisgaard and Z. Sasvári, *Characteristic Functions and Moment Sequences. Positive Definiteness in Probability*, Nova Science Publishers, Huntington, 2000.

[7] T. M. Bisgaard and Z. Sasvári, When does $\mathbb{E}(X^k \cdot X^j) = \mathbb{E}(X^k) \cdot E(X^j)$ imply independence?, *Statist. Probab. Lett.* **76** (2006), 1111–1116.

[8] L. E. Blumenson, A derivation of n-dimensional spherical coordinates, *Amer. Math. Monthly* **67** (1960), 63–66.

[9] J.-P. Chilès and P. Delfiner, *Geostatistics. Modelling Spatial Uncertainty*, Wiley, New York, 1999.

[10] B. De Finetti, *Theory of Probability. Vol. 2*, Wiley, New York, 1975.

[11] K. de Leeuw and I. Glicksberg, The decomposition of certain group representations, *J. Anal. Math.* **15** (1965), 135–192.

[12] F. Derrien, *Strictly positive definite functions on the real line*, HAL: Hyper Articles en Ligne, hal-00519325, version 1 (2010). http://hal-univ-artois.archives-ouvertes.fr/docs/00/51/93/25/PDF/SPD.pdf.

[13] R. E. Edwards, *Fourier Series. A Modern Introduction. Vol. 1*, Springer, New York-Heidelberg-Berlin, 1979.

[14] A. V. Efimov, An analogue of Rudin's theorem for continuous radial positive definite functions of several variables (in Russian), *Trudi Instituta Matematiki i Mehaniki UrO RAN* **18** (2012), 1–8.

[15] W. Ehm, T. Gneiting, and D. Richards, Convolution roots of radial positive definite functions with compact support, *Trans. Amer. Math. Soc.* **356** (2004), 4655–4685.

[16] W. Feller, *An Introduction to Probability Theory and its Applications. Vol. 1*, 3rd edn., Wiley, New York, 1968.

[17] W. Feller, *An Introduction to Probability Theory and its Applications, Vol. 2*, 2nd edn., Wiley, New York, 1971.

[18] G. B. Folland, *Real Analysis*, Wiley, New York, 1984.

[19] I. I. Gikhman and I. V. Skorokhod, *The Theory of Stochastic Processes. I*, translated from the Russian by S. Kotz. Corrected printing of the 1st edn. Classics in Mathematics, Springer, Berlin, 2004.

[20] T. Gneiting, Radial positive definite functions generated by Euclid's hat, *J. Multivar. Anal.* **69** (1999), 88–119.

[21] T. Gneiting, Decomposition theorems for α-symmetric positive definite functions, *Math. Nachr.* **208** (1999), 117–120.

[22] T. Gneiting, Criteria of Pólya type for radial positive definite functions, *Proc. Amer. Math. Soc.* **129** (2001), 2309–2318.

[23] T. Gneiting, Strictly and non-strictly positive definite functions on spheres, Cornell University Library, arXiv:1111.7077v4 [math.PR] (2012), pp. 1–30.

[24] N. Goloshchapova, M. Malamud, and V. Zastavnyi, Radial positive definite functions and spectral theory of the Schrödinger operators with point interactions, *Math. Nachr.* **285** (2012), 1839–1859.

[25] V. I. Gorbachuk and M. L. Gorbachuk, *M.G. Krein's Lectures on Entire Operators*, Birkhäuser, Basel-Boston-Berlin, 1997.

[26] L. Grafakos and G. Teschl, On Fourier transforms of radial functions and distributions, Cornell University Library, arXiv:1112.5469 [math.CA] (2012), pp. 1–12.

[27] E. Hewitt and K. A. Ross, *Abstract Harmonic Analysis. I*, Springer, Berlin, 1963.

[28] E. Hewitt and K. Stromberg, *Real and Abstract Analysis*, Springer, New York-Heidelberg-Berlin, 1965.

[29] N. J. Higham, *Functions of Matrices. Theory and Computation*, SIAM, Philadelphia, 2008.

[30] B. Jessen, *Abstrakt Maalog Integraltheori*, Ch. Johansens Bogtrykkeri, Kobenhavn, 1947.

[31] B. Jessen and A. Wintner, Distribution functions and the Riemann zeta function, *Trans. Amer. Math. Soc.* **38** (1935), 48–88.

[32] Johnson, W. P., The curious history of Faà di Bruno's formula, *Amer. Math. Monthly* **109** (2002), 217–234.

[33] A. M. Kagan, Yu. V. Linnik, and C. R. Rao, *Characterization Problems in Mathematical Statistics*, Wiley, New York-London-Sydney-Toronto, 1973.

[34] K. Karhunen, Über lineare Methoden in der Wahrscheinlichkeitsrechnung, *Ann. Acad. Sci. Fenn. Ser. A I* **37** (1947), 1–79.

[35] C. Kleiber and J. Stoyanov, Multivariate distributions and the moment problem, *J. Multivar. Anal.* **113** (2013), 7–18.

[36] M. G. Krein, On the integral representation of a continuous Hermitian-indefinite function with a finite number of negative squares. (Russian), *Doklady Akademii Nauk SSSR, Serija matematika* **125** (1959), 31–34.

[37] R. G. Laha and V. K. Rohatgi, *Probability Theory*, Wiley, New York–Chichester–Brisbane–Toronto, 1979.

[38] H. Langer, M. Langer, and Z. Sasvári, Continuations of Hermitian indefinite functions and corresponding canonical systems: an example, *Meth. Func. Anal. Top.* **10** (2004), 39–53.

[39] G. Letac and Q. I. Rahman, A factorisation of the Askey's characteristic function $(1-\|t\|_{2n+1})_+^{n+1}$, *Ann. Inst. Henri Poincaré, Sect. B* **22** (1986), 169–174.

[40] P. Lévy, Extensions d'un théorème de D. Dugué et M. Girault, *Z. Wahrscheinlichkeitstheorie Verw. Geb.* **1** (1962), 159–173.

[41] W. Maak, *Fastperiodische Funktionen*, Springer, Berlin, 1950.

[42] G. Matheron, The intrinsic random functions and their applications, *Adv. Appl. Prob.* **5** (1973), 439–468.

[43] G. Pólya, Remarks on characteristic functions, *Proc. Berkeley Symp. Math. Stat. Probab.* **1** (1949), 115–123.

[44] G. Pólya and G. Szegö, *Problems and Theorems in Analysis*, Springer, Berlin-Heidelberg-New York, 1972.

[45] A. Rényi, *Foundations of Probability*, Holden-Day, San Francisco, 1970.

[46] W. Rudin, *Real and Complex Analysis*, McGraw-Hill, New York, 1966.

[47] W. Rudin, *Principles of Mathematical Analysis,* 3rd edn., McGraw-Hill, New York, 1976.

[48] Z. Sasvári, On common fixed points of linear contractions, *Proc. Amer. Math. Soc.* **108** (1990), 565–566.

[49] Z. Sasvári, *Positive Definite and Definitizable Functions*, Akademie Verlag, Berlin, 1994.

[50] Z. Sasvári, On norm dependent positive definite functions, *Monatsh. Math.* **120** (1995), 319–325.

[51] Z. Sasvári, On the number of negative squares of certain functions, in: *Contributions to operator theory in spaces with an indefinite metric: The Heinz Langer anniversary volume*, Operator Theory: Advances and Applications 106, pp. 337–353, Birkhäuser, Basel, 1998.

[52] Z. Sasvári, On generalized correlation functions of intrinsically stationary processes of order k, *Statist. Probab. Lett.* **78** (2008), 3381–3387.

[53] Z. Sasvári, Correlation functions of intrinsically stationary random fields, in: *Modern Analysis and Applications: The Mark Krein Centenary Conference. Vol. 1*, Operator Theory: Advances and Applications 190, pp. 451–470, Birkhäuser, Basel–Berlin, 2009.

[54] R. Schaback and Z. Wu, Operators on radial functions, *J. Comput. Appl. Math.* **73** (1996), 257–270.

[55] J. A. Shohat and J. D. Tamarkin, *The Problem of Moments*, American Math. Soc., Providence, 1943, published in 1950.

[56] E. C. Titchmarsh, *The Theory of Functions,* 2nd edn., Oxford University Press, London, 1939.

[57] N. G. Ushakov, *Selected Topics in Characteristic Functions*, VSP, Utrecht, 1999.

[58] H. Wendland, *Scattered Data Approximation*, Cambridge University Press, Cambridge, 2005.

[59] E. T. Whittaker and G. N. Watson, *A Course of Modern Analysis,* 4th edn. (repr. 1927), Cambridge University Press, Cambridge, 1999.

[60] D. V. Widder, *The Laplace Transform*, Princeton University Press, Princeton, 1946.

[61] J. Yeh, *Lectures on Real Analysis*, World Scientific Publishing, Singapore, 2000.

Index

Bei Fragen zur Produktsicherheit wenden Sie sich bitte an:
If you have any questions regarding product safety,
please contact:

Walter de Gruyter GmbH
Genthiner Straße 13
10785 Berlin
productsafety@degruyterbrill.com